NANOMATERIALS

Physical, Chemical, and Biological Applications

NANOMATERIALS

Physical, Chemical, and Biological Applications

Edited by
Nandakumar Kalarikkal, PhD
Sabu Thomas, PhD
Obey Koshy, PhD

Apple Academic Press Inc.	Apple Academic Press Inc.
3333 Mistwell Crescent	9 Spinnaker Way
Oakville, ON L6L 0A2 Canada	Waretown, NJ 08758 USA

© 2018 by Apple Academic Press, Inc.

First issued in paperback 2021

Exclusive worldwide distribution by CRC Press, a member of Taylor & Francis Group

No claim to original U.S. Government works

ISBN-13: 978-1-77463-051-8 (pbk)
ISBN-13: 978-1-77188-461-7 (hbk)

Library and Archives Canada Cataloguing in Publication

Nanomaterials (Kalarikkal)

Nanomaterials : physical, chemical, and biological applications / edited by Nandakumar Kalarikkal, PhD, Sabu Thomas, PhD, Obey Koshy, PhD.

This book is published as a result of the Second International Conference on Nanostructured Materials and Nanocomposites (ICNM 2014), held at Mahatma Gandhi University, Kottayam, Kerala, India, December 19-21, 2014.
Includes bibliographical references and index.
Issued in print and electronic formats.
ISBN 978-1-77188-461-7 (hardcover).--ISBN 978-1-315-36590-9 (PDF)

1. Nanostructured materials--Congresses. I. Kalarikkal, Nandakumar, editor II. Thomas, Sabu, editor III. Koshy, Obey, editor IV. International Conference on Nanostructured Materials and Nanocomposites (2nd : Kottayam, India : 2014) V. Title.

| TA418.9.N35N36 2017 | 620.1'15 | C2017-907076-2 | C2017-907077-0 |

Library of Congress Cataloging-in-Publication Data

Names: Kalarikkal, Nandakumar, editor. | Thomas, Sabu, editor. | Koshy, Obey, editor.
Title: Nanomaterials : physical, chemical, and biological applications/edited by, Nandakumar Kalarikkal, PhD, Sabu Thomas, PhD, Obey Koshy, PhD.
Description: Oakville, ON ; Waretown, NJ : Apple Academic Press, [2018] | Includes bibliographical references and index.
Identifiers: LCCN 2017052456 (print) | LCCN 2017059261 (ebook) | ISBN 9781315365909 (ebook) | ISBN 9781771884617 (hardcover : alk. paper) | ISBN 1771884614 (hardcover : alk. paper) | ISBN 1315365901 (eBook)
Subjects: LCSH: Nanostructured materials. | Nanocomposites (Materials)
Classification: LCC TA418.9.N35 (ebook) | LCC TA418.9.N35 N2596 2018 (print) | DDC 620.1/15--dc23
LC record available at https://lccn.loc.gov/2017052456

Apple Academic Press also publishes its books in a variety of electronic formats. Some content that appears in print may not be available in electronic format. For information about Apple Academic Press products, visit our website at **www.appleacademicpress.com** and the CRC Press website at **www.crcpress.com**

ABOUT THE EDITORS

Nandakumar Kalarikkal, PhD

Nandakumar Kalarikkal, PhD, is the Director of the International and Inter University Centre for Nanoscience and Nanotechnology as well as an Associate Professor in the School of Pure and Applied Physics at Mahatma Gandhi University, Kerala, India. His current research interests include synthesis, characterization, and applications of various nanostructured materials, laser plasma, and phase transitions. He has published more than 80 research articles in peer-reviewed journals and has co-edited seven books. Dr. Kalarikkal obtained his Master's degree in physics with a specialization in industrial physics and his PhD in Semiconductor Physics from Cochin University of Science and Technology, Kerala, India. He was a postdoctoral fellow at NIIST, Trivandrum, and later joined Mahatma Gandhi University, Kerala, India.

Sabu Thomas, PhD

Sabu Thomas, PhD, is the Founder and Director of the International and Inter University Centre for Nanoscience and Nanotechnology and a professor of polymer science and engineering at the School of Chemical Sciences of Mahatma Gandhi University, Kottayam, Kerala, India. Professor Thomas has received a number of national and international awards, including a Fellowship of the Royal Society of Chemistry; Distinguished Professorship from Josef Stefan Institute, Slovenia; an MRSI medal; a CRSI medal; and the Sukumar Maithy award. He is on the list of most productive researchers in India and holds the fifth position. Professor Thomas has published over 600 peer-reviewed research papers, reviews, and book chapters. He has delivered over 300 plenary/inaugural and invited lectures at national/international meetings in over 30 countries. He has established a state-of-the-art laboratory at Mahatma Gandhi University in the area of polymer science and engineering and nanoscience and nanotechnology through external funding from various organizations and institutes, including DST, CSIR, UGC, DBT, DRDO, AICTE, ISRO, TWAS, KSCSTE, BRNS, UGC-DAE, Du Pont, USA, General Cables, USA, Surface Treat Czech Republic, and Apollo Tyres. The h index of Professor Thomas is 75, and his work has been cited more than 23,850 times. He also has four patents to his credit.

Obey Koshy, PhD

Obey Koshy, PhD, is currently working as a postdoctoral fellow under the UGC-DS Kothari Scheme at the International and Interuniversity Centre for Nanoscience and Nanotechnology at Mahatma Gandhi University, Kerala, India. He completed his PhD in nanoscience and nanotechnology from the University of Kerala, India, after which he joined the Utah State University, USA, as a visiting scholar. His research interests include flexible sensors and photovoltaics among other related topics.

CONTENTS

LIST OF CONTRIBUTORS

Amitava Mukherjee
Centre for Nanobiotechnology, VIT University, Vellore, Tamil Nadu, India

Anushka Gupta
Quantum and Molecular Engineering Laboratory, Indian Institute of Technology Kharagpur, Kharagpur 721302, India

Arum Kumar Chakraborty
Department of Civil Engineering, Indian Institute of Engineering Science and Technology, Shibpur, Howrah, West Bengal, India

Bharat A. Bhanvase
Department of Chemical Engineering Laxminarayan Institute of Technology, Rashtrasant Tukadoji Maharaj Nagpur University, Nagpur 440033, Maharashtra, India

Bhaskar Bethi
Department of Chemical Engineering, National Institute of Technology, Warangal 506004, Telangana, India

Bibin John
Lithium Ion and Fuel Cell Division, Energy Systems Group, Vikram Sarabhai Space Centre, Thiruvananthapuram 695022, India

Bincy Rose Vergis
Department of chemistry, M.S. Ramaiah Institute of Technology, Bangalore 560054, Karnataka, India

C. Gouri
Lithium Ion and Fuel Cell Division, Energy Systems Group, Vikram Sarabhai Space Centre, Thiruvananthapuram 695022, India

C. Joseph Kennady
Department of Chemistry, School of Science and Humanities, Karunya University, Coimbatore 641114, India

C.P. Sandhya
Lithium Ion and Fuel Cell Division, Energy Systems Group, Vikram Sarabhai Space Centre, Thiruvananthapuram 695022, India

E. Sivakumar
Bioprocess Laboratory, Department of Microbial Biotechnology, Bharathiar University, Coimbatore 641046, Tamil Nadu, India

G. Ramkumar
Bioprocess Laboratory, Department of Microbial Biotechnology, Bharathiar University, Coimbatore 641046, Tamil Nadu, India

G. S. Shahane
Department of Electronics, DBF Dayanand College of Arts and Science, Solapur 413002, Maharashtra, India. E-mail: shahanegs@yahoo.com

H. Muktha
Department of Biotechnology, M.S. Ramaiah Institute of Technology, Bangalore 560054, Karnataka, India

John Thomas
Centre for Nanobiotechnology, VIT University, Vellore, Tamil Nadu, India. E-mail: john.thomas@vit. ac.in; th_john28@yahoo.co.in

K. B. Sravan Kumar
Quantum and Molecular Engineering Laboratory, Indian Institute of Technology Kharagpur, Kharagpur 721302, India

K. Parameswari
Department of Chemistry, School of Science and Humanities, Karunya University, Coimbatore 641114, India. E-mail: parameswari@karunya.edu

K. Samrat
Department of Biotechnology, M.S. Ramaiah Institute of Technology, Bangalore 560054, Karnataka, India

K. Swaminathan
Bioprocess Laboratory, Department of Microbial Biotechnology, Bharathiar University, Coimbatore 641046, Tamil Nadu, India

K. Venkatesan
Bioprocess Laboratory, Department of Microbial Biotechnology, Bharathiar University, Coimbatore 641046, Tamil Nadu, India

M. Suresh Kumar
Department of Chemical Engineering, National Institute of Technology, Warangal 506004, Telangana, India

M.N. Chandraprabha
Department of Biotechnology, M.S. Ramaiah Institute of Technology, Bangalore 560054, Karnataka, India. E-mail: chandraprabhamn@yahoo.com

Mainak Ghosal
Indian Institute of Engineering Science and Technology, Shibpur, Howrah, West Bengal, India. E-mail: mainakghosal2010@gmail.com

Mallika Saharia
Quantum and Molecular Engineering Laboratory, Indian Institute of Technology Kharagpur, Kharagpur 721302, India

Manjusha Padole
Quantum and Molecular Engineering Laboratory, Indian Institute of Technology Kharagpur, Kharagpur 721302, India

Meifang Zhou
Department of Chemistry, University of Melbourne, VIC 3010, Australia

Muthupandian Ashokkumar
Department of Chemistry University of Melbourne, VIC 3010, Australia

Nagaraju Kottam
Department of Chemistry, M.S. Ramaiah Institute of Technology, Bangalore 560054, Karnataka, India

Natarajan Chandrasekaran
Centre for Nanobiotechnology, VIT University, Vellore, Tamil Nadu, India

P. M. G. Nambissan
Applied Nuclear Physics Division, Saha Institute of Nuclear Physics, Kolkata 700064, India. E-mail: pmg.nambissan@saha.ac.in, pmgnambissan@gmail.com

P.A. Mahanwar
Department of Polymer and Surface Engineering, Institute of Chemical Technology, Mumbai, India. E-mail: pa.mahanwar@ictmumbai.edu.in

P.R. Deepu Krishnan
Bioprocess Laboratory, Department of Microbial Biotechnology, Bharathiar University, Coimbatore 641046, Tamil Nadu, India

Parag A. Deshpande
Quantum and Molecular Engineering Laboratory, Indian Institute of Technology Kharagpur, Kharagpur 721302, India. Email: parag@che.iitkgp.ernet.in

Praveenkumar Upadhyay
Basic Sciences: Chemistry, NIIT University, NH-8 Jaipur/Delhi Highway, Neemrana 301705, Rajasthan, India

R. Harikrishna
Department of chemistry, M.S. Ramaiah Institute of Technology, Bangalore 560054, Karnataka, India

R. Sankar Ganesh
Bioprocess Laboratory, Department of Microbial Biotechnology, Bharathiar University, Coimbatore 641046, Tamil Nadu, India

R. Sharath
Department of Biotechnology, M.S. Ramaiah Institute of Technology, Bangalore 560054, Karnataka, India. E-mail: sharathsarathi@gmail.com; knrmsr@gmail.com

R.H. Dongre
Department of Polymer and Surface Engineering, Institute of Chemical Technology, Mumbai, India

S. R. Murthy
Department of Mechanical Engineering, Geethanjali College of Engineering and Technology, Cheeryal (V), Keesara (M), RR Dist., Hyderabad 501301, Telangana, India
Department of Physics, Osmania University, Hyderabad 500007, Telangana, India.
E-mail: ramanasarabu@gmail.com

S. Sabarathinam
Bioprocess Laboratory, Department of Microbial Biotechnology, Bharathiar University, Coimbatore 641046, Tamil Nadu, India

S. Sadhasivam
Bioprocess Laboratory, Department of Microbial Biotechnology, Bharathiar University, Coimbatore 641046, Tamil Nadu, India

S. Thanigaivel
Centre for Nanobiotechnology, VIT University, Vellore, Tamil Nadu, India

Shabana Shaik
Department of Chemical Engineering, National Institute of Technology, Warangal 506004, Telangana, India

Shirish, H. Sonawane
Department of Chemical Engineering, National Institute of Technology, Warangal 506004, Telangana, India. E-mail: shirish@nitw.ac.in; shirishsonawane09@gmail.com

Shraddha Shah
Department of Biotechnology, M.S. Ramaiah Institute of Technology, Bangalore 560054, Karnataka, India

V. Jayavignesh
Bioprocess Laboratory, Department of Microbial Biotechnology, Bharathiar University, Coimbatore 641046, Tamil Nadu, India

V. Sai Phani Kumar
Quantum and Molecular Engineering Laboratory, Indian Institute of Technology Kharagpur, Kharagpur 721302, India

Vikas Mittal
Department of Chemical Engineering, Petroleum Institute, Abu Dhabi, UAE

Vivek Srivastava
Basic Sciences: Chemistry, NIIT University, NH-8 Jaipur/Delhi Highway, Neemrana 301705, Rajasthan, India. E-mail: vivek.shrivastava@niituniversity.in

Y. M. Purushothaman
Bioprocess Laboratory, Department of Microbial Biotechnology, Bharathiar University, Coimbatore 641046, Tamil Nadu, India

LIST OF ABBREVIATIONS

AC	alternating current
AF	acriflavin
AFM	atomic force microscopy
AILs	acidic ionic liquid
Al_2O_3	aluminum oxide
AOPs	advanced oxidation processes
BET	Brunauer–Emmett–Teller
BIL	basic ionic liquids
BSE	backscattered electrons
BVS	bond valence sum
CB	conduction band
Ce^{4+}	cerium ion
CHC	catalytic hydrogen combustion
CLEAs	cross-linked enzyme aggregates
CMC	carboxy methyl cellulose
CNT	carbon nanotubes
Co–Fe	cobalt–ferrite
CTAB	cetyl trimethyl ammonium bromide
Cu Kα	copper Kα
Cv	cavitation number
CV	crystal violet
CVD	chemical vapor deposition
DLS	dynamic light scattering
DMSO	dimethyl sulfoxide
EDAX	energy dispersive analysis of X-rays
EDC	ethylcarbodiimide
EDTA	ethylenediaminetetraacetic acid
fcc	face centered cubic
Fe_2O_3	hematite
Fe_3O_4	magnetite
$FeCl_3$	ferric chloride
FESEM	field emission scanning electron microscopy
FTIR	Fourier transform infrared
FWHM	full width at half maximum

G-TiO$_2$	nanocomposites
GO	graphene oxide
GO-ZnO	graphene oxide-zinc oxide
H$_2$O$_2$	hydrogen peroxide
HC	hydrodynamic cavitation
HiPco	high-pressure carbon monoxide
HPC	high performance concrete
HPGe	high pure germanium
HREM	high-resolution electron microscopy
IC	indigo carmine
ILs	ionic liquids
LDA	local density approximation
MB	methylene blue
mCLEAs	magnetic cross-linked enzyme aggregates
MF	melamine formaldehyde
MG	malachite green
MgFeO4	magnesium ferrite
MMA	methyl methacrylate
Mn(CH3COO) 2·4H2O	manganese acetate
MNPs	metal nanoparticles
MO	methyl orange
MRI	magnetic resonance imaging
MTCC	microbial type culture collection
MV	methyl violet
MWCNT	multi-walled carbon nanotubes
NaBH$_4$	sodium borohydride
NaOH	sodium hydroxide
NC	nanocement
NFA	nanoflyash
NHS	N-hydroxysuccinimide
NILs	neutral ionic liquid
NMR	nuclear magnetic resonance
NPs	nanoparticles
NSF	nano-silica fume
OCP	open-circuit potential
ODH	oxalyl dihydrazide
OG	orange G
OP	organic pollutant
OPC	ordinary portland cement
OSC	oxygen storage capacity

PAS	positron annihilation spectroscopy
PBS	phosphate-buffered saline
PCE	poly-carboxylate ether
PCs	photocatalysts
Pd	palladium
PE	primary electrons
PEG	polyethylene glycol
PEI	polyethyleneimine
PMMA	poly methyl methacrylate
PQP	premium quality paint
PSD	particle size distribution
PVA	polyvinyl alcohol
PVP	poly-vinyl pyrrolidone
PWscf	Plane-wave self-consistent field
R6G	rhodamine 6G
RBBR	remazol brilliant blue R
RGO	reduced graphene oxide
RhB	rhodamine B
ROS	reactive oxygen species
RR2	reactive red 2
RSDT	reactive spray deposition technology
RTIL	room temperature ionic liquid
SCE	standard calomel electrode
SCR	selective catalytic reduction
SCS	solution combustion synthesis
SDS	sodium dodecyl sulfate
SE	secondary electrons
SEM	scanning electron microscope
SF	silica fume
SiO_2	nano-silica
SOFC	solid oxide fuel cells
SPIONs	super paramagnetic iron oxide nanoparticles
SPY	standard putty powder sample
TEM	transmission electron microscopy
Ti/TiO_2	titanium dioxide-coated titanium
$Ti_{0.75}Ce_{0.25}O_2$	titanium into ceria
TiO_2	titanium dioxide
TiO_2-SiO_2	titania–silica
TPR	temperature-programmed reduction
TSIL	task specific ionic liquid

UHPC	ultra-high performance concrete
VB	valence band
WGS	water–gas shift
XPS	X-ray photoelectron spectroscopy
XRD	X-ray diffraction
$Zn(NO3)2 \cdot 6H2O$	zinc nitrate
ZnO	zinc oxide
ZnS	zinc sulfide
Zr^{4+}	zirconium ion
ZVI	zero valent iron
$\gamma\text{-}Fe_2O_3$	maghemite

PREFACE

Nanoscience and nanotechnology is emerging as a leading frontier of research and development in natural sciences, engineering as well as the anticipated pillars of the high-tech industry of this century. Presently, nanoscience and nanotechnology represents the most active discipline all around the world and is considered as the fastest growing technological revolution the human history has ever seen. The immense interest in the science of materials confined within the atomic scales stems from the fact that these nanomaterials exhibit fundamentally unique properties with great potential of bringing about revolutionary advancements of next-generation technologies in electronics, computing, optics, biotechnology, medical imaging, medicine, drug delivery, structural materials, automotives, aerospace, food, and energy.

Nanomaterials: Physical, Chemical, and Biological Applications is a book about the recent developments in the area of nanoscience and nanotechnology. This book is published as a result of the International Conference held on the topic Nanostructured Materials and Nanocomposites in Mahatma Gandhi University, Kottayam, Kerala, India. This book deals with the recent developments in the synthesis and characterization of nanomaterial and also its incorporation into polymer matrixes. The biological applications of nanomaterials have also been discussed in detail.

Chapter 1 deals with the synthesis of Al-doped SnO_2 thin films by spin coating. The XRD studies show the phase of thin films are pure tetragonal SnO_2. Gas sensing characteristics showed that SnO_2 films are sensitive as well as fast responding to CO gas at 275°C. A high sensitivity for CO gas indicates that the SnO_2 films are selective for this gas. The mechanism of gas sensing is explained adequately.

Chapter 2 is a review on the synthesis, surface functionalization, and immobilization of lipase enzyme on magnetic nanoparticles and the application of immobilized enzyme in biodiesel production. This deals with the potential of nanoparticles as carriers for enzymes.

Chapter 3 deals with the application of GO-ZnO as nanophotocatalyst for the degradation of malachite green (MG) dye in aqueous solution using a hybrid advanced oxidation process consisted of hydrodynamic cavitation.

Chapter 4 is a review which deals with the role of nanosized materials on the performance of lithium ion batteries. The potential advantages of nano-materials with a focus on the recent advances in cathode, anode, and elec-trolyte are reviewed. The challenges in the area of nanomaterials for lithium ion batteries are also summarized.

Chapter 5 aims to investigate the use of silver nanoparticles and their antibacterial activity against the disease-causing bacterial pathogens, which are most predominantly found in the cultivation of aquaculture fish farming. The green synthesis of silver nanoparticles using *Caulerpa scalpelliformis,* which is reported to have antibacterial activity against infection of Gram-negative bacterium such as *Pseudomonas aeruginosa* and is identified as a fish pathogen in this study, is described. The in vivo pathogenicity was performed with *Catla catla* fish and treatment of AgNPs was demonstrated both in vitro and in vivo.

Chapter 6 deals with the synthesis and characterization of nanoparticles of copper metal and $Ni_{0.5}Zn_{0.5}Fe_2O_4$ by using electroplating and microwave-hydrothermal (M-H) methods, respectively, and they were further used to prepare $Cu/Ni_{0.5}Zn_{0.5}Fe_2O_4$ nanocomposites by mechanical milling method. The phase identification, crystallinity, and morphology of the prepared nanoparticles were characterized by using X-ray diffraction (XRD), trans-mission electron microscopy (TEM), and FTIR. The magnetic hysteresis measurements were carried out to obtain important magnetic properties such as saturation magnetization (Ms) and coercivity (Hc) of the samples. The applicability of the present nanocomposites for the use of electromagnetic microwave absorption was examined in terms of their dielectric, magnetic properties, and shielding effectiveness.

Chapter 7 deals with the introduction of positron annihilation spectroscopy as a versatile nuclear experimental spectroscopic probe that can reflect the properties of the defects in nanomaterials and it is shown that the technique can offer solutions to a number of issues related to the exotic features nanomaterials exhibit compared to their bulk counterparts. Apart from these, the success of coincidence Doppler broadening measurements in the identification of the momentum distribution of electrons in a complex nanomaterial system is also briefly discussed. The aim of this chapter is to highlight the sensitivity and success of a non-conventional defect-investigative probe for understanding nanomaterials and following their evolution under varying experimental conditions such as crystallite size variation, heat treatment, and doping.

Chapter 8 deals with the synthesis and characterization of GO and graphene-TiO_2 by using TEM and XRD. Decolorization of MG has been achieved using hydrodynamic cavitation and in combination with other advanced oxidation processes. The hydrodynamic cavitation was first optimized in terms of different operating parameters such as operating inlet pressure, and pH of the operating medium to get the maximum degradation of MG. In the hybrid techniques, the combination of hydrodynamic cavitation with H_2O_2, G-TiO_2, and G-TiO_2/H_2O_2 has been used to get the enhanced degradation efficiency through hydrodynamic cavitation device.

Chapter 9 summarizes the recent developments in the field of synthesis, characterization, and activity of ceria-based catalysts. A detailed report on the activity of the catalysts and mechanism of the reactions over these catalysts is presented here.

Chapter 10 deals with the studies related to the effect of concentration and mode of addition of various components on properties of paint for decorative applications. These powder putty and powder paint formulations have been evaluated and compared with conventional decorative paints. The powder paint formulations have also been evaluated for exterior applications, namely, mechanical, thermally insulating, and chemical performance properties. The effects of concentration of glass microspheres on properties of powder paints have also been studied.

Chapter 11 deals with the phase manipulation of TiO_2 by coating TiO_2 on titanium substrate by the thermal decomposition and anodization methods using different precursors. The electrodes were studied for their surface morphology, phase formation, and corrosion resistance. Phase of TiO_2 was found to be anatase and efficiency of electrocatalytic activity was high in the electrode obtained by the thermal decomposition of $TiCl_3$/HNO_3 compared to the other electrodes prepared by other methods and precursors.

Chapter 12 deals with the antimicrobial activity of zero valent iron nanoparticles (nZVI) against Gram-negative (*Escherichia coli, Klebsiella pneumonia, Pseudomonas aeruginosa*) bacteria, and fungal (*Candida albicans, and Trichophyton rubrum*) strains.

Chapter 13 reports the simple method for in situ synthesis of Fe_2O_3 and Al_2O_3 with poly methyl methacrylate (PMMA) and its characterization. The conversion of MMA to PMMA is approximately 90%.

Chapter 14 discusses the recent as well as significant developments in the synthesis of ionic liquid mediated nanoparticles.

Chapter 15 deals with the synthesis and biological application of spinel ferrites such as antibacterial and nanomedicine applications.

Chapter 16 gives a brief discussion on the change of properties of cement when nanomaterials such as nanosilica, carbon nanotube, etc. are added to it.

Nano-additions on ordinary Portland cement (OPC) influence its hardening properties appreciably in both short term and medium term.

We hope this book will act as a reference book on a wide range of topics in nanoscience including physics, chemistry, biology, and polymer science.

Editors
Obey Koshy
Sabu Thomas
Nandakumar Kalarikkal

PART I
Synthesis and Characterization of Nanomaterials/Nanocomposites

CHAPTER 1

SYNTHESIS AND CHARACTERIZATION OF Al-DOPED SnO_2 THIN FILMS

G. S. SHAHANE[*]

Department of Electronics, DBF Dayanand College of Arts and Science, Solapur 413002, Maharashtra, India
[*]Corresponding author. E-mail: shahanegs@yahoo.com

CONTENTS

ABSTRACT

Aluminum-doped tin oxide thin films are synthesized by a sol–gel spin coating method. The synthesized samples are thin, uniform, and adherent to the substrate support. The XRD studies show that the phase of thin films is pure tetragonal SnO_2. For pure SnO_2, the lattice parameters are $a = 4.7284$ Å and $c = 3.2673$ Å. Not much change is observed in lattice parameters with Al-doping concentration. The average crystallite size is 6.6 nm and decreases with the increase in Al-doping concentration. SEM studies show that growth of the film takes place with porous structure embedded with nanogranules, increasing the open surface area of the film. Optical study revealed that band gap of SnO_2 is 3.96 eV with direct band-to-band transitions and increases with Al-doping concentration. Gas-sensing characteristics showed that SnO_2 films are sensitive as well as fast responding to CO gas at 275°C. The sensor response increases with Al-doping concentration up to 1 wt% and decreases afterwards. A high sensitivity for CO gas indicates that the SnO_2 films are selective for this gas. The rise time and recovery time are 27 and 80 s, respectively. The mechanism of gas sensing is explained adequately.

1.1 INTRODUCTION

Gas sensors based on metal oxide sensitive layers are playing an important role in the detection of toxic pollutants and inflammable gases. Since the gas-sensing properties are strongly dependent on the surface of the materials exposed to gases, the sensor based on thin film nanostructures of semiconducting metal oxides are expected to exhibit better sensitive properties than bulk.[1] Microstructure control of the sensing materials, especially the grain size and porosity, hence, become fundamental for the enhancement of gas-sensing performance. Tin oxide with a wide band gap ($E_g = 3.6$ eV) is one of the most important and extensively used metal oxide semiconductor materials for gas sensors.[2–4] The major drawback of undoped SnO_2 gas sensor material is its low sensitivity and selectivity.[4] It is important to note that n-type semiconductor materials such as SnO_2 have few oxygen adsorption sites available due to the development of potential barriers on the particle surface. Consequently, incorporating species which have a comparably higher number of adsorption sites with high fractional occupancy in the SnO_2 sensing material can have a significant effect on the sensor performance. The effect of the addition of dopants, such as Au, Pd, Pt, Sb, Zn,

Ni, and Al in tin oxide thin films on the gas-sensing properties has been previously studied.[5-13] The implantation of the dopants increased the sensitivity and the selectivity towards many gases.[5-13] The method and preparative conditions of SnO$_2$ nanoparticles are also very important to control the microstructure of sensing bodies and thus expected to influence the electrical properties. Variety of SnO$_2$ nanoparticles synthesizing and thin film coating methods such as solid state reaction, hydrothermal, sol–gel, spray pyrolysis, sonochemical, thermal evaporation, CVD, spin coating, etc. have been widely investigated.[2,14-22] Among the various methods, the sol–gel process is a wet-chemical technique widely used in the fields of materials science and ceramic engineering. The sol–gel approach is a cheap and low-temperature technique that allows for the fine control of the product's chemical composition. This chapter describes the structural, morphological, optical, and gas-sensing properties of nanocrystalline tin oxide thin films synthesized by sol–gel spin coating method.

1.2 EXPERIMENTAL DETAILS

The thin films of Al-doped SnO$_2$ were deposited onto glass substrates by sol–gel method.[18,23] For synthesis, AR grade SnCl$_4$.5H$_2$O was dissolved in double distilled water to get 8.9 M solution. For aluminum doping, Al$_2$(SO$_4$)$_3$.16H$_2$O was dissolved in water and added to the host solution in the appropriate proportion. The solution was stirred for 20 min at 70°C using magnetic stirrer to get clear and homogeneous solution. This was then added drop wise into 40 mL isopropyl alcohol, with continuous stirring to form a gel. The films were deposited from this gel using a spin coating unit (MILMAN-XT56) and heated in the furnace at 400°C for one hour. The thin films were then characterized through various characterization techniques. X-ray diffraction technique was used for its structural analysis. A Philips PW-3710 X-ray diffractometer with CuKα radiation (λ = 1.54056 Å) was used for this purpose. The surface morphology of the films was recorded on SEM (JEOL JSM 6360) scanning electron microscope operating at 20 kV. The topography of the films was recorded by an atomic force microscope (INNOVA 1B3BE). The UV–Vis spectra of these films were recorded using Shimadzu UV–VIS-NIR spectrophotometer (UV-3600). For gas-sensing characterization, the sensor element was kept in an air tight chamber of volume 250 mL. Two silver electrodes, separated by 1 mm, were deposited on SnO$_2$ film and silver wires were attached using silver paint. A predefined concentration of gas was introduced in this chamber by a syringe. The response of the sensor was measured using ARM processor based data acquisition system, designed and developed in the laboratory. The sensor was tested for H_2S, NH_3, CO, H_2 and NO_2 gases.

1.3   RESULTS AND DISCUSSION

1.3.1   STRUCTURAL CHARACTERIZATION

The deposited films are thin, uniform, and strongly adherent to the substrate support. The X-ray diffractograms of Al-doped SnO_2 thin films are shown in Figure 1.1. The X-ray diffraction patterns show that broad peaks indicating nanocrystallite particles are in the thin films. The d-values and intensities of the observed diffraction peaks match with the single crystalline phase of tetragonal SnO_2 (JCPD card no. 41–1445) with (1 1 0) plane as a preferred orientation. No diffraction peaks from any other impurities are detected indicating that the synthesized samples are pure tin oxide with tetragonal structure and dopant atoms are homogeneously incorporated into the tin oxide matrix. For pure SnO_2, the lattice parameters are a = 4.7284 Å and c = 3.2673 Å. The lattice parameters are slightly different from the standard JCPD values (a = 4.738 Å and c = 3.187 Å) indicating that the samples are under strain. The reason is that the sample is ultrafine in nature and it is possible that the lattice may be under strain-induced forces. When referred to the Al-doping, not much change is observed in the lattice parameters with Al-doping concentration. The average crystallite size was determined by the Scherer's relation:

$$D = 0.89\lambda/\beta\cos\theta \qquad (1.1)$$

where D is the crystallite size, λ is wavelength of X-ray, β is full width at half maximum (FWHM) measured in radians, and θ is the Bragg angle. The crystallite size was calculated and the values are listed in Table 1.1. The average crystallite size for pure SnO_2 is 6.6 nm and decreases with increase in Al-doping concentration. The decrease in crystallite size with doping can be explained as follows. Because of the interaction on the boundaries between host and dopant crystallite, the motion of crystallites is restricted. In other words, the advancing of grain boundaries which is required for crystal growth is stunted. As a result, the size of crystallites is decreased by the doping of impurities (Fig. 1.1).

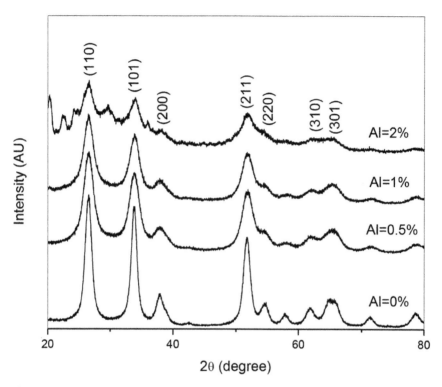

FIGURE 1.1   XRD patterns of SnO$_2$: Al thin films.

TABLE 1.1   Structural and Optical Analysis of SnO$_2$: Al Thin Films.

Al-doping Concentration (wt%)	Lattice Parameter		Crystallite Size (nm)	Band Gap (eV)
	a (Å)	*c* (Å)		
0	4.7284	3.2673	6.6	3.96
0.5	4.7366	3.2702	5.3	4.00
1	4.7284	3.2673	4.6	4.03
2	4.7397	3.2693	4.2	4.04

1.3.2   SURFACE MORPHOLOGICAL STUDIES

The SEM images of SnO$_2$ thin films with different Al-doping concentrations are shown in Figure 1.2. It is seen that the growth of the film takes place with the porous structure embedded with nanogranules, increasing the open surface area of the film. The results are in close agreement with XRD studies (Fig. 1.2).

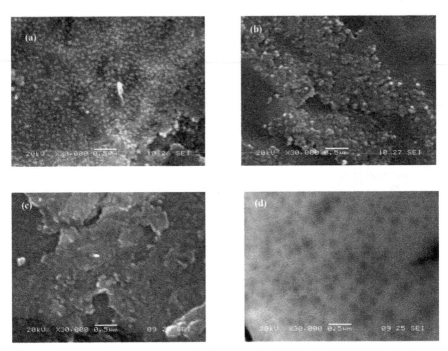

FIGURE 1.2   SEM images of SnO$_2$ thin films with (a) 0% Al, (b) 0.5% Al, (c) 1% Al, and (d) 2% Al.

Atomic force microscopy was used to record the topography of the Al-doped SnO$_2$ thin films. Figure 1.3 shows the surface topography of pure and doped SnO$_2$ films observed by AFM in two dimensional and three dimensional views. The surface roughness of the film over a 5 μm × 5 μm area was measured. Two dimensional view shows that the SnO$_2$ film surface is covered uniformly by fine grains. The three dimensional view shows that the growth of film takes place with closely placed sharp peaks and valleys. The maximum surface roughness of the film is 144 nm, which demonstrates that the surface morphology of SnO$_2$ film is rough and the film has a large open surface area. With Al-doping, the peaks become broad with a maximum surface roughness of 112 nm. This type of growth supports the application of these films as the gas sensor, since the film has a large open surface area to be exposed for the gas (Fig. 1.3).

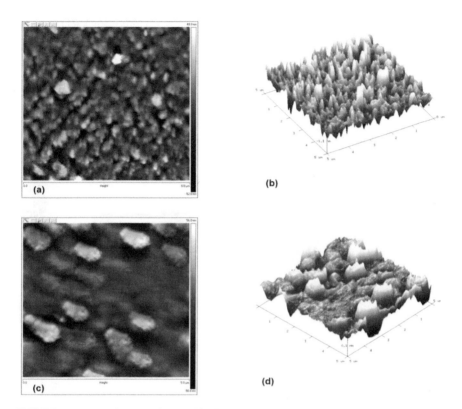

FIGURE 1.3 AFM images of SnO$_2$ thin films: (a and b) 0% Al and (c and d) 1% Al.

1.3.3 OPTICAL PROPERTIES

The optical absorption spectra of Al-doped SnO$_2$ thin films recorded over the wavelength range 280–900 nm at room temperature is shown in Figure 1.4. A sharp ultraviolet absorption edge at approximately 290 nm is observed. The spectra was studied to evaluate the absorption coefficient (α) energy gap (E_g) and the nature of transition involved. It is found that the optical absorption coefficient is higher for all the composites. The absorption coefficient (α), energy gap (E_g), and photon energy (hv) are related as:[23]

$$\alpha hv = A \ (hv - E_g)^{n/2} \tag{1.2}$$

Assuming the mode of transition to be of the direct allowed type ($n = 1$), the band gap energies have been calculated from the variation of $(\alpha hv)^2$ versus hv. Figure 1.5 shows the variation of $(\alpha hv)^2$ versus hv, which shows straight

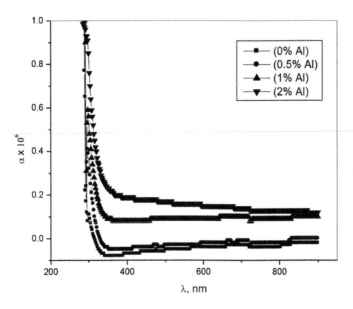

FIGURE 1.4 Optical absorption spectra of SnO_2:Al thin films.

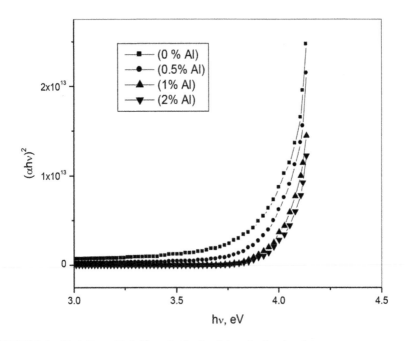

FIGURE 1.5 Variation of $(\alpha h\nu)^2$ vs. $h\nu$ for the determination band gap.

line behavior on the higher energy side, that confirm the direct type of transitions involved in these films.[24] The values of band gap energy for all the compositions are listed in Table 1.1. The band gap (E_g) of the pure SnO$_2$ thin film is 3.96 eV, which is higher than the reported value. The higher value of band gap can be ascribed to the nanocrystalline nature of the samples. With increase in Al-doping concentration, a blue shift in band gap is observed and can be attributed to the size quantization effects due to decreased grain size as revealed from the XRD studies. The similar values of the optical band gaps are reported by Gürakar et al. for doped SnO$_2$ thin films (Figs. 1.4 and 1.5, Table 1.1).[25]

1.3.4 GAS RESPONSE

The gas-sensing properties of SnO$_2$ thin films were studied in terms of its sensitivity, selectivity, temperature dependence, and speed of response. The sensitivity of a gas sensor is defined as $S = R_a/R_g$, where R_a and R_g are the resistances of the sensing element in air and in presence of the target gas.[13,26] The gas-sensing properties of SnO$_2$ film were carried out for the CO gas. The sensitivity is dependent on film porosity, film thickness, operating temperature, presence of additives, and crystallite size.[1] It is observed that resistance of the SnO$_2$ film decreases with the CO gas concentration and the film is highly sensitive to CO gas. The gas-sensing mechanism can be explained in terms of conductance modulation by adsorption of atmospheric oxygen on the surface and subsequent reaction of the adsorbed oxygen with the target gas. The atmospheric oxygen molecules are adsorbed on the surface of the sensor. These molecules take electrons from the conduction band of n-type SnO$_2$ to be adsorbed as $O_{2,ads}^-$ and O_{ads}^-, leading to the formation of an electron depleted region near the surface of SnO$_2$ particle. The form of the adsorbed oxygen (either molecule or atom) depends on the temperature of the sensor, where $O_{2,ads}^-$ species have been observed at lower temperatures (below 175°C) and O_{ads}^-, species have been observed at higher temperatures (above 175°C). Due to the formation of the depletion region, the material shows high resistance state in ambient air. On exposure, the CO gas molecules react with negatively charged oxygen adsorbates and the trapped electrons are given back to the conduction band of SnO$_2$. The energy released during this process would be sufficient for electrons to jump up into the conduction band of tin oxide, causing an increase in the conductivity of the sensor.[27] So, the steady state value of the resistance depends on the concentration of the CO gas. The overall reactions are as follows:

$$2CO + O_{2,ads}^- \rightarrow 2CO_2 + e^- \tag{1.3}$$

$$CO + O_{ads}^- \rightarrow CO_2 + e^- \tag{1.4}$$

Response of the sensor to a target gas depends on the working temperature of the sensor element. Hence, it is essential to optimize the working temperature of the SnO_2 sensor to give maximum sensitivity and selectivity towards CO gas. For this, the sensor element was exposed to 100 ppm concentration of CO gas and its response was recorded in 100–350°C temperature range. Figure 1.6 shows the response of the sensor as a function of temperature. It is seen that the response increases with temperature, reaches maximum at 275°C and further decreases. Thus, in the present case, the optimum operating temperature for SnO_2 films is 275°C, at which the sensor response attained its peak value. This behavior can be explained as follows. At low operation temperatures, the low response can be expected because the gas molecules do not have enough thermal energy to react with the surface adsorbed oxygen species. With increase in temperature, the thermal energy becomes high enough to overcome the potential barrier, and a significant increase in electron concentration was resulted from the sensing reaction. At higher temperatures, the sensor response is restricted by the speed of diffusion of gas molecules. At some intermediate temperature, the speed values of the two processes become equal, and at that point, the sensor response reaches its maximum (Fig. 1.6).[23]

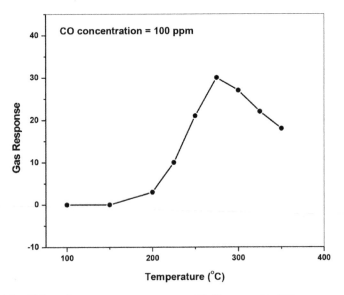

FIGURE 1.6   Effect of temperature on response of SnO_2 sensor for CO gas.

1.3.5 SELECTIVITY

The ability of a sensor to respond to a certain gas in presence of other gases is known as selectivity. The selectivity of a sensor in relation to a definite gas is closely associated with its operating temperature. Here, the selectivity measured in terms of selectivity coefficient/factor of a target gas to another gas is defined as $K = S_A/S_B$, where S_A and S_B are the responses of a sensor to a target gas and an interference gas, respectively. To check the selectivity of the sensor, it was exposed to H$_2$S, NH$_3$, CO, H$_2$, and NO$_2$ gases. Figure 1.7 shows the histogram of gas response for different gases at a fixed concentration of 100 ppm. The histogram revealed that the SnO$_2$ sensor offered response to H$_2$S (3), NH$_3$ (6), CO (30), H$_2$ (4), and NO$_2$ (8). The SnO$_2$ film showed more selectivity for CO over other gases. The selectivity coefficient for CO over other gases has the values 10, 5, 7.5, and 3.75 against H$_2$S, NH$_3$, H$_2$, and NO$_2$, respectively. This may be due to the fact that different gases have different energies for reaction to occur on the surface of SnO$_2$ thin film (Fig. 1.7).

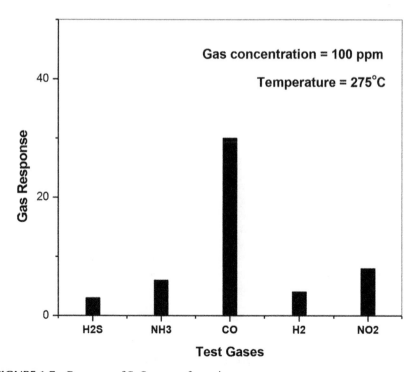

FIGURE 1.7 Response of SnO$_2$ sensor for various gases.

1.3.6   EFFECT OF AL-DOPING ON GAS RESPONSE

Figure 1.8 shows the effect of Al-doping concentration on the response of sensor operated at 275°C. It is observed that the sensor response increases with Al-doping concentration up to 1 wt% and decreases afterwards. The effect of doping can be explained as follows. Al-doping can have several effects on the properties important to gas-sensing applications, including inhibiting SnO_2 grain growth, modifying the electron Debye length and modifying the gas-surface interactions.[1] Another important role of Al-doping is that the substitution of Sn^{4+} by Al^{3+} acts as an acceptor impurity, forming localized p-regions in n-type SnO_2, forming the p–n junctions.[25,27] This may increase the depletion barrier height due to the electron transfer from n-type to p-type materials. When the sensor is exposed to CO gas, the electrons trapped by adsorbed oxygen species and p-type materials are fed back to the n-type material through surface interactions, resulting in a decrease in sensor resistance, improving the sensor response significantly (Fig. 1.8).[1]

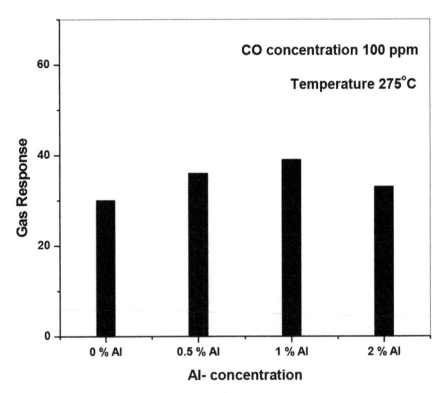

FIGURE 1.8   Effect of Al-doping concentration on sensor response.

1.3.7   LINEARITY

Figure 1.9 shows the variation of sensitivity with the concentration of CO gas. The gas concentration was varied from 20 to 250 ppm. It is observed that sensitivity increases almost linearly with the gas concentration up to 200 ppm and deviates afterwards. At lower gas concentrations, a mono-layer of gas molecules would be formed on the surface of the sensor which interacts more actively, giving a linear response. At higher gas concentrations, the multilayers of the gas molecules would result into deviation from linearity (Fig. 1.9).[28]

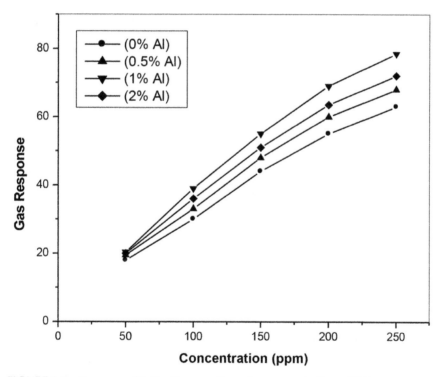

FIGURE 1.9   Response of SnO$_2$: Al sensor for various concentrations of CO gas.

1.3.8   SPEED OF RESPONSE

Figure 1.10 depicts the response of the sensor as a function of time under the exposure to CO gas at 100 ppm concentration. The conductivity increased rapidly by exposing the sensor to CO gas and recovered toward the original value after introducing clean air. This demonstrates a high potential for nanostructured metal oxide sensors with superior sensitivity. The response and recovery times of the sensor were calculated and found to be 27 and 80 s, respectively (Fig. 1.10).

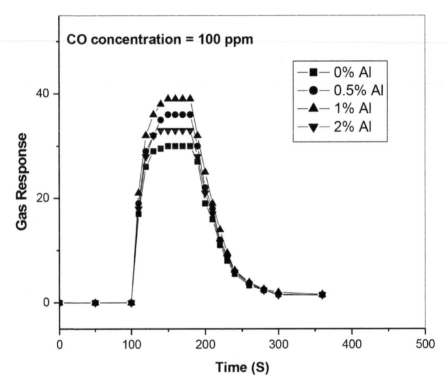

FIGURE 1.10   Speed response of SnO_2: Al sensor.

1.4   CONCLUSION

In summary, Al-doped SnO_2 thin films were prepared using sol–gel spin coating technique. Gas-sensing properties of the Al-doped SnO_2 thin films were systematically investigated and compared with those of the pristine SnO_2 thin films. Obtained results showed that the Al-doped sensor has a good selectivity to CO with higher responses, compared with the undoped sensor. The enhanced sensor response is due to the small crystallite size, modifications in the electron Debye length and modifying the gas-surface interactions due to the formation of localized p–n junctions between the p-type Al-SnO$_2$ and n-type SnO$_2$. This work also suggests that the Al-dopant can be a promising substitute for the noble metal additives such as Ag, Au, Pt, and Pd to fabricate chemical sensors at a much lower cost.

ACKNOWLEDGMENTS

The author (GSS) is thankful to University Grants Commission (WRO), Pune, for the financial assistance under the MRP scheme no F.47-031/12.

KEYWORDS

- **SnO$_2$ thin films**
- **structural characterization**
- **optical studies**
- **gas sensor**
- **nanocrystalline**

REFERENCES

1. Sun, Y. F.; Liu, S. B.; Meng, F. L.; Liu, J. Y.; Jin, Z.; Kong, L. T.; Liu, J. H. Metal Oxide Nanostructures and Their Gas Sensing Properties: A Review. *Sensors.* **2012,** *12,* 2610–2631.
2. Firooz, A. A.; Mahjoub, A. R.; Khodadadi, A. A. Effects of Flower-like, Sheet-Like and Granular SnO$_2$ Nanostructures Prepared by Solid-State Reactions on CO Sensing. *Mat. Chem. Phys.* **2009,** *115,* 196–199.
3. Hieu, N. V. Highly Reproducible Synthesis of Very Large-Scale Tin Oxide Nanowires Used for Screen-Printed Gas Sensor. *Sens. Actuators B.* **2010,** *144,* 425–431.
4. Berger, F.; Sanchez, J. B.; Heintz, O. Detection of Hydrogen Fluoride Using SnO$_2$-Based Gas Sensors: Understanding of the Reaction Mechanism. *Sens. Actuators B.* **2009,** *143,* 152–157.
5. Fasaki, I.; Suchea. M.; Mousdis, G.; Kiriakidis. G.; Kompitsas. M. The Effect of Au and Pt Nanoclusters on the Structural and Hydrogen Sensing Properties of SnO$_2$ Thin Films. *Thin Solid Films.* **2009,** *518,* 1109–1113.
6. Majula, P.; Arunkumar, S.; Manorama, V. Au/SnO$_2$ An Excellent Material for Room Temperature Carbon Monoxide Sensing. *Sens Actuators B.* **2011,** *152,* 168–175.
7. Srivastava, J. K.; Pandey. P.; Mishra, V. N.; Dwivedi, R. Sensing Mechanism of Pd-Doped SnO$_2$ Sensor for LPG Detection. *Solid State Sci.* **2009,** *11,* 1602–1605.

8. Khun, K. K.; Mahajan, A.; Bedi, R. K. Nanostructured Sb Doped SnO_2 Thick Films for Room Temperature NH_3 Sensing. *Chem. Phys. Lett.* **2010,** *492,* 119–122.
9. Ding, X.; Zeng, D.; Xie, C. Controlled Growth of SnO_2 Nanorods Clusters vie Zn Doping and Its Influence on Gas-Sensing Properties. *Sensors Actuators B.* **2010,** *149,* 336–344.
10. Li, H.; Xu, J.; Zhu, Y.; Chen, X.; Xiang, Q. Enhanced Gas Sensing by Assembling Pd Nanoparticles onto the Surface of SnO_2 Nanowires. *Talanta* **2010,** *82,* 458–463.
11. Liu, X.; Zhang, J.; Guo, X.; Wu, S.; Wang, S. Enhanced Sensor Response of Ni-Doped SnO_2 Hollow Spheres. *Sensors Actuators B.* **2011,** *152,* 162–167.
12. Zheng, Y.; Wang, J.; Yao, O. Formaldehyde Sensing Properties of Electrospun NiO-Doped SnO_2 Nanofibers. *Sensors Actuators B.* **2011,** *156,* 723–730.
13. Sinha, S. K. Synthesis of Novel Al-Doped SnO_2 Nanobelts with Enhanced Ammonia Sensing Characteristics. *Adv. Mater. Lett.* **2015,** *6 (2),* 133–138.
14. Wang, L.; Lou, Z.; Zhang, T.; Fan, H.; Xu, X. Facile Synthesis of Hierarchical SnO_2 Semiconductor Microspheres for Gas Sensor Application. *Sensors Actuators B.* **2011,** *155,* 285–289.
15. Jajarmi, P.; Barzegar, S.; Ebrahimi, G. R.; Varahram, N. Production of SnO_2 Nano-Particles by Hydrogel Thermal Decomposition Method. *Mater. Lett.* **2011,** *65,* 1249–1251.
16. Wang, Y.; Mu, Q.; Wang, G.; Zhou, Z. Sensing Characterization to NH_3 of Nanocrystalline Sb-Doped SnO_2 Synthesized by a Nanoqueous Sol–Gel Route. *Sensors Actuators B.* **2010,** *145,* 847–853.
17. Patil, L. A.; Shinde, M. D.; Bari, A. R.; Deo, V. V. Highly Sensitive and Quickly Responding Ultrasonically Sprayed Nanostructured SnO_2 Thin Films for Hydrogen Gas Sensing. *Sensors Actuators B.* **2009,** *143,* 270–277.
18. Sedghi, S. M.; Mortazavi, Y.; Khodadadi, A. Low Temperature CO and CH_4 Dual Selective Gas Sensor Using SnO_2 Quantum Dots Prepared by Sonochemical Method. *Sensor Actuators B.* **2010,** *145,* 7–12.
19. Thong, L. V.; Loan, L. T. N.; Hieu, N. V.; Comparative Study of Gas Sensor Performance of SnO_2 Nanowires and Their Hierarchical Nanostructures. *Sensors Actuators B.* **2010,** *150,* 112–119.
20. Zhao, J.; Wu, S.; Liu, J.; Gong, S.; Zhou, D. Tin Oxide Thin Films Prepared by Aerosol-Assisted Chemical Vapour Deposition and the Characteristics on Gas Detection. *Sensors Actuators B.* **2010,** *145,* 788–793.
21. Izydorczyk, W.; Wackzynski, K.; Izydorczyk, J.; Karansinski, P.; Mazurkiewick, J.; Magnuski, M.; Ulianow, J.; Niemiec, N. M.; Filipowski, W. Electrical and Optical Properties of Spin Coated SnO_2 Nanofilms. *Mater. Sci.* **2014,** *32* (4), 729–736.
22. Prdhan, U. U.; Naveen Kumar, S. K. Humidity Sensing Properties of TiO_2-SnO_2 Thin Films Fabricated Using Spin Coating Technique. *Arch. Phys. Res.* **2011,** *2* (3), 56–59.
23. Vidap, R. V.; Mathe, V. L.; Shahane, G. S. Structural, Optical and Gas Sensing Properties of Spin Coated Nanocrystalline Zinc Oxide Thin Films. *J. Mater. Sci. Mater. Electron.* **2013,** *24,* 3170–3174.
24. Baber, A. R.; Shinde, S. S.; Moholkar, A. V.; Bhosle, C. H.; Rajpure, K. Y. Structural and Optoelectronic Properties of Sprayed Sb: SnO_2 Thin Films: Effects of Substrate Temperature and Nozzle to Substrate Distance. *J. Semicond.* **2011,** *32* (10), 1–9.
25. Gürakar, S.; Serin, T.; Serin, N. Electrical and Microstructural Properties of (Cu, Al, In)-Doped SnO_2 Films Deposited by Spray Pyrolysis. *Adv. Mat. Lett.* **2014,** *5* (6), 309–314.

26. Young, J. J.; Lee, C. S. Characteristics of the TiO$_2$/SnO$_2$ Thick Film Semiconductor Gas Sensor to Determine Fish Freshness. *J. Ind. Engi. Chem.* **2011,** *17,* 237–242.
27. Sriram, S.; Thayumanavan, A. Effect of Al concentration on the Optical and Electrical Properties of SnO$_2$ Thin Films Prepared by Low Cost Spray Pyrolysis Technique. *Int. J. ChemTech Res.* **2013,** *5* (5), 2204–2209.
28. Shinde, S. D.; Patil, G. E.; Kajale, D. D.; Gaikwad, V. B.; Jain, G. H. Synthesis of ZnO Nanorods by Spray Pyrolysis for H$_2$S Gas Sensor. *J. Alloys Compounds.* **2012,** *528,* 109–114.

CHAPTER 2

SYNTHESIS, SURFACE MODIFICATION, AND APPLICATION OF IRON OXIDE NANOPARTICLES IN LIPASE IMMOBILIZATION FOR BIODIESEL PRODUCTION: A REVIEW

P. R. DEEPU KRISHNAN*, V. JAYAVIGNESH, G. RAMKUMAR, Y. M. PURUSHOTHAMAN, S. SADHASIVAM, S. SABARATHINAM, K. VENKATESAN, E. SIVAKUMAR, R. SANKAR GANESH, and K. SWAMINATHAN

Department of Microbial Biotechnology, Bioprocess Laboratory, Bharathiar University, Coimbatore 641046, Tamil Nadu, India

Corresponding author. E-mail: deepukrishnan7@gmail.com

CONTENTS

ABSTRACT

Nanoparticles are submicron moieties with at least one dimension that is less than 100 nm. They exhibit certain novel physical and chemical properties that are distinct from the characteristics of their bulk counterparts. Fabrication of nanostructured materials such as metal nanoparticles, nanofibers, nanotubes, nanosheets, and nanocomposites for biological applications has evoked great interest in researchers around the globe. During the past few years, a significant number of reports have been published on exploring the potential of nanoparticles as carriers for enzymes. The large surface area to volume ratio of nanoparticles is one of the most advantageous characteristics over conventional materials used in immobilization and thus they offer high enzyme loading during the immobilization process. Enzymes immobilized onto nanoparticles also exhibit higher stability in a wide range of temperature and pH, compared to free enzymes. In recent years, magnetic nanoparticles particularly iron oxides (magnetite, Fe_3O_4 and maghemite, $\gamma\text{-}Fe_2O_3$) have received significant importance due to their strong magnetic properties, chemical stability, and decreased toxicity. They find broad spectrum of biological applications such as contrast agents in magnetic resonance imaging (MRI), separation of biomolecules or cells, agents for drug delivery, tumor diagnosis, tissue engineering, in hyperthermia therapy, etc. Surface functionalized magnetic iron oxide nanoparticles serve as excellent candidates for immobilization of lipase enzyme as their magnetic property facilitates easy and effective recovery of the enzyme bound particles by the application of a magnetic field. This review focuses on the synthesis, surface functionalization, and immobilization of lipase enzyme on magnetic nanoparticles and the application of immobilized enzyme in biodiesel production.

2.1 MAGNETIC IRON OXIDE NANOPARTICLES

Among all forms of iron oxides, magnetite (Fe_3O_4), and maghemite ($\gamma\text{-}Fe_2O_3$) are gaining much attention as suitable substrates for enzyme immobilization in the process of biodiesel production. Magnetite, Fe_3O_4, has a cubic inverse spinel structure in which the oxygen atoms form a face centered cubic (fcc) closed packing arrangement and the Fe cations are occupied in the interstitial tetrahedral and octahedral sites. The iron cations exist in two valence states, Fe^{2+} and Fe^{3+} in the octahedral sites.[1] Maghemite, $\gamma\text{-}Fe_2O_3$, has a spinel structure with cationic vacancies in the octahedral sites and the vacant sites are occupied by Fe(III) ions in such a way that two filled sites are followed by one vacant site.[2] Due to the presence of unpaired electrons in 3*d* orbitals, iron atoms have a strong magnetic moment and they exist in different magnetic states such as paramagnetic, ferromagnetic, and ferrimagnetic forms. Magnetite and maghemite are ferromagnetic at room temperature in its bulk form. However, when the a core size of these particles is <10 nm, they exhibit a unique phenomenon called as superparamagnetism where these materials can be magnetized when exposed to an externally applied magnetic field, but exhibit zero magnetization at an applied field of zero. These superparamagnetic iron oxide nanoparticles (SPIONs) coated with biocompatible materials can be surface functionalized to immobilize enzymes and they can be easily separated from the biodiesel reaction mixture by the application of an external magnetic field.

2.2   SYNTHESIS OF IRON OXIDE NANOPARTICLES

Over the past few decades, several methods have been developed to synthesize iron oxide nanoparticles. However, synthesis of magnetic nanoparticles with customized size, specific shape, stability, and good monodispersity still remains as a challenge. Among the numerous methods that can be employed in the synthesis of iron oxide nanoparticles, the most common methods are co-precipitation, thermal decomposition, microemulsion, and hydrothermal synthesis. The following sections compile the most popular methods of iron oxide nanoparticle synthesis.

2.2.1   CO-PRECIPITATION

The most common method for synthesizing Fe_3O_4 or γ-Fe_2O_3 nanoparticles is co-precipitation. Synthesis of iron oxide nanoparticles through this route is simpler, more tractable, and more efficient with appreciable control over size, composition and sometimes even the shape of the nanoparticles.[3] In this method, ferric (Fe^{3+}) and ferrous (Fe^{2+}) ions are mixed in 2:1 molar ratio in the presence of a base at a specific temperature to synthesize magnetite (Fe_3O_4) which is obtained as a black precipitate. The chemical reaction of magnetite (Fe_3O_4) formation may be represented as follows:

$$Fe^{2+} + 2Fe^{3+} + 8OH^- \rightarrow Fe_3O_4 + 4H_2O$$

However, magnetite (Fe_3O_4) is not very stable and is sensitive to oxidation. Thus, this reaction is performed under an inert atmosphere to prevent the oxidation reaction. The size and shape of nanoparticles depend on various reaction parameters such as type of the salts used (e.g., chlorides, sulfates, nitrates, perchlorates, etc.), ratio of Fe^{2+} and Fe^{3+} ions, rate of stirring, the pH value, ionic strength of the media, and reaction temperature.[4] Wu et al.[5] have reported the synthesis of magnetite (Fe_3O_4) nanoparticles by co-precipitation method and they have also studied the influence of different reaction temperature on the morphology, structure, and magnetic properties of the synthesized nanoparticles. Kang et al.[6] reported the synthesis of Fe_3O_4 nanoparticles with a mean diameter of 8.5 ± 1.3 nm through co-precipitation method and the particles were found to be monodispersed and uniformly distributed without using any surfactant during the synthesis. Maghemite (γ-Fe_2O_3) nanoparticles can be synthesized by chemical co-precipitation method, followed by a temperate oxidation process. Yoon[7] reported the synthesis of superparamagnetic maghemite nanoparticles by this method and the average particle size was 13.9 nm. Co-precipitation process has the advantage of synthesizing large amount of nanoparticles but the wide particle size distribution, formation agglomerated particles, and difficulties in the purification of the final product due to high pH value of the reaction mixture and generation of highly basic wastewater remain as a major challenge in this method. Recently, many researchers reported the synthesis of monodispersed iron oxide nanoparticles by the addition of surfactants (e.g., polyvinyl alcohol (PVA), chitosan, dextran, etc.). The role of surfactants on the size, shape, and monodispersity of the iron oxide nanoparticles is discussed in the following section of this review. Synthesis of nanoparticles by co-precipitation method followed by a hydrothermal or solvothermal treatment method is proposed by several authors to obtain iron oxide nanoparticles with good monodispersity.[8]

2.2.2 THERMAL DECOMPOSITION

Iron oxide nanoparticles can be prepared by high temperature decomposition of organometallic precursors [e.g., Fe(acac)$_3$ where "acac" is acetylacetonates, Fe(cup)$_3$ where "cup" is N-nitrosophenylhydroxylamine, or Fe(CO)$_5$] in high-boiling non-aqueous environment containing surfactants. Monodispersed magnetic nanoparticles with control over size and shape can be synthesized by this method.[9,10] Sun and Zeng[11] have reported the synthesis of monodisperse magnetite nanoparticles with controlled size (3–20 nm) by high-temperature (265°C) reaction of iron (III) acetylacetonate, Fe(acac)$_3$, in phenyl ether in the presence of alcohol, oleic acid, and oleylamine. They have also reported that the as-synthesized Fe_3O_4 nanoparticle assemblies can be easily transformed into γ-Fe$_2$O$_3$ nanoparticle assemblies by annealing at oxygen rich environment. William et al.[12] demonstrated the synthesis of magnetite nanocrystals by thermal decomposition of iron carboxylate salts and the particles were reported to have a spherical shape and narrow size distribution (6–30 nm). The size and morphology of the nanoparticles are influenced by the reaction time, temperature, ratio of the reactants, nature of the solvent, and precursors.

2.2.3 MICROEMULSION

Microemulsions are thermodynamically stable dispersions of two immiscible phases stabilized by an interfacial layer of surfactant molecules. Microemulsion routes, particularly water-in-oil (w/o), are employed in the shape and size-controlled synthesis of iron oxide nanoparticles. In principle, the mixing of two identical w/o microemulsions containing the precursors results in the continuous collision, coalescence, and breaking of microdroplets to form precipitates in the micelles. Nanoparticles formed in the miscelles can be recovered by the addition of solvents such as acetone or ethanol. Liu and Wang[13] have reported a single step microemulsion method for the synthesis of magnetite particles having an average diameter of 10 nm and shape of quasisphere. Liang et al.[14] have demonstrated a two-step microemulsion method to prepare cubic Fe$_3$O$_4$ nanocrystalline particles with size ranging from 100 to 400 nm. They have also studied the influence of various factors such as the molar ratio of water to surfactant, alkali concentration, temperature, initial iron concentration, and aging time on the morphology and the size of the nanoparticles. Vidal-Vidal et al.[15] have reported the synthesis of monodisperse maghemite nanoparticles by microemulsion method using oleylamine as precipitating agent. The synthesized particles were spherical in shape, capped with a monolayer coating of oleylamine, and exhibited a narrow size distribution (3.5 ± 0.6 nm) with high saturation magnetization values. Despite the size control and monodispersity of the magnetic particles, the microemulsion method has the drawback of a relatively low yield.

2.2.4 HYDROTHERMAL SYNTHESIS

Hydrothermal processing involves wet chemical reaction in the presence of solvents (aqueous or non-aqueous) in a closed container under high pressure and temperature conditions. Iron oxide nanoparticles with controlled size, shape, and better crystallinity than those from other processes can be synthesized by this method (Fig. 2.1). Hydrothermal synthesis of Fe_3O_4 nanoparticles, particularly nano-sized crystals, have been reported by several authors. Zheng et al.[16] have proposed the synthesis of ferromagnetic Fe_3O_4 nanoparticles with diameter of ~27 nm by hydrothermal route in the presence of a surfactant, sodium bis(2-ethylhexyl)sulfosuccinate (AOT). Wan et al.[17] have prepared single crystal Fe_3O_4 nanorods with an average diameter of 25 nm and length of 200 nm by a soft-template-assisted hydrothermal process. Zhang et al.[18] have developed a simple hydrothermal route to prepare necklace-shaped superstructures self-assembled by magnetite (Fe_3O_4) nanoparticles coated with polyvinyl pyrrolidone (PVP). Daou et al.[19] have demonstrated the synthesis of magnetite nanoparticles with an average size of 39 nm and good monodispersity by co-precipitation method, followed by hydrothermal treatment at 250°C. They have also reported the synthesis of maghemite nanoparticles by heating magnetite at 300°C for 12 h. In the hydrothermal process, the size and morphology of the nanoparticles can be manipulated by controlling the reaction time, temperature, concentration, and ratios of the reactants, nature of the solvent, and type of precursors.

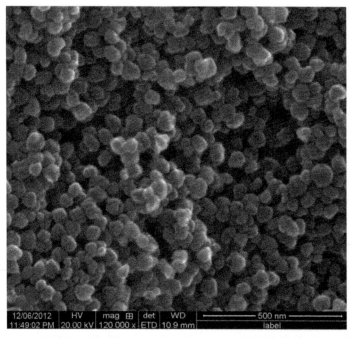

FIGURE 2.1   FE-SEM image of iron oxide (Fe_3O_4) nanoparticles synthesized by hydrothermal method.

The above-mentioned methods have been widely used for the past few decades in the synthesis of iron oxide nanoparticles. Co-precipitation method is found to be the simplest and most efficient route to obtain iron oxide particles, whereas thermal decomposition and hydrothermal methods have the advantage of producing nanoparticles with desired size and morphology. Electrochemical method,[20–22] sonochemical synthesis,[23–26] aerosol/vapor synthesis,[27,28] etc., are also employed in the synthesis of iron oxide nanoparticles.

2.3 SURFACE FUNCTIONALIZATION AND LIPASE IMMOBILIZATION

Magnetic nanoparticles, in the absence of any surface coating, exhibit hydrophobic interactions between the particles and they form large clusters with increased particle size. In the presence of an external magnetic field, these clusters acquire dipole moment that is proportional to the strength of the magnetic field.[29,30] The dipole–dipole interaction between two large particle clusters causes a mutual magnetization, resulting in the formation of aggregated magnetic nanoparticles. Therefore, these nanoparticles are stabilized with surfactants to obtain well-dispersed particles. Inclusion of surfactants during nanoparticle synthesis also helps in avoiding the surface oxidation, manipulates the surface energy, and prevents phase transformations. The surfactants used in the encapsulation of nanoparticles can be polymeric or inorganic materials. Stabilization of iron oxide nanoparticles by polymeric surfactants is achieved by two different ways, that is, the coating can be done after precipitation or it can be added directly to the reaction mixture. Polymeric coating will create steric repulsive forces between the nanoparticles that in turn balance the magnetic and the van der Waals attractive forces acting on them. Thus, the magnetic particles achieve stabilization. Both synthetic and natural polymers are used as surfactants. Natural polymeric materials have attracted attention of researchers as they are renewable, stable, biocompatible, non-toxic, and biodegradable.[31–33]. The most commonly used natural polymeric surfactants include dextran, starch, chitosan, and gelatin. Dextran, a polysaccharide polymer of glucose molecule, is widely used in the polymer of iron oxide nanoparticles, due to its biocompatibility and stability.[34–37] Bautista et al.[38] has demonstrated the synthesis of dextran coated iron oxide (maghemite, γ-Fe_2O_3) nanoparticles by laser pyrolysis and by co-precipitation method. Several authors have reported the synthesis of iron oxide nanoparticles-chitosan composites and its application in lipase immobilization.[39–42] Mikhaylova et al.[33] have reported the synthesis of starch coated SPIONs with an average particle diameter of 6 nm by controlled chemical co-precipitation method. The synthetic polymeric materials include polyethylene glycol (PEG), PVA, PVP, polystyrene, polymethyl methacrylate, etc., and the nanoparticles coated with these materials are reported to have sizes ranging from 10 to 50 nm.[30] Encapsulation with silica is one of the most prominent methods of inorganic coating, as it is chemically inert, biocompatible, and non-toxic. Jang and Lim[29] have demonstrated the synthesis of well-dispersed silica coated Fe_3O_4 nanoparticles with nearly 10 nm diameter magnetic cores and about 2 nm thick silica shells using co-precipitation method. Coating with metallic or non-metallic shells such as gold, silver, platinum, carbon, etc., is another facile method of encapsulating magnetic nanoparticles. For instance, iron nanoparticles possessing a well-defined core-shell structure of gold and silver with size ranging from 18 to 30 nm were successfully synthesized by Mandal et al.[43] However, the single metallic coating of gold, silver, or carbon is reported to have a decreased saturation magnetization (Ms) value.[44] Hence, at most attention should be given in choosing a suitable surfactant for the synthesis and surface modification of any nanoparticle. Encapsulation of nanoparticles with surfactants also offers the possibility of binding various biomolecules to the surface modified iron nanoparticles. Surfactants contain various functional groups, which can be covalently linked to the biomolecules through crosslinking agents such as N-(3-dimethylaminopropyl)-N-ethylcarbodiimide (EDC), N-hydroxysuccinimide (NHS), glutaraldehyde, etc. Kuo et al.[45] have studied the covalent immobilization of *Candida rugosa* lipase on chitosan-coated Fe_3O_4 nanoparticles using EDC and NHS as crosslinking agents. It was hypothesized in their study that the carboxyl groups of lipase could be activated by EDC to form an unstable enzyme—EDC complex. In the presence of NHS, an ester between lipase and NHS would form and it reacts with the amino groups on the Fe_3O_4—chitosan nanoparticles, facilitating the immobilization of the enzyme.

2.4   BIODIESEL PRODUCTION USING LIPASE IMMOBILIZED ON MAGNETIC NANOPARTICLES

Biodiesel, chemically defined as the fatty acid alkyl esters derived from vegetable oils and animal fats, has received considerable attention as an alternative to petroleum based fuels, due to its renewable, biodegradable, non-toxic, and reduced emission properties. A schematic representation of life cycle of biodiesel is provided in Figure 2.2.

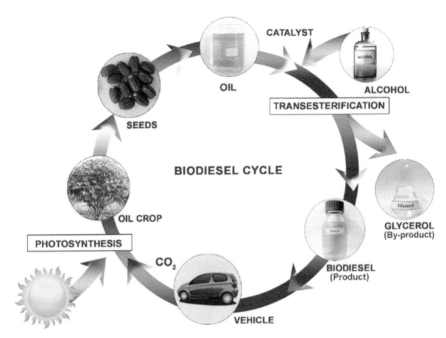

FIGURE 2.2 Schematic representation of life cycle of biodiesel.

Lipases (EC 3.1.1.3) are versatile tools that can catalyze the transesterification reaction between triglycerides and alcohols to produce alkyl esters of long chain fatty acids (biodiesel) and the byproduct, glycerol. Lipase mediated transesterification offers the advantages of increased biodiesel yield, easy recovery of products and by products and environmental friendly method of biodiesel production. Immobilization of lipase onto various supports facilitates enhanced yield of biodiesel along with the cost effectiveness of the biodiesel production through reusability of the enzyme. Recently, researchers have been focusing on the application of SPIONs in lipase immobilization, and several reports have been published on the successful synthesis of biodiesel through this method. Among the various commercially available lipases, *C. rugosa* lipase has been widely used in immobilization onto iron nanoparticles for biodiesel production. Dyal et al.[46] have investigated the stability and activity of *C. rugosa* lipase immobilized on γ-Fe_2O_3 nanoparticles. They have prepared maghemite nanoparticles by sonication of $Fe(CO)$ in decalin and the subsequent annealing of the amorphous Fe_2O_3

nanoparticles. The average size of the synthesized nanoparticles is was 20 ± 10 nm with saturation magnetization value of 61 emu/g. *C. rugosa* lipase was then immobilized on the surface functionalized nanoparticles using glutaraldehyde as the cross-linking agent. They have reported that the immobilized enzyme can be easily separated from the reaction medium, stored, and reused with consistent results. Solanki and Gupta[47] have demonstrated the purification and immobilization of *C. rugosa* lipase on superparamagnetic Fe_3O_4 nanoparticles for transesterification reaction. They have synthesized iron oxide nanoparticles by co-precipitation method followed by surface modification using Tween 80 and polyethyleneimine (PEI). The surface coated nanoparticles were reported to have 10 ± 2 nm size, and there was no aggregation when lipase was immobilized onto the surface modified nanoparticles. Moreover, the immobilized lipase showed 110 times increase in initial rates of transesterification. Xie and Ma[48] have investigated the enzymatic transesterification of soybean oil using lipase immobilized on magnetic iron nanoparticles. In their study, magnetic Fe_3O_4 particles were prepared by co-precipitation of Fe^{2+} and Fe^{3+} ions in an ammonia solution followed by a treatment under hydrothermal conditions. The synthesized particles were reported to have a spherical shape with a mean particle size of 11.2 nm. Lipase from *Thermomyces lanuginosa* was covalently bound on Fe_3O_4 magnetic nano-particles via 1-ethyl-3-(3-dimethylaminopropyl) carbodiimide (EDAC) activation method and binding of lipase to magnetic particles was confirmed by enzyme assays, transmission electron microscopy (TEM), and Fourier transform infrared (FT-IR) spectra. The bound lipase was then used as a biocatalyst for the transesterification soybean oil with methanol to produce biodiesel. The effects of various reaction parameters on the transesterification of soybean oil were also estimated and they have reported a maximal conversion of 94% under optimum conditions. Wang et al.[49] have reported the immobilization of porcine pancreas lipase, *C. rugosa* lipase, and *Pseudomonas cepacia* lipase onto amino-functionalized Fe_3O_4 nanoparticles. Magnetic nanoparticles of 20 nm size and a saturation magnetization of 63 emu g^{-1} was synthesized by chemical co-precipitation method followed by immobilization of lipase enzyme using glutaraldehyde as cross linking agent. Higher percentage of lipase immobilization (80.3–85.2%) and higher enzyme activity (82.6–95.7% of those of the free lipases) was reported in their study. However, they have chosen immobilized lipase from *P. cepacia* for their further transesterification studies in soybean oil as lipase from *P. cepacia* has the greatest methanol resistance among different lipases. Biodiesel production, when performed in a reactor, was reported to be 100% in the first three cycles, and even after nine cycles, the conversion rate remained 44.3%. This higher conversion rate when compared to the 47.60% conversion using free lipase is an indication of the immense potential of magnetic iron oxide nanoparticles in immobilized lipase mediated biodiesel production in industrial scale. Raita et al.[50] have investigated the immobilization of *Thermomyces lanuginosus* lipase on 3-aminopropyl triethyoxysilane modified Fe_3O_4 nanoparticles for biodiesel production from refined palm oil. When the reaction variables, namely, enzyme loading, methanol to FFA molar ratio, water content, and co-solvent to oil ratio, were optimized by central composite design, 97.2% FAME yield was reported in their study and the immobilized enzyme was separated by magnetization and recycled for at least five consecutive batches with > 80% residual activity. Microalgae have been emerged as potential candidates for biodiesel production, due to its high productivity and increased growth rate when compared with other feedstock. Tran et al.[51] have reported biodiesel production from microalgal oil (*Chlorella vulgaris* ESP-31) using *Burkholderia* lipase, immobilized on alkyl-grafted $Fe_3O_4–SiO_2$ nanoparticles. They have stated that the immobilized lipase was tolerant to a higher methanol to oil molar ratio and it can be repeatedly used for six cycles without significant loss of its original activity. Recently, nanoparticles are associated with enzymes using a coupling agent, glutaraldehyde, and the resultant product is called as cross-linked enzyme aggregates (CLEAs). Glutaraldehyde is inexpensive, readily available in commercial quantities, has a low impact on the enzyme activity and greatly improves the enzyme stability.[52,53] Cruz-Izquierdo et al.[54] have developed a method for the preparation of magnetic cross-linked enzyme aggregates (mCLEAs) of *Candida antarctica* lipase B using glutaraldehyde as the cross-linking agent. The resulting mCLEAs were used as biocatalysts in the transesterification of waste frying olive oil, jatropha oil, cameline oil, and crude soybean oil.

Recent advances in synthesis and functionalization of iron oxide nanoparticles have opened wide avenues in enzyme immobilization. High susceptibility to magnetic field, superparamagnetic properties, biocompatibility, and decreased toxicity favors the use of iron oxide nanoparticles as an incredibly potential substrate for enzyme immobilization. Robustness and reusability of these nanobiocatalysts make it advantageous over other enzyme carriers. Most of the above mentioned reports affirm that lipase bound iron oxide nanoparticles can be considered as the most effective catalyst in transesterification reactions and thus, the green synthesis of biodiesel in a cost-effective way is going to become a reality in the near future.

KEYWORDS

- nanoparticles
- iron oxides
- surface functionalization
- enzyme immobilization
- lipase
- biodiesel

REFERENCES

1. Cornelis, K; Hurlburt, C. S. *Manual of Mineralogy;* Wiley: New York, 1977.
2. Cornell, R. M.; Schwertmann, U. *The Iron Oxides: Structure, Properties, Reactions, Occurrences and Uses;* 2nd ed. Wiley-VCH: Weinheim, 2003.
3. Gupta, A. K.; Gupta, M. Synthesis and Surface Engineering of Iron Oxide Nanoparticles for Biomedical Applications. *Biomaterials* **2005,** *26* (18), 3995–4021.
4. Lu, A. H.; Salabas, E. E.; Schüth, F. Magnetic Nanoparticles: Synthesis, Protection, Functionalization, and Application. *Angew. Chem. Int. Ed.* **2007,** *46* (8), 1222–1244.
5. Wu, W.; He, Q.; Chen, H.; Tang, J.; Nie, L. Sonochemical Synthesis, Structure and Magnetic Properties of Air-Stable Fe_3O_4/Au Nanoparticles. *Nanotechnology* **2007,** *18* (14), 145609.
6. Kang, Y. S.; Risbud, S.; Rabolt, J. F.; Stroeve, P. Synthesis and Characterization of Nanometer-size Fe_3O_4 and γ-Fe_2O_3 Particles. *Chem. Mater.* **1996,** *8* (9), 2209–2211.
7. Yoon, S. Preparation and Physical Characterizations of Superparamagnetic Maghemite Nanoparticles. *J. Magn.* **2014,** *19* (4), 323–326.
8. Daou, T. J.; Pourroy, G.; Begin-Colin, S.; Greneche, J. M.; Ulhaq-Bouillet, C.; Legare, P.; Rogez, G. Hydrothermal Synthesis of Monodisperse Magnetite Nanoparticles. *Chem. Mater.* **2006,** *18* (18), 4399–4404.
9. Dumestre, F.; Chaudret, B.; Amiens, C.; Renaud, P.; Fejes, P. Superlattices of Iron Nanocubes Synthesized from Fe $[N (SiMe_3)_2]_2$. *Science.* **2004,** *303* (5659), 821–823.
10. Li, Z.; Sun, Q.; Gao, M. Preparation of Water-Soluble Magnetite Nanocrystals from Hydrated Ferric Salts in 2-Pyrrolidone: Mechanism Leading to Fe_3O_4. *Angew. Chem. Int. Ed.* **2005,** *44* (1), 123–126.
11. Sun, S.; Zeng, H. Size-Controlled Synthesis of Magnetite Nanoparticles. *J. Am. Chem. Soc.* **2002,** *124* (28), 8204–8205.
12. William, W. Y.; Falkner, J. C.; Yavuz, C. T.; Colvin, V. L. Synthesis of Monodisperse Iron Oxide Nanocrystals by Thermal Decomposition of Iron Carboxylate Salts. *Chem. Commun.* **2004,** *20*, 2306–2307.
13. Liu, Z. L.; Wang, X.; Yao, K. L.; Du, G. H.; Lu, Q. H.; Ding, Z. H.; Xi, D. Synthesis of Magnetite Nanoparticles in W/O Microemulsion. *J. Mater. Sci.* **2004,** *39* (7), 2633–2636.

14. Liang, X.; Jia, X.; Cao, L.; Sun, J.; Yang, Y. Microemulsion Synthesis and Characterization of Nano-Fe_3O_4 Particles and Fe_3O_4 Nanocrystalline. *J. Dispers. Sci. Technol.* **2010,** *31* (8), 1043–1049.

15. Vidal-Vidal, J.; Rivas, J.; Lopez-Quintela, M. A. Synthesis of Monodisperse Maghemite Nanoparticles by the Microemulsion Method. *Colloids. Surf A. Physicochem. Eng. Asp.* **2006,** *288* (1), 44–51.

16. Zheng, Y. H.; Cheng, Y.; Bao, F; Wang, Y. S. Synthesis and Magnetic Properties of Fe_3O_4 Nanoparticles. *Mater. Res. Bul.* **2005,** *41* (3), 525–529.

17. Wan, J.; Chen, X.; Wang, Z.; Yang, X.; Qian, Y. A Soft-Template-Assisted Hydrothermal Approach to Single-Crystal Fe_3O_4 Nanorods. *J. Cryst. Growth.* **2005,** *276* (3), 571–576.

18. Zhang, Z. J.; Chen, X. Y.; Wang, B. N.; Shi, C. W. Hydrothermal Synthesis and Self-Assembly of Magnetite (Fe_3O_4) Nanoparticles with the Magnetic and Electrochemical Properties. *J. Cryst. Growth.* **2008,** *310* (24), 5453–5457.

19. Daou, T. J.; Pourroy, G.; Begin-Colin, S.; Greneche, J. M.; Ulhaq-Bouillet, C.; Legaré, P.; Rogez, G. Hydrothermal Synthesis of Monodisperse Magnetite Nanoparticles. *Chem. Mater.* **2006,** *18* (18), 4399–4404.

20. Franger, S.; Berthet, P.; Berthon, J. Electrochemical Synthesis of Fe_3O_4 Nanoparticles in Alkaline Aqueous Solutions Containing Complexing Agents. *J. Solid State. Electrochem.* **2004,** *8* (4), 218–223.

21. Martinez, L.; Leinen, D.; Martin, F.; Gabas, M.; Ramos-Barrado, J. R.; Quagliata, E.; Dalchiele, E. A. Electrochemical Growth of Diverse Iron Oxide (Fe_3O_4, α-FeOOH, and γ-FeOOH) Thin Films by Electrodeposition Potential Tuning. *J. Electrochem. Soc.* **2007,** *154* (3), D126–D133.

22. Pascal, C.; Pascal, J. L.; Favier, F.; Elidrissi Moubtassim, M. L.; Payen, C. Electrochemical Synthesis for the Control of γ-Fe_2O_3 Nanoparticle Size. Morphology, Microstructure, and Magnetic Behavior. *Chem. Mater.* **1999,** *11* (1), 141–147.

23. Kim, E. H.; Lee, H. S.; Kwak, B. K.; Kim, B. K. Synthesis of Ferrofluid with Magnetic Nanoparticles by Sonochemical Method for MRI Contrast Agent. *J. Magn. Magn. Mater.* **2005,** *289*, 328–330.

24. Shafi, K. V.; Koltypin, Y.; Gedanken, A.; Prozorov, R.; Balogh, J.; Lendvai, J.; Felner, I. Sonochemical Preparation of Nanosized Amorphous $NiFe_2O_4$ Particles. *J. Phys. Chem. B.* **1997,** *101* (33), 6409–6414.

25. Kumar, R. V.; Koltypin, Y.; Xu, X. N.; Yeshurun, Y.; Gedanken, A.; Felner, I. Fabrication of Magnetite Nanorods by Ultrasound Irradiation. *J. Appl. Phys.* **2001,** *89* (11), 6324–6328.

26. Vijayakumar, R.; Koltypin, Y.; Felner, I..; Gedanken, A. Sonochemical Synthesis and Characterization of Pure Nanometer-Sized Fe_3O_4 Particles. *Mater. Sci. Eng.* **2000,** *286* (1), 101–105.

27. Alexandrescu, R.; Morjan, I.; Voicu, I.; Dumitrache, F.; Albu, L.; Soare, I.; Prodan, G. Combining Resonant/Non-Resonant Processes: Nanometer-Scale iron-Based Material Preparation via CO_2 Laser Pyrolysis. *App. Surf. Sci.* **2005,** *248* (1), 138–146.

28. Morales, M. P.; Bomati-Miguel, O.; De Alejo, R. P.; Ruiz-Cabello, J.; Veintemillas-Verdaguer, S.; O'Grady, K. Contrast Agents for MRI Based on Iron Oxide Nanoparticles Prepared by Laser Pyrolysis. *J. Magn. Magn. Mater.* **2003,** *266* (1), 102–109.

29. Jang, J. H.; Lim, H. B. Characterization and Analytical Application of Surface Modified Magnetic Nanoparticles. *Microchem. J.* **2010,** *94* (2), 148–158.

30. Teja, A. S.; Koh, P. Y. Synthesis, Properties, and Applications of Magnetic Iron Oxide Nanoparticles. *Prog. Cryst. Growth. Ch.* **2009,** *55* (1), 22–45.

31. Berry, C. C.; Wells, S.; Charles, S.; Aitchison, G.; Curtis, A. S. Cell Response to Dextran-Derivatised Iron Oxide Nanoparticles Post Internalisation. *Biomaterials* **2004,** *25* (23), 5405–5413.
32. Kim, D. K.; Mikhaylova, M.; Wang, F. H.; Kehr, J.; Bjelke, B.; Zhang, Y.; Muhammed, M. Starch-Coated Superparamagnetic Nanoparticles as MR Contrast Agents. *Chem. Mater.* **2003,** *15* (23), 4343–4351.
33. Mikhaylova, M.; Kim, D. K.; Bobrysheva, N.; Osmolowsky, M.; Semenov, V.; Tsaka-lakos, T.; Muhammed, M. Superparamagnetism of Magnetite Nanoparticles: Dependence on Surface Modification. *Langmuir.* **2004,** *20* (6), 2472–2477.
34. Gamarra, L. F.; Brito, G. E. S.; Pontuschka, W. M.; Amaro, E.; Parma, A. H. C.; Goya, G. F. Biocompatible Superparamagnetic Iron Oxide Nanoparticles Used for Contrast Agents: A Structural and Magnetic Study. *J. Magn. Magn. Mater.* **2005,** *289,* 439–441.
35. Laurent, S.; Nicotra, C.; Gossuin, Y.; Roch, A.; Ouakssim, A.; Vander Elst,; Muller, R. N. Influence of the Length of the Coating Molecules on the Nuclear Magnetic Relaxivity of Superparamagnetic Colloids. *Phys. Status. Solid.* **2004c,** *1* (12), 3644–3650.
36. Lee, K. M.; Kim, S. G.; Kim, W. S.; Kim, S. S. Properties of Iron Oxide Particles Prepared in the Presence of Dextran. *Korean. J. Chem. Eng.* **2002,** *19* (3), 480–485.
37. Paul, K. G.; Frigo, T. B.; Groman, J. Y.; Groman, E. V. Synthesis of Ultrasmall Superparamagnetic Iron Oxides Using Reduced Polysaccharides. *Bioconjug. Chem.* **2004,** *15* (2), 394–401.
38. Bautista, M. C.; Bomati-Miguel, O.; Zhao, X.; Morales, M. P.; Gonzalez-Carreno, T.; de Alejo, R. P.; Veintemillas-Verdaguer, S. Comparative Study of Ferrofluids Based on Dextran-Coated Iron Oxide and Metal Nanoparticles for Contrast Agents in Magnetic Resonance Imaging. *Nanotechnology* **2004,** *15* (4), S154.
39. Ren, Y.; Rivera, J. G.; He, L.; Kulkarni, H.; Lee, D. K.; Messersmith; Facile, P. B. High Efficiency Immobilization of Lipase Enzyme on Magnetic Iron Oxide Nanoparticles via a Biomimetic Coating. *BMC Biotechnol.* **2011,** *11* (1), 63.
40. Wu, Y.; Wang, Y.; Luo, G.; Dai, Y. In situ Preparation of Magnetic Fe_3O_4-Chitosan Nanoparticles for Lipase Immobilization by Cross-Linking and Oxidation in Aqueous Solution. *Bioresour. Technol.* **2009,** *100* (14), 3459–3464.
41. Kuo, C. H.; Liu, Y. C.; Chang, C. M. J.; Chen, J. H.; Chang, C.; Shieh, C. J. Optimum Conditions for Lipase Immobilization on Chitosan-Coated Fe_3O_4 Nanoparticles. *Carbohydr. Polym.* **2012,** *87* (4), 2538–2545.
42. Liu, T.; Zhao, L.; Sun, D.; Tan, X. Entrapment of Nanoscale zero-Valent iron in Chitosan Beads for Hexavalent Chromium Removal from Wastewater. *J. Hazard. Mater.* **2010,** *184* (1), 724–730.
43. Mandal, M.; Kundu, S.; Ghosh, S. K.; Panigrahi, S.; Sau, T. K.; Yusuf, S. M.; Pal, T. Magnetite Nanoparticles with Tunable Gold or Silver Shell. *J. Colloid. Interface. Sci.* **2005,** *286* (1), 187–194.
44. Wu, W.; He, Q.; Jiang, C. Magnetic Iron Oxide Nanoparticles: Synthesis and Surface Functionalization Strategies. *ChemInform.* **2009,** *40* (24).
45. Kuo, C. H.; Liu, Y. C.; Chang, C. M. J.; Chen, J. H.; Chang, C.; Shieh, C. J. Optimum Conditions for Lipase Immobilization on Chitosan-Coated Fe_3O_4 Nanoparticles. *Carbohydr. Polym.* **2012,** *87* (4), 2538–2545.
46. Dyal, A.; Loos, K.; Noto, M.; Chang, S. W.; Spagnoli, C.; Shafi, K. V.; Ulman, A.; Cowman, M.; Gross, R. A. Activity of Candida Rugosa Lipase Immobilized on $\gamma\text{-}Fe_2O_3$ Magnetic Nanoparticles. *J. Am. Chem. Soc.* **2003,** *125* (7), 1684–1685.

47. Solanki, K.; Gupta, M. N. Simultaneous Purification and Immobilization of Candida Rugosa Lipase on Superparamagnetic Fe$_3$O$_4$ Nanoparticles for Catalyzing Transesterification Reactions. *New J. Chem.* **2011,** *35* (11), 2551–2556.

48. Xie, W.; Ma, N. Enzymatic Transesterification of Soybean Oil by Using Immobilized Lipase on Magnetic Nano-Particles. *Biomas. Bioener.* **2010,** *34* (6), 890–896.

49. Wang, X.; Dou, P.; Zhao, P.; Zhao, C.; Ding, Y.; Xu, P. Immobilization of Lipases onto Magnetic Fe$_3$O$_4$ Nanoparticles for Application in Biodiesel Production. *Chem. Sus. Chem.* **2009,** *2* (10), 947–950.

50. Raita, M.; Arnthong, J.; Champreda, V.; Laosiripojana, N. Modification of Magnetic Nanoparticle Lipase Designs for Biodiesel Production from Palm Oil. *Fuel Process. Technol.* **2015,** *134,* 189–197.

51. Tran, D. T.; Yeh, K. L.; Chen, C. L.; Chang, J. S. Enzymatic Transesterification of Microalgal Oil from *Chlorella vulgaris* ESP-31 for Biodiesel Synthesis Using Immobilized *Burkholderia* Lipase. *Bioresour. Technol.* **2012,** *108,* 119–127.

52. Cipolatti, E. P.; Silva, M. J. A.; Klein, M.; Feddern, V.; Feltes, M. M. C.; Oliveira, J. V.; de Oliveira, D. Current Status and Trends in Enzymatic Nanoimmobilization. *J. Mol. Catal. B Enzym.* **2014,** *99,* 56–67.

53. Sheldon, R. A.; Van Pelt, S. Enzyme Immobilisation in Biocatalysis: Why, What and How. *Chem. Soc. Rev.* **2013,** *42* (15), 6223–6235.

54. Cruz-Izquierdo, Á.; Picó, E. A.; López, C.; Serra, J. L.; Llama, M. J. Magnetic Cross-Linked Enzyme Aggregates (mCLEAs) of *Candida Antarctica* Lipase: An Efficient and Stable Biocatalyst for Biodiesel Synthesis. *PLoS One.* **2014,** *9* (12), e115202.

CHAPTER 3

DOPING EFFECTS IN WIDE BAND GAP SEMICONDUCTOR NANOPARTICLES: LATTICE VARIATIONS, SIZE CHANGES, WIDENING BAND GAPS BUT NO STRUCTURAL TRANSFORMATIONS!

P. M. G. NAMBISSAN*

Applied Nuclear Physics Division, Saha Institute of Nuclear Physics, Kolkata 700064, India

Corresponding author. E-mail: pmg.nambissan@saha.ac.in; pmgnambissan@gmail.com

CONTENTS

ABSTRACT

The conventional experimental techniques leave a number of lacunae in their limited success while trying to explore the fascinating properties of the nanomaterials and nanosystems. The defects are an area that is often untouched for discussion as if their role in governing the structure and properties of nanomaterials is inconsequential and it is the ultimate outcome which is of sole importance. Here positron annihilation spectroscopy is being introduced as a versatile nuclear experimental spectroscopic probe that can reflect about the properties of the defects in nanomaterials and it is shown that the technique can offer solutions to a number of issues related to the exotic features nanomaterials exhibit compared to their bulk counterparts. For example, the negative vacancies in wide band gap semiconductor nanomaterials can be precisely sensed by the positrons and, by monitoring their transformation in terms of size and concentration, can provide information on their structure, phase, and composition. Positrons can also sensitively probe the structural and phase transformations in spinel nanomaterials and their success is established through a number of experimental data illustrating the smooth and qualitative variation of the positron lifetime or its intensity. Apart from these, the success of coincidence Doppler broadening measurements in the identification of the momentum distribution of electrons in a complex nanomaterial system is also briefly discussed. The aim of this chapter is to highlight the sensitivity and success of a non-conventional defect-investigative probe for understanding nanomaterials and following their evolution under varying experimental conditions such as crystallite size variation, heat treatment and doping.

3.1   INTRODUCTION

The studies of nanomaterials should encompass the four important and fundamental aspects, namely, their synthesis, characterization, property analysis, and applications. There are innumerable techniques also that are nowadays easily accessible to users. The judicious selection of the most desirable one depends on one's interest and the particular issue concerning the nanomaterial objects that the investigator wants to address. For example, the popular methods of synthesis of nanomaterials such as ball-milling, sol–gel method, hydrothermal precipitation, solvothermal route, ion implantation, etc., each has advantages and disadvantages that can be partly overcome by a different choice but still not fully and exactly complementing.

The techniques normally in use for characterizing the nanoparticles are X-ray diffraction (XRD), high-resolution electron microscopy (HREM), atomic force microscopy (AFM), etc. However, bigger challenges come when one has to focus upon a particular property of the nanomaterial system concerned. Techniques like optical absorption, photoluminescence studies, resistivity, magnetic susceptibility measurements, etc., often tend to focus upon the properties that mutually exclude one another. The issues related to the applications of nanomaterials like for device fabrication in technological and industrial areas are just kept outside the scope of this chapter.

Despite whichever method of synthesis is adopted, several properties of nanomaterials are also related to the high concentration of defects at the grain surfaces and interfaces. Defects in material systems, especially of nanoscale dimensions, are to be extensively studied for want of information concerning their correlation to the measured properties. For example, vacancy type defects have been proved to have a direct bearing on the magnetic properties of several semiconducting nanocrystalline systems.[1,2] Defects in fact play a crucial role in modifying (favorably or adversely) their properties. While the extent of favor or damage defects can offer is significant, the size and concentration of these defects in many cases are too small to be investigated with the available sensitivity of the conventional techniques. Positron annihilation spectroscopy (PAS) is a very useful experimental probe in this context and its reliability has been proved beyond ambiguity in a number of studies recently.[3-5] Positrons by virtue of their ability to get trapped in specific defects and to give the corresponding information about the electronic structure and properties of the solid through measurable quantities of the outgoing annihilation gamma radiation have been in the eye of attraction of physicists and material scientists alike for several decades and recently the potential of the techniques to explore the properties of nanomaterials has been realized and appreciated. An extensive review article is also available in literature on the potential and applicability of this technique for the defect characterization in nanomaterial systems.[6] However, for the sake of completion, a brief outline of the fundamental principles of these techniques is given below.

3.2   POSITRON ANNIHILATION EXPERIMENTS

The existence of a positively charged electron (later known as the positron) was first predicted by P.A.M. Dirac in 1928. It was conceived as the antiparticle of the electron. In 1932, it was discovered by C. D. Andersen in a cloud chamber experiment. Positron has got a mass same as that of the electron but a unit positive charge. When the positron comes into physical contact with an electron, they annihilate and two gamma rays are emitted. Each gamma ray will have energy close to 0.511 MeV, which is the rest mass equivalent of either the electron or the positron. Dirac himself had formulated the basic equation connecting the lifetime of the positron and the density of electrons that eventually cause its annihilation. The popular equation has the simplified form for a defect-free homogeneous solid as[7]

$$\tau = \left(\pi r_0^2 c \xi n_e \right)^{-1} \tag{3.1}$$

where $r0$ is the classical electron radius, c is the velocity of light and ne is the density of electrons at the annihilation site. ξ ($=2$ for Al) is an enhancement factor which accounts for the enhancement of the local electron density due to Coulomb interaction with the positron. In metals, for example, it is given by Ref. [8].

$$\xi(r) = 1 + 1.23\,r_s + 0.8295\,r_s^{3/2} - 1.26\,r_s^2 + 0.3826\,r_s^{5/2} + 0.167\,r_s^3 \tag{3.2}$$

where $r_s = (4\,\pi\,n(r)/3)^{-1/3}$ is the radius of a fictitious sphere containing one conduction electron.

In a solid containing a given type of positron trapping sites (such as defects or interfaces), the annihilation of positrons will take place according to the rate equations

$$\frac{dn_b}{dt} = -y_b n_b - k_d n_b \tag{3.3a}$$

and

$$\frac{dn_d}{dt} = -y_d n_d + k_d n_b \tag{3.3b}$$

where nb and nd are the number of positrons present in the bulk (free positrons) and in the defects (trapped positrons), respectively. The positron lifetime spectrum is the addition of the two solutions. The final equation is

$$N(t) = n_b + n_d = I_1 e^{-t/\tau 2} + I_2 e^{-t/\tau_2} \tag{3.4}$$

Here, τs represent the positron lifetimes and Is the respective intensities. Additional terms will be present in the case of solids containing more kinds of defects such as vacancy clusters, large surfaces, and cavities. The positron lifetimes and the relative intensities in a sense reflect information on the sizes and concentrations of the defects present in the solid. Physical inter- pretations of the experimental results often become real challenges as there are only a finite number of variables (i.e., the lifetimes and their intensities), which can increase, decrease or remain unchanged under variations of the experimental conditions.

While the positron lifetime experiments directly map the electron density profiles of the solid, there are two other characteristic features of the electron–positron annihilation process that can throw light on the elec- tron momentum distribution around the positron trapping sites. The angular correlation and Doppler shift in the energies of the annihilation radiation are directly related to the momentum of the electrons. The positron emitted from the radioactive nuclei, after entering the solid, will lose its kinetic energy by dissipative processes like ionization, electronic excitation, elec- tron-hole pair creation, phonon interaction, etc. The probability of annihi- lation with an electron is higher at very low velocities of the positron, as given by the equation,

$$\sigma(v) = \frac{\pi \gamma_0^2 c}{v}$$
(3.5)

Hence, prior to annihilation, the kinetic energy and linear momentum of the thermalized positron are negligible compared to those of the electron. Conservation of linear momentum and energy then will imply that the two gamma rays deviate from anticollinearity by an angle

$$\theta = \frac{p_z}{m_c c}$$
(3.6a)

and suffer a Doppler shift in their energies given by

$$\Delta E = \frac{p_x c}{2}$$
(3.6b)

In the angular correlation spectrum, the annihilations due to the low momentum valence electrons give an inverted parabolic shape to the central region of the spectrum, whereas the result of the annihilations with the high momentum core electrons is a Gaussian at the wings.[7] The cut-off between the two marks the Fermi momentum and is accurately determinable owing to the extremely good spatial resolution one can obtain in this case. However, the source-to-detector distance needs to be large leading to very much reduced count rate, which will be then necessitating the use of very large detectors and considerably large measuring times. These measurements are therefore not carried out in the works being reported here. Compared to the angular correlation studies, Doppler broadening measurements yield rapid information within relatively shorter measuring times and with good statistics. The resolution is, however, severely limited by that of the solid state detector, almost by an order of magnitude poorer than that in angular correlation, but is still reliable where qualitative comparisons are required and sufficient.

3.2.1 POSITRON ANNIHILATION IN NANOMATERIALS

Apart from its known and proven advantages over conventional defect spectroscopic probes, the ability of thermalized positrons to diffuse out and annihilate from the surfaces of crystallites smaller than their thermal diffusion lengths (lth ~20–100 nm in typical solids) make them suitable for the studies of nanocrystalline materials of different sizes, shapes, crystallinity, and morphology. The sensitivity of positrons to the charge state of vacancy type defects is also appropriate for the distinction between anionic and cationic defects in ionic solids and semiconductors. The oxygen vacancies in wide band gap semiconductors like ZnO, MgO, NiO, CdO, etc., are positively charged and hence repel positrons from entering into them. The metallic vacancies, on the other hand, attract them and do so with a trapping rate stronger than that of neutral vacancies. It should be mentioned that neutral vacancies also trap the positrons since the absence of a positive ion core provides an attractive potential well. A divacancy formed by the absence of a neighboring pair of cation and anion, for example, is a case in point and is a trapping site for the positron. In the case of multiple vacancy clusters, the overall charge will be the decisive factor, that is, the number of cationic vacancies should be in excess. Repulsion is the result otherwise.

3.3 EXPERIMENTAL DETAILS

3.3.1 SYNTHESIS OF THE SAMPLES

The samples are basically synthesized through the bottom-up approach, that is, by the normal solvothermal route. The top-down method of ball-milling of the samples is discouraged in the positron annihilation studiesowing to the sensitivity of the technique to the possible contamination effects and the presence of impurities in the samples. On the other hand, purity and a better control on the nanocrystallite size distribution can be achieved by the chemical process of synthesis followed by annealing at the desired temperatures for the required durations. Appropriate amounts of the proper reagents were chosen and mixed in the right proportion followed by rigorous stirring till the final precipitate is formed. It was filtered, washed several times in water or ethanol and dried in vacuum in an oven at ambient temperature. Specific details of synthesis pertaining to the respective samples can be found in the references cited.

3.3.2 CHARACTERIZATION

The products were characterized by XRD and the crystallite sizes were calculated by using the standard Scherrer equation[9]

$$d_c = \frac{0.9\lambda}{\beta \cos \theta} \tag{3.7}$$

where β is the full width at half maximum of the peak and θ the Bragg angle. In some cases, the sizes were also determined from HRTEM images.

Further, the lattice constant is an important parameter that needs to be estimated for samples of different crystallite sizes. This is done by combining the Bragg's law of XRD $2\,dhkl \sin \theta = n\lambda$ with the interplanar separation $dhkl = a/(h2 + k2 + l2)1/2$ where h, k, and l are the Miller indices. A relation between the average crystallite sizes and the corresponding lattice constants is often discussed in nanomaterial studies as it bears signature of the strain and relaxation processes that ultimately stabilizes the nanocrystallite sizes within a very narrow distribution.[10–21]

3.3.3 POSITRON ANNIHILATION MEASUREMENTS

Since the samples are in powder form, the normal course of action followed in many studies had been to make pellets and use them to sandwich the positron source. This, however, is not followed here in positron annihilation studies as we need the pellets in thickness sufficient enough to stop all the positrons and the degree of compaction is not found uniform throughout this thickness. Second, there are possibilities of introduction of dislocations that may also trap the positrons, thereby rendering the physical interpretation of the results difficult and cumbersome. The powdered sample is therefore taken in a glass tube with the positron source (see below) embedded at the geometrical center of the column of it. Care is taken to cover the source from all the sides in sufficient thickness so that all the positrons are stopped in their ranges within the samples and none of them reach the glass walls. The glass tube is continuously evacuated through a needle valve using a rotary vane pump to take out the air trapped within the powder and the particles thereafter settle down under their own weight. This also ensures that the source and sample are always under dry and moisture-free conditions.

3.3.4   THE POSITRON SOURCE AND THE EXPERIMENTAL SETUP

The positron lifetime in principle is measured as the time interval between the emission of the positron from a radioactive source and its subsequent annihilation with an electron of the sample of interest. While (one of) the annihilation gamma rays will serve as the end signal, the start signal needs to be derived from a concomitant process so that the positron can enter and interact with the sample electrons. Fortunately, a versatile radioactive source[22] Na exists which decays to the first exited state of its daughter nucleus[22] Ne (Fig. 3.1) and the latter immediately (within ~1–2 ps) de-excites to the ground state by releasing the excess energy of 1.276 MeV as a gamma ray. The detection of this and one of the annihilation gamma rays enables the measurement of the positron lifetime using a slow-fast gamma–gamma coincidence spectrometer. The schematic diagram of the setup is shown in Figure 3.2. It consists of two BaF2 scintillators coupled with XP2020Q photomultiplier tubes serving as the radiation detectors and the associated nuclear electronics. The time resolution of this set up is ~170 ps (full width at half maximum). For coincidence Doppler broadening measurements, two high pure Germanium (HPGe) detectors are used. They had resolutions 1.27 and 1.33 keV at 0.511 MeV. The schematic diagram of the experimental setup is as shown in Figure 3.3. About a million coincidence counts are collected in each positron lifetime spectrum whereas the number of coincidence events generated in each CDB spectrum was about 20 million.

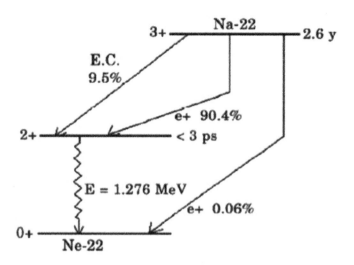

FIGURE 3.1 The decay of the ^{22}Na nucleus.

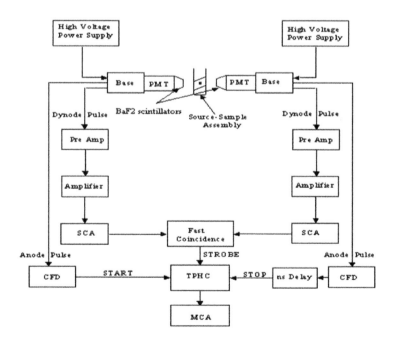

FIGURE 3.2 Schematic diagram of the positron lifetime spectrometer.

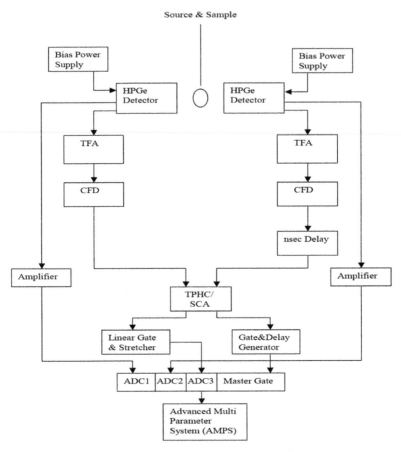

FIGURE 3.3 Schematic diagram of the experimental setup for coincidence Doppler broadening measurements.

3.4 DATA ANALYSIS

The positron lifetime spectrum is a sum of decaying exponentials, the slopes, and intensities of which is resolved using a universally accepted computer program PALSfit.[10] The analysis requires the deconvolution of the instrumental resolution function, which is obtained by fitting the rising part of the spectrum (Fig. 3.4) with a Gaussian function (consisting of one, two, or three Gaussians) and the subtraction of the background. Besides, the contributions coming from the positron source is to be accurately determined and corrected for in all the lifetime data analysis.

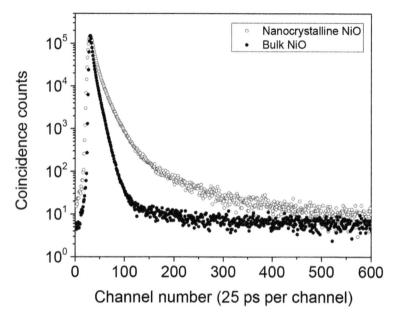

FIGURE 3.4   Peak-normalized positron lifetime spectra of bulk and nanocrystalline nickel oxide.

The source is normally prepared in the form of a deposition of[22] NaCl dissolved in dilute HCl on to a thin (~2 mg.cm^{-2}) Ni foil. The extended portion of the foil is folded to cover the source, the line of folding being well outside the source-deposited area. Inevitably, a fraction of the emitted positrons, while penetrating through the Ni foil, will get annihilated by the Ni electrons with a lifetime ~110 ps.

This fraction is calculated using the equation derived by Bertolaccini and Zappa[11] as given below.

$$I_{foil}(\%) = 0.324Z^{0.93}x^{3.45/z^{0.41}} \tag{3.8}$$

Here Z is the atomic number of the sample and x is the thickness of the foil in mg.cm^{-2}. The lifetimes and intensities of the other two source contributions, that is, from the annihilation of positrons with the electrons of the source material (^{22}NaCl+HCl) and that of the positrons backscattered from the foil-sample interface, are obtained from the positron lifetime spectrum of a reference sample. The reference sample (Al, in this case) should be single crystalline and of purity 99.999% and annealed at a temperature (625°C, in this case) close to its melting point so as to make it completely defect-free and such that, during the analysis of its positron lifetime spectrum, its bulk lifetime (166 ps, in this case) can be fixed and the other components can be treated as the positron lifetimes from the source. These values are to be fed at the appropriate places in the input when the data is analyzed using the PALSfit program.[10]

A typical CDB spectrum is shown in Figure 3.5. The two detectors and amplifiers give the Doppler-shifted gamma ray spectra in the form of counts versus energies $E1 = E0 +/-\Delta E$ and $E2 = E0 -/+ \Delta E$ of the electron-positron annihilation gamma rays in the two opposite directions. The master gate signal generated from the coincidence of the two gamma rays will ensure the rejection of uncorrelated gamma rays being recorded. Still, each of these spectra will inevitably have a high nuclear gamma ray background coming from the Compton absorption of high energy gamma rays by the detectors. There will be also the finite broadening due to the energy resolutions associated with the detectors. $E0$ $0.511 \neq$ MeV corresponds to the annihilation gamma ray events which got recorded in channels on either side of the peak due to the resolution effects, that is, incomplete charge collection (<0.511 MeV) and summation (>0.511 MeV).12 Therefore $E1 + E2 = 2E0 = 1.022$ MeV ensures that there is no loss and gain in amplitude of the signals due to these effects and the corresponding distribution of $E1 - E2 = 2\Delta E$ along $E1 + E2 = 1.022$ MeV will give the exact distribution of the electron momentum around the positron annihilation sites, according to eq 3.6(b), free of the effects of background and resolutions. A two-parameter spectrum is therefore generated with $E1 + E2$ and $E1 - E2$ in the two coplanar perpendicular axes and the coincidence events accordingly distributed (Fig. 3.5). The analysis of the CDB spectra is done by collecting the events within a window of a finite width (1.022 ± 0.0024 MeV) in order to improve the spectral statistics and also to the account for the binding energies of the electrons.[12] Further, in order to magnify the differences in the electron momentum distribution that may be too small to reflect with precision, it has been a practice routinely followed by many researchers to generate ratio curves, also known as quotient spectra, by dividing the (2) ΔE distribution spectra with a spectrum similarly obtained for a reference sample. An example is presented in Figure 3.6. The same Al sample mentioned in the previous paragraph has been used as reference for analyzing the CDB spectrum of ZnO and the characteristic peak at $10 \times 10-3$ m0c is identified as due to positron annihilation with the $2p$ electrons of oxygen ions. This means positrons are getting trapped into the zinc vacancies, which are surrounded by the nearest neighbor oxygen ions.

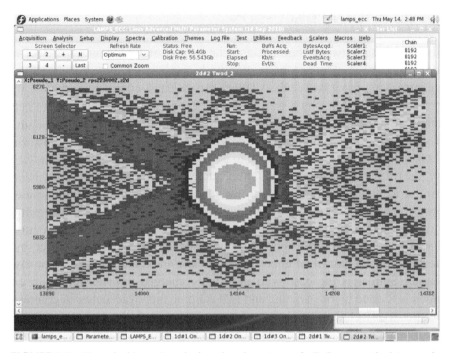

FIGURE 3.5 The coincidence Doppler broadened spectrum of a ZnO nanoparticulate sample.

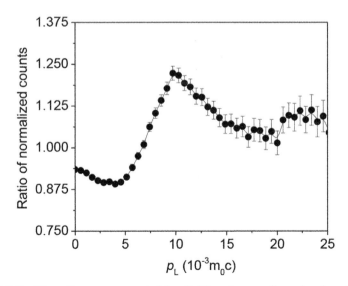

FIGURE 3.6 The ratio curve generated by dividing the one-dimensional projection of the CDB spectrum of a ZnO sample by that of pure and well annealed Al single crystalline sample.

3.5 RESULTS AND DISCUSSION

Distinguishing between the bulk and the nanocrystalline counterparts of a given material is a task to be treated with caution, for there can be other accompanying processes such as lattice expansion or contraction (mainly in metals), structural or phase transformation (common in spinels), and other finite size effects (typical in semiconductors) that can camouflage the nature of variation of physical parameters at very small sizes of the nanoparticles. There are also additional facts need to be kept in mind when one has to interpret the results of positron annihilation studies on semiconductors in the bulk form or as nanocrystallites. Due to the repulsion from anionic monovacancies, which are positively charged, the positrons cannot be trapped by them and hence they are insensitive to the effects of cationic substitutions. Still changes can be reflected if positrons are trapped in divacancies formed by the absence of a pair of neighboring cation and anion and hence electrically neutral. Similar is the case with negatively charged trivacancies, each of which is formed by one anionic and two cationic monovacancies, in contrast to a positively charged trivacancy formed by one cationic and two anionic monovacancies. A cationic monovacancy is negatively charged and positrons will be strongly trapped and hence the elemental surroundings can be accurately mapped through CDBS. A distinguishable peak characteristic of the $2p$ electrons of oxygen ions is very commonly observed in the case of wide band gap metal oxide semiconductors. See the illustrations presented later.

3.5.1 THE SENSITIVITY OF POSITRON LIFETIME SPECTRA TO NANOCRYSTALLINITY AND ACCOMPANYING EFFECTS

To begin the discussion with, we present the peak-normalized positron lifetime spectra of two sets of samples—Figure 3.7a depicts the spectra of bulk and the nanocrystalline cadmium sulfide samples and Figure 3.7b shows the spectra of two samples of zinc sulfide which are nanocrystalline but one is pure and the other is with zinc substituted by 30 at% manganese. These are raw experimental data and no smoothening or approximation has been carried out. They thus reveal the extreme sensitivity of the positron annihilation lifetime spectroscopic probe for unraveling the effects of nanocrystallization and physical or chemical changes taking place inside the samples. The distinction between the bulk and the nanocrystalline samples is obvious, for the reasons of positrons encountering with more number of open volume structures such as nanocrystallite surfaces and interfaces. The positrons which are not trapped in any of the defects within the nanocrystal-lites will diffuse over to the crystallite surfaces when the crystallite sizes are less than the thermal diffusion length (lth) of the positrons. We shall discuss about lth a little later. The effects of substitution also get vividly displayed in the positron annihilation characteristics. A very unique reason for positrons to sense the effects of substitution is also related to their diffusion to the surfaces. The ionic radius of Mn^{2+} is 0.8 Å whereas that of Zn^{2+} is 0.73 Å.[13] Thus the increasing substitution of Zn^{2+} ions by Mn^{2+} ions will introduce increasing strain within the nanocrystallites and the strain can be minimized if the substitution process is restricted to the unit cells on the few surface atomic layers. Thus increasing the surface to volume ratio, or in other words decreasing the crystallite sizes, is a direct consequence of the substitution process and the changed electronic environment is displayed in the positron life time characteristics of the spectrum (Fig.3.7b). The size-induced anoma-lous features of NiO nanoparticles and the effects of Mn-substitution have been discussed in detail in two recently published papers.[14,15]

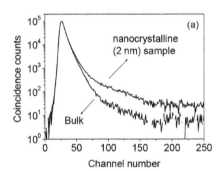 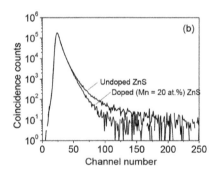

FIGURE 3.7   Peak-normalized positron lifetime spectra of (a) bulk and nanocrystalline cadmium sulfide and (b) undoped and 20 at% manganese-doped zinc oxide.

3.5.2   THE SIZE-DEPENDENT LATTICE EXPANSION OR CONTRACTION EFFECTS

The changes in the lattice constants of solids when their compositional building blocks are reduced to nanometric dimensions constitute a case in point worth examining in detail. In the case of elements and especially in metals, normally a contraction is observed although it is not very uncommon to come across a lattice expansion in very small nanocrystals of some of them. Nanocrystalline niobium, titanium, zirconium, etc., are examples.[16–18] In spinels too, usually the nanocrystallite size reduction has been found accompanied by a lattice contraction,[19,20] although an exception in the case of magnesium ferrite ($MgFeO_4$) has been observed recently.[21] Semiconductor nanocrystals, in particular those of wide band gap metal sulfides and oxides, present distinct but diverse features as far as their lattice responses to nanocrystallization are concerned. Recently this has been the subject of an interesting review by Manuel Diehm et al.[22] and the origin for the lattice expansion or contraction has been examined. It has been mentioned that the negative surface stress is the key reason for lattice expansion and the excess of lattice sums or point defects of various charge states can be excluded from the list of possible explanations. But vacancies can enhance the lattice expansion caused by negative surface stress.

According to the authors of the reference cited above, the fact that ions are surrounded by ions of the opposite charge needs special consideration. If a vacancy is created, the ions on either side can now repel each other. But this has been reported not to cause any substantial increase in the inter–ionic distance but only local relaxation around the vacancy.

Chattopadhyay et al.[16] have treated this problem in an engineering perspective. According to them, the very large network of intercrystallite boundaries in nanomaterials make them favorably moldable for a number of practical applications. In particular, the atoms or ions situated at the surfaces of the nanocrystals get an additional degree of vibration in a direction perpendicular to the surface and can be conceived to have excess free volume associated with them. The fractional excess free volume can be written in terms of the crystallite diameter dc as

$$\Delta V_F = \frac{\left(d_c + \Delta/2\right)^2 - d_c^2}{d_c^2}$$

(3.9)

where Δ is the thickness of the intercrystallite boundary, usually taken as 1 nm. The plot shown in Figure 3.8 shows an asymptotic increase of this quantity at very small sizes of the nanocrystallites.

The authors of the above mentioned paper have further studied the pattern of behavior of the bulk modulus and the negative hydrostatic pressure associated with the nanocrystallites when their sizes are reduced to typically as small as 2 nm.[16] They showed that the bulk modulus varied directly with the crystallite size and decreased very sharply below about 10 nm. They argued that the mechanical stability of the intercrystallite boundaries decreased as the free volume per atom increased. In other words, the crystal lattice no more maintained the rigidity of atomic packing within the nanoparticles and especially toward the surfaces. The variation of the negative hydrostatic pressure—Ph is more interesting. It increases and finally levels off for crystallites of very small sizes, less than ~5 nm. It is treated in the following way. The saturated value of Ph at very small particle sizes correspond to an overall gain in the fractional free volume and the same can be treated to correspond to a lattice expansion. In addition to this, the amount of energy involved in such an expansion is only slightly different from that required for structural transformations to be initiated especially in metallic nanoparticles. The combined effect of a decreasing grain size and an increasing fractional free volume at the grain boundaries would suggest an increase in free energy of the particle or crystallite (for metals), the immediate consequence of which will be a structural or phase transformation to counter the effect so that the structure and compositional stability can be retained.

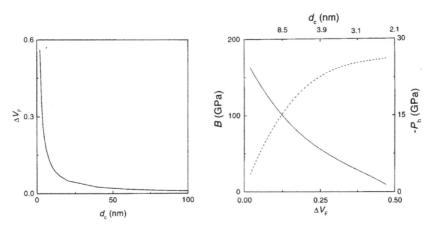

FIGURE 3.8 The variation of the excess free volume (ΔV_F) of the atoms at the grain boundaries as a function of the grain size (d_c). From Ref. [16] figure on the right hand side shows the variation of the bulk modulus B (solid line) and the negative hydrostatic pressure Ph (broken line) as functions of the excess free volume ΔV_F and the grain size d_c.

3.5.3 THE SIZE-DEPENDENCY CAUSED BY SUBSTITUTION AND DOPING

Although the discussion in the previous section was mainly focused on the effect of size variation on the lattice constants of the nanocrystallites, the changes occurring when ions of a certain species are substituted by ions of a different species but with physically interpretable similarities need to be considered. An example of single crystalline zinc oxide (ZnO) nanocones under Mn substitution is discussed here. The ZnO nanocones were synthesized through a solvothermal route. Appropriate amounts of zinc nitrate ($Zn(NO_3)_2 \cdot 6H_2O$), manganese acetate ($Mn(CH_3COO)_2 \cdot 4H_2O$), and sodium hydroxide (NaOH) were dispersed in an ethanol–water solvent. The mixture is then heated in a closed Teflon-lined stainless steel chamber for 12 h at 413 K.[23] A series of samples with varying Mn^{2+} content were synthesized by introducing different concentrations (0.1, 0.5, 1, 6, 10, and 15 at%) of the Mn source. Energy dispersive analysis of X-rays (EDAX) is used in order to determine the actual intake and incorporation of the doped element(s) into the structure of the host material. The amount of Mn actually got substituted into the lattice was almost the entire amount that was initially estimated and added.[23] However, the most striking observation was a gradual decrease of the size of the nanocones with the increase in concentration of Mn. The explanation we could offer was in terms of the mismatch between the replaced Zn^{2+} ion (radius 0.74 Å) the replacing Mn^{2+} ion (radius 0.80 Å). The incorporation of Mn^{2+} in the place of Zn^{2+} will introduce strain in the lattice. To overcome the strain, the lattice needs additional space. As explained earlier, excess free volume is available with the ions residing on the surface layer. Hence, it is easy for the Mn^{2+} ions to replace the Zn^{2+} ions residing in the unit cells of the surfaces of the nanoparticles. As the concentration of Mn^{2+} ions is increased, more such surfaces are required and the material reorganizes itself to have larger surface to volume ratio or, in other words, smaller nanocones.[23]

An interesting comment to be made here is with respect to a similar experiment of Mn-substitution in zinc sulfide (ZnS) nanorods.[24] The original wurtzite (hexagonal) structure immediately transformed into a cubic one on substitution of even 0.1 at% Zn by Mn. At the same time, a similar structural transformation was not observed in the case of ZnS nanoparticles. It appears that even the morphology of the nanostructural material has a deterministic role to play in molding its destiny under external treatments like doping. Surprisingly, in the case of Mn-doped ZnO nanoparticles, no structural transformation till $x = 35$–40 at% was observed. Beyond this level of concentration, phases like $ZnMn_2O_4$ are found to be formed in tracer amounts.[25] XRD has confirmed these findings.

3.5.4 QUANTUM CONFINEMENT EFFECTS AND POSITRON ANNIHILATION STUDIES

Quantum confinement effects manifest in the form of a nearly discrete energy band structure in the case of metal nanoparticles.[14,26] In a paper published by Wood and Ashcroft as early as in 1982 (before the word nanoparticle caught the attention of scientists and researchers),[27] it has been shown that a nanoparticle can behave like a "big atom" with its energy band becoming nearly discrete in nature at a temperature depending purely on the size of the nanoparticle or, in other words, the number N of the atoms or molecules constituting the nanoparticle. The temperature (T) corresponding to the onset of quantum confinement is directly related to N as

$$k_B T = \frac{\varepsilon_F}{N} \qquad (3.10)$$

where ε_F is the Fermi energy of the metal and kB is the Boltzmann constant. This theory has been able to explain the finite size effects in Ag nanoparticles, which exhibit semiconductor-like behavior at extreme low nanometric composition.[28,29] In semiconductor nanoparticles, the effect of quantum confinement is often observed as a characteristic blue shift in the UV or optical absorption spectra of samples of decreasing nanoparticle sizes.[26–30] Our observation in the case of CdS nanoparticles is shown in Figure 3.9. The band gap energies obtained for bulk CdS and samples of crystallite sizes 7.5, 4.5, and 2.5 nm are 2.4, 2.5, 2.6, and 3.1 eV.

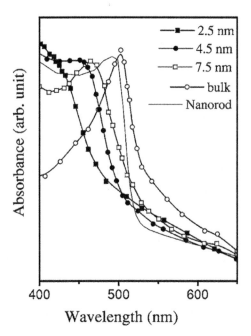

FIGURE 3.9 The optical absorption spectra of the CdS samples. From Ref. [26].

The criterion for observation of such blue shifts in semiconductor nanoparticles is slightly different from that for metal nanoparticles. The exciton Bohr radius is the determining parameter and the nanoparticle sizes (here their radii) have to be below this for a distinct increase in the band gap value to be seen when measured. Brus et al.[31] have given the following expression for the exciton Bohr radius which has been nowadays popularly used.

$$E_{g(nano)} - E_{g(bulk)} = \frac{\hbar^2 \pi^2}{2er^2}\left(\frac{1}{m_e m_0} + \frac{1}{m_h m_0}\right) - \frac{1.8e}{4\pi\varepsilon\varepsilon_0 \gamma} \tag{3.11}$$

where r represents the corresponding radius and m_0 is the electron mass. In our calculation for NiO nanoparticles, we have used the relative effective masses of the electron (m_e) and hole (m_h) as 0.7 and 0.8[32,33] and it led to a particle size of 9.3 nm for the onset of quantum confinement effects. Significantly it agreed with the widening of band gaps in particles below this size (Fig. 3.10). We also observed a further increase of the band gap below a particle size of about 5 nm. Fernandez-Garcia and Rodriguez[34] had categorized such effects as weak, intermediate and strong depending on whether the particle size of the sample under investigation is greater than, equal or less than the exciton Bohr radius r. They suggested a model, according to which, the exciton radius should be related to the reduced mass of the electron–hole pair through the relation

$$r = \frac{\varepsilon \hbar^2}{\mu e^2} \tag{3.12}$$

where ε is the dielectric constant of the metal oxide semiconductor. However, in many cases, this model gives an extremely low value for the excitonic Bohr radius and more refinements are required to satisfactorily explain the onset of quantum confinement effects in such materials. Our recent studies on NiO nanoparticles and discussions on the effects of Mn-substitution are published elsewhere.[14,15]

Investigations based on positron annihilation measurements had been carried out recently on titanium dioxide (TiO_2) nanoparticles. The sensitivity of the techniques to the anatase to rutile structural transformation had been clearly brought out in this work.[35] Li-doping effects in ZnO have shown the retention of zinc vacancies in excess which in turn, helped to sustain long range high T_c ferromagnetism over a wide range of concentrations.[1,2]

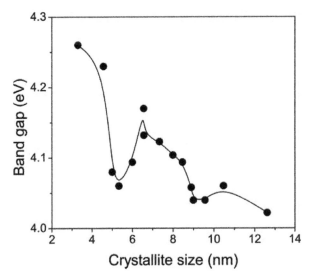

FIGURE 3.10 The band gap estimated from the optical absorption spectra versus average particle sizes of the crystallites of the different samples. (Reprinted with permission from Das Anjan; Mandal,; Atis Chandra; Roy, Soma; Nambissan P. M. G. Positron Annihilation Studies of Defects and Fine Size Effects in Nanocrystalline Nickel Oxide. *J. Exp. Nanosci.* **2015**, *10*, 622–639. © 2015 Taylor & Francis.).

The positron annihilation studies on several such semiconductor nanocrystalline samples did indicate the presence of vacancy type defects in all of them and the following question naturally is then to be answered. What prompts the nanocrystalline materials to choose a structurally imperfect configuration than a perfect one? The answer has two parts. Is stability an utmost criterion for the independent existence of a nanomaterial system? If so, why do stable nanocrystalline materials contain a variety of imperfections such as defects, surfaces and interfaces? A perfect solid need not necessarily be the most stable one and as a corollary, the most stable configuration of a material especially at nanoscale may not be a perfect one. To put it more straightforward, defects are essentially to be considered as an inherent and integral aspect of any nanocrystalline material and they may have consequential effects on its numerous physical properties. Talking about the promises held by positron annihilation studies in this context, we could point out its success in a number of avenues like the investigations of structural, phase and morphological transformations, doping effects, quantum size effects, amorphization, and intermetallic compound formation and even in the understanding of novel physical systems like graphene and its composite derivatives. Some of them are discussed in the references.[36–40]

3.5.5   COMPARISON WITH NANOSPINELS

Before concluding this paper, a striking contrast is to be pointed out with regard to the studies on nanospinels. The behavior of compounds with structure of the form $A(D_{1-x}T_x)B(D_xT_{2-x})O_4$ and especially of nanocrystalline composition has been the subject of several investigations in the recent past.[41,42] The unit cell of this structure is basically a cubic one with lattice constant a but has alternative octants (one-eighth with side $a/2$) occupied by divalent (D) and trivalent (T) cations at tetrahedral (A—) and octahedral (B—) sites with coordinated oxygen (O) ions, four and six, respectively. There are altogether eight tetrahedral and 16 octahedral sites in a unit cell and if they are occupied by the respective number of divalent and trivalent cations, it is termed as a normal spinel. (The number of coordinated oxygen ions is 32 in a unit cell, which means they share their bonds with both the types of ions.)[43,44]

Under certain very specific experimental conditions, a normal spinel can undergo inversion and this is a process in which the eight D ions from the A-sites move over to the B-sites in exchange of eight of the 16 T ions moving to the A-sites. The rest eight T ions continue to be at their original B-sites. The inverted spinels exhibit properties markedly different from their normal counterparts and this has made experiments involving such inversions interesting from both the physical and technological viewpoints.

The factors prompting such inversions include the nanoparticle size, temperature, doping concentration, etc. As an example, we have illustrated in Figure 3.11 the results of positron lifetime measurements in nanocrystalline $Zn_{1-x}Ni_xFe_2O_4$.[45] Bulk (coarse-grained) $ZnFe_2O_4$ is normal spinel while $NiFe_2O_4$ is an inverse one. When reduced to nanocrystalline composition, they get inverted. When Zn^{2+} of $ZnFe_2O_4$ is gradually replaced by Ni^{2+}, a clear and distinct transformation to the spinel configuration of the opposite kind is observed. XRD and Mossbauer spectroscopy normally help in supporting these findings.[46] The flow-chart shown in Figure 3.12 summarizes these processes and also ascertains that these processes are reversible and interchangeable to a great degree of success.

The observation from a more recent work on Ca-substituted $MgFe_2O_4$ is shown in Figure 3.13. Here the mean lifetime of positrons plotted as a function of the concentration of Ca^{2+} ions steadily decreases indicating that the structure gradually transforms from its mixed spinel configuration to a denser lattice (which can be proved to be orthorhombic) as a result of substitution and by virtue of its nanocrystalline character.[47] In one of our earlier studies, we had come across a conspicuous decrease of the positron lifetime when

γ-Fe_2O_3 (spinel) nanocrystals changed over to the α-phase (orthorhombic)[48] and the present observation in Ca-doped $MgFe_2O_4$ is consistent with it.

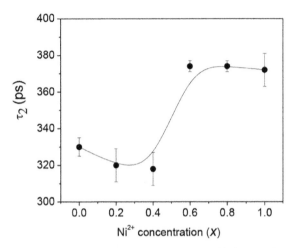

FIGURE 3.11   The variation of the defect-specific positron lifetime τ_2 with the concentration (x) of substituted Ni^{2+} ions in $Zn_{1-x}Ni_xFe_2O_4$ nanocrystals.

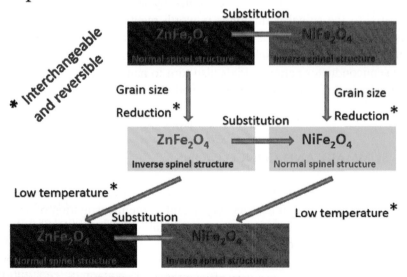

FIGURE 3.12   A schematic diagram to explain the interchangeability of the physical processes leading to the structural inversion in $ZnFe_2O_4$ and $NiFe_2O_4$.

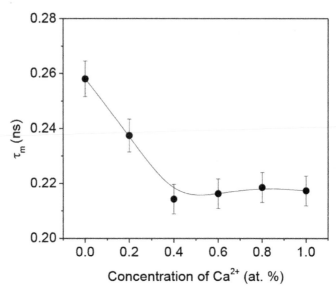

FIGURE 3.13 The positron mean lifetime τ_m versus the concentration (x) of substituted Ca^{2+} ions in $mg_{1-x}Ca_xFe_2O_4$ nanocrystals.

The striking similarity in composition between the nanocrystals of metal oxide semiconductors and spinel-structured mixed oxides is not to be missed. Still the better stability exhibited by semiconductor nanocrystals compared to the spinel compounds is worth mentioning. While structural transformation is almost the accepted method of strain energy release in spinels, the same option is seldom utilized by the semiconductors. However, these semiconductor nanomaterials also turn to nanospinel configuration at higher concentrations of doping, as we had observed in some of our earlier studies.[15–25]

3.6 SUMMARY AND CONCLUSIONS

PAS had been an established tool for the study of vacancy type defects in solids. It is demonstrated here as very useful in the investigation of several interesting physical properties of nanomaterials. A few highlights of this attempt are summarized as follows.

- Doping by cations is a very efficient way of manipulating the properties of semiconductor nanoparticles, although one has to find methods to cope with the changes in the lattice parameters and crystallite sizes that may lead to other finite size effects. Positron annihilation parameters are sensitive not only to the changes in elemental composition but also these concomitant effects.

- Defects, as in the case of Li-substituted ZnO, are found to significantly modify the optical properties, an indirect influence of the finite sizes of nanoparticles.

- In our studies on Eu-doped CeO_2, not discussed in this article, another observation was made in the form of reduction of Ce^{4+} to Ce^{3+} leading to the release of oxygen vacancies and the formation of Ce-vacancy aggregates and the effect is reflected in the variations of the lattice constant and the crystallite sizes too.[49,50]

- As a sensitive and reliable probe to explore defects on an atomic scale, PAS can give valuable complementary support to findings on nanoparticles using the other conventional and more popular experimental techniques such as XRD, TEM, AFM, UV–Vis absorption, etc.

ACKNOWLEDGMENTS

The results reported in this work are taken from the papers published by the author based on collaborative research programs with scientists and researchers of various institutions and it is a pleasure to thank each one of them for their invaluable contributions. Any unintentional omissions are sincerely regretted. Mentions are to be made of S. Chakrabarti, Soumitra Kar, Subhajit Biswas, Tandra Ghoshal, S.K. De (late) Subhadra Chaudhuri, U. Rana and S. Malik (IACS, Kolkata), B. Roy, B. Karmakar, O. Mondal, S. Chakrabarty and M. Pal (CSIR-CMERI, Durgapur), Anjan Das, A.C. Mandal (University of Burdwan, Durgapur), Soma Roy and K. Chakrabarti (SINP, Kolkata), Atul V. Thorat, J.D. Holmes and M.A. Morris (University of Cork, Ireland), S. Chakraverty, Subarna Mitra, S. Sinha, S. Ghosh, Gobinda Gopal Khan and K. Mandal (SNBNCBS, Kolkata), C. Upadhyay and H.C. Verma (IIT, Kanpur) and Ann Rose Abraham, B. Raneesh and Nandakumar Kalarikkal of M.G. University, Kottayam, Kerala.

The author also takes this opportunity to place his deep sense of gratitude to the organizers of ICNM-2014 at Mahatma Gandhi University, Kottayam, Kerala for the excellent hospitality and academic atmosphere extended to him by them during the days of the event.

KEYWORDS

- defects
- nanomaterials
- phase transformation
- positron annihilation
- vacancies

REFERENCES

1. Ghosh, S.; Khan,; Gobinda Gopal,; Mandal, K.; Thapa,; Samudrajit and Nambissan, P. M. G. Positron Annihilation Studies of Vacancy-Type Defects and Room Temperature Ferromagnetism in Chemically Synthesized Li-doped ZnO nanocrystals. *J. Alloys Comp.* **2014,** *590,* 396–405.

2. Ghosh, S.; Nambissan, P. M. G.; Thapa, S.; Mandal, K. Defect Dynamics in Li Substituted Nanocrystalline ZnO: A Spectroscopic Analysis. *Physica B.* **2014,** *454,* 102–109.

3. Brandt, W.; Dupasquier, A. In Positron Solid State Physics. *Proceedings of the 83rd International School of Physics "Enrico Fermi";* July 14–24, 1981, Lake Como., Villa Monastero., Eds.; Italy: North- Holland, Amsterdam, The Netherlands, 1983; p 710.

4. Dupasquier, A.; Mills, Jr., A. P. Positron Spectroscopy of Solids. Proceedings of the 125th International School of Physics "Enrico Fermi"; Italy, July 6–16, 1993, Lake Como., Villa Monastero., Eds.; IOS Press: Amsterdam, The Netherlands, 1995; p 780.

5. Krause-Rehberg, R.; Leipner, H. S. Positron Annihilation in Semiconductors—Defect Studies; Springer: Berlin; Germany, 1999; p 379.

6. Nambissan, P.M.G. Defects Characterization in Nanomaterials Through Positron Annihilation Spectroscopy, In *Nanotechnology: Synthesis Characterization;* Shishir Sinha, N. K., Navani, J. N., Govil., Eds.; Studium Press LLC: Houston, 2013; Vol. 2, p 455–491.

7. Siegel, R. W. Positron Annihilation Spectroscopy. *Ann. Rev. Mater. Sci.* **1980,** *10,* 393–425.

8. Boronski, E.; Nieminen, R. M. Electron-Positron Density-Functional Theory. *Phys. Rev. B.* **1986,** *34,* 3820–3831.

9. Cullity, B. D. Elements of X-Ray Diffraction; Addison-Wesley: Boston, MA,1956; pp 1–531.

10. Olsen, J. V.; Kirkegaard, P.; Pedersen, N. J.; Eldrup, M. PALSfit: A New Program for the Evaluation of Positron Lifetime Spectra. *Phys. Status Solid* C. **2007,** *4,* 4004–4006.

11. Bertolaccini, M.; Zappa, L. Source-Supporting Foil Effect on the Shape of Positron Time Annihilation Spectra. *Nuovo Cimento.* 1967, *52,* 487–494.

12. Asoka-Kumar, P., Alatalo, M., Ghosh, V. J., Kruseman, A. C., Nielsen, B.; Lynn, K. G. Increased Elemental Specificity of Positron Annihilation Spectra. *Phys. Rev. Lett.* **1996,** *77,* 2097–2100.

13. Burns, R. G. Mineralogical Applications of Crystal Field Theory; Cambridge University Press: Cambridge, 1993; 1–551.

14. Das Anjun; Mandal,; Atis Chandra; Roy, Soma; Nambissan P. M. G. Positron Annihilation Studies of Defects and Fine Size Effects in Nanocrystalline Nickel Oxide. *J. Exp. Nanosci.* **2015**, *10,* 622–639.

15. Das Anjun; Mandal, Atis Chandra; Roy, Soma; Nambissan P. M. G. Mn-Doping in NiO Nanoparticles: Defects-Modifications and Associated Effects Investigated Through Positron Annihilation Spectroscopy. *J. Nanosci. Nanotech.* **2015**, *15,* in press. doi. : 10.1166/jnn.2015.10963.

16. Chattopadhyay, P. P.; Nambissan, P. M. G.; Pabi, S. K.; Manna, I. Polymorphic Bcc to Fcc Transformation of Nanocrystalline Niobium Studied by Positron Annihilation. *Phys. Rev. B.* **2001,** *63,* 054107.

17. Manna, I.; Chattopadhyay, P. P.; Nandi, P.; Nambissan, P. M. G. Positron Lifetime Studies of the Hcp to Fcc Transformation Induced by Mechanical Attrition of Elemental Titanium. *Phys. Lett. A.* **2004,** *328,* 246–254.

18. Manna, I.; Chattopadhyay, P. P.; Banhart, F.; Fecht, H.-J. Formation of Face-Centered-Cubic Zirconium by Mechanical Attrition. *Appl. Phys. Lett.* **2002,** *81,* 4136–4138.

19. Nambissan, P. M. G.; Upadhyay, C.; Verma, H. C. Positron Lifetime Spectroscopic Studies of Nanocrystalline $ZnFe_2O_4$. *J. Appl. Phys.* **2003,** *93,* 6320–6326.

20. Chakraverty, S.; Mitra, Subarna, Mandal, K.; Nambissan, P. M. G.; Chattopadhyay, S. Positron Annihilation Studies of Some Anomalous Features of $NiFe_2O_4$ Nanocrystals Grown in SiO_2. *Phys. Rev. B.* **2005,** *71,* 024115.

21. Abraham, Ann Rose; Raneesh, B.; Das D. Kalarikkal,; Nandakumar,; Magnetic Response of Superparamagnetic Multiferroic Core-Shell Nanostructures, AIP Conference Proceedings, 1731, 050151, 2016.

22. Manuel Diehm, P.; Ágoston, Péter; Albe, Karsten. Size-Dependent Lattice Expansion in Nanoparticles: Reality or Anomaly? *Chem. Phys. Chem.* (Special Issue: Nanomaterials). **2012,** *13,* 2443–2454.

23. Ghoshal, Tandra; Kar, Soumitra; Biswas, Subhajit; De, S. K.; Nambissan, P. M. G. Vacancy-Type Defects and Their Evolution Under Mn Substitution in Single Crystalline ZnO Nanocones Studied by Positron Annihilation. *J. Phys. Chem. C.* **2009,** *113* (9), 3419–3425.

24. Kar, S.; Biswas, S.; Chaudhuri, S.; Nambissan, P.M. G. Substitution-Induced Structural Transformation in Mn-doped ZnS Nanorods Studied by Positron Annihilation Spectroscopy. *Nanotechnology* **2007,** *18,* 225606.

25. Roy, B.; Karmakar, B., Nambissan, P. M. G.; Pal, M. Mn Substitution Effects and Associated Defects in ZnO Nanoparticles Studied by Positron Annihilation. *NANO: Brief Rep. Rev.* **2011,** *6* (2), 173–183.

26. Kar, Soumitra; Biswas, Subhajit; Chaudhuri, Subhadra; Nambissan, P. M. G. Finite-Size Effects on Band Structure of CdS Nanocrystallites Studied by Positron Annihilation. *Phys. Rev. B.* **2005,** *72,* 075338.

27. Wood, D. M.; Ashcroft, N. W. Quantum Size Effects in the Optical Properties of Small Metallic Particles. *Phys. Rev. B.* **1982,** *25,* 6255–6274.

28. Roy, B.; Chakravorty, D. Anomalous Resistance Change in Nanocrystalline Metallic Systems. *Solid State Commun.* **1993,** *87,* 71–75.

29. Mukherjee, M.; Chakravorty, D.; Nambissan, P. M. G., Annihilation Characteristics of Positrons in a Polymer Containing Silver Nanoparticles. *Phys. Rev. B.* **1998,** *57,* 848–856.

30. Nanda, J.; Kuruvilla, Beena Annie; Sarma, D. D. Photoelectron Spectroscopic Study of CdS Nanocrystallites. *Phys. Rev. B.* **1999,** *59,* 7473–7479.

31. Brus, L. E. A Simple Model for the Ionization Potential, Electron Affinity, and Aqueous Redox Potentials of Small Semiconductor Crystallites. *J. Chem. Phys.* **1983,** *79,* 5566–5571.

32. Choi, Seung Chul; Koumoto, Kunihito; Yanagida, Hiroaki. Electrical Conduction and Effective Mass of a Hole in Single-Crystal NiO. *J. Mater. Sci.* **1986,** *21,* 1947–1950.

33. Irwin, Michael D.; Servaites, Jonathan D.; Bruce Buchholz, D.; Leever, Benjamin J.; Liu, Jun; Emery, Jonathan D.; Zhang, Ming; Song, Jung-Hwan; Durstock, Michael F.; Freeman, Arthur J.; Bedzyk, Michael J.; Hersam, Mark C.; Chang, Robert P. H.; Ratner, Mark A.; Marks, Tobin J. Structural and Electrical Functionality of NiO Interfacial Films in Bulk Heterojunction Organic Solar Cells. *Chem. Mater.* **2011,** *23,* 2218–2226.

34. Fernández-Garcia, Marcos; Rodriguez, José A. In *Nanomaterials.* Inorganic and Bioinorganic Perspectives; Lukehart, Charles M.; Scott, Robert A. Eds.; Wiley: New York, 2008; 453–512.

35. Ghosh, S.; Khan, Gobinda Gopal; Mandal, K.; Samanta, Anirban; Nambissan, P. M. G. Evolution of Vacancy-Type Defects, Phase Transition and Intrinsic Ferromagnetism during Annealing of Nanocrystalline TiO2 Studied by Positron Annihilation Spectroscopy. *J. Phys. Chem. C.* **2013,** *117,* 8458–8467.

36. Nambissan, P. M. G. Probing the Defects in Nano-Semiconductors Using Positrons. *J. Phys. Conf. Series.* **2011,** *265,* 012019.

37. Nambissan, P. M. G. Nano Sulfide and Oxide Semiconductors as Promising Materials for Studies by Positron Annihilation. *J. Phys. Conf. Series.* **2013,** *443,* 012040.

38. Nandi, P.; Chattopadhyay, P. P.; Nambissan, P. M. G.; Banhart, F.; Fecht, H. J.; Manna, I. Microstructural Aspects and Positron Annihilation Study on Solid State Synthesis of Amorphous and Nanocrystalline Al60_xTi40Six Aalloys Prepared by Mechanical Alloying. *J. Non-Cryst. Solids.* **2005,** *351,* 2485–2492.

39. Mukherjee, A.; Banerjee, M.; Basu, S.; Nambissan. P. M. G.; Pal, M. Gadolinium Substitution Induced Defect Restructuring in Multiferroic BiFeO3: Case Study by Positron Annihilation Spectroscopy. *J. Phys. D. Appl. Phys.* **2013,** *46,* 495309.

40. Rana, Utpal; Nambissan, P. M. G.; Malika, Sudip ; Chakrabarti, Kuntal. Effects of Process Parameters on the Defects in Graphene Oxide–Polyaniline Composites Investigated by Positron Annihilation Spectroscopy. *Phys. Chem. Chem. Phys.* **2014,** *16,* 3292–3298.

41. Nambissan, P. M. G.; Upadhyay, C.; Verma, H. C. Identification of Positron Trapping Sites in Nanocrystalline ZnFe$_2$O$_4$ by Coincidence Doppler Broadening Measurements. *Mater. Sci. Forum.* **2004,** *445–446,* 162–164.

42. Mitra, S.; Mandal, K.; Sinha, S.; Nambissan, P. M. G.; Kumar, S. Size and Temperature Dependent Cationic Redistribution in NiFe2O4(SiO2) Nanocomposites: Positron Annihilation and Mossbauer Studies. *J. Phys. D.* **2006,** *39,* 4228–4235.

43. Smit, J.; Wijn, H. P. J. *Ferrites – Physical Properties of Ferrimagnetic Oxides in Relation to Their Technical Applications;* N. V. Philip's Gloeilampenfabrieken: Eindhoven, Holland, 1959; pp 136–175.

44. Scordari, F. Ionic Crystals. in Fundamentals of Crystallography; Giacovazzo, C. Ed.; Oxford University Press: New York, 1992; pp 403–464.

45. Nambissan, P. M. G.; Mondal, O.; Chakrabarty, S.; Pal, M. Ni-Substitution Induced Inversion in ZnFe$_2$O$_4$ Seen by Positron Annihilation. *Mater. Sci. Forum.* **2013,** *733,* 219–223.

46. Mohan, H.; Shaikh, I. A.; Kulkarni, R. G. Magnetic Properties of the Mixed Spinel $CoFe_{2-x}Cr_xO_4$. *Physica B.* **1996,** *217,* 292–298.

47. Khot, V. M.; Salunke, A. B.; Phadatare, M. R.; Pawar, S. H. Formation, Microstructure and Magnetic Properties of Nanocrystalline MgFe2O4, Mater. *Chem. Phys.* **2012,** *132,* 782– 787.

48. Chakrabarti, S.; Chaudhuri, S.; Nambissan, P. M. G. Positron Annihilation Lifetime Changes Across the Structural Phase Transition in Nanocrystalline Fe2O3, *Phys. Rev. B.* **2005,** *71,* 064105.

49. Thorat, Atul V.; Ghoshal, Tandra; Holmes, Justin D.; Nambissan, P.M.G.; Morris, Michael A. A Positron Annihilation Spectroscopic Investigation of Europium-Doped Cerium Oxide Nanoparticles, *Nanoscale* **2014,** *6,* 608–615.

50. Thorat, A. V.; Ghoshal, T.; Morris, M. A.; Nambissan, P. M. G. Eu-doped Cerium Oxide Nanoparticles Studied by Positron Annihilation. *Acta Phys. Polonica A.* **2014,** *125,* 756–759.

CHAPTER 4

SYNTHESIS OF IONIC LIQUID MEDIATED NANOPARTICLE SYNTHESIS

PRAVEENKUMAR UPADHYAY and VIVEK SRIVASTAVA*

Basic Sciences: Chemistry, NIIT University, NH-8 Jaipur/Delhi Highway, Neemrana 301705, Rajasthan, India

Corresponding author. E-mail: vivek.shrivastava@niituniversity.in

CONTENTS

ABSTRACT

In this chapter, we discuss the recent as well as significant developments in the synthesis of ionic liquid (IL) mediated nanoparticles (NPs). Apart from that, we also discuss a small background of IL considering types, properties, and effect of IL's counterparts on NPs synthesis. There are promising consequences proposing that the IL route can lead the design and fabrication of NPs through the variety of size, nature, configuration, and functionality. All discussed examples in this chapter clearly depicted that ILs, add great value to the area of NP synthesis.

4.1 INTRODUCTION

In concern with greenhouse effect, researchers are looking to create or search less volatile compounds which can replace the conventional, toxic, and volatile organic solvents. The curiosity of researcher had increased over last few decades in the area of ionic liquids (ILs). The chemical and physical properties of ILs make them more interesting as an area of research. Mainly researches are looking to apply ILs as a suitable solvent or as a catalyst in different type's organic transformations as they show non-detectable vapor pressure along with other unique properties like high stability, high density, large liquid range (i.e., more than 300°C), less vapor pressure, high viscosity, and good solubility makes them over conventional solvents. The less volatile nature under ambient conditions like high temperature and pressure, decreases the chances of harmful organic substance emission in the environment.[1]

ILs are entirely made up of poorly coordinated ions (cations/anions). Such chemical nature of IL makes them able to facilitate various fascinating properties which make them a new class of material which can be modified as per the desired requirment.[2] In Table 4.1 some examples of typical cations and anions are given which are commonly used for preparation of ILs.[1]

Imidazolium, pyrollidinium, pyridinium, ammonium, phosphonium, and sulfonium ions are the most commonly used typical cations, also known as heteroatomic cations, while trifluoromethanesulfonate, *Bis* (trifluoromethanesulfonyl) amide, hexafluorophosphate, tetrafluoroburate, and dicyanimide are the most commonly used typical anions. The most common synthetic process for IL synthesis is divided into two parts (1) alkylation and (2) metathesis (Scheme 4.1).[1]

TABLE 4.1 Most Commonly Used Typical Cations and Anions for ILs.[1]

Cations		Anions		
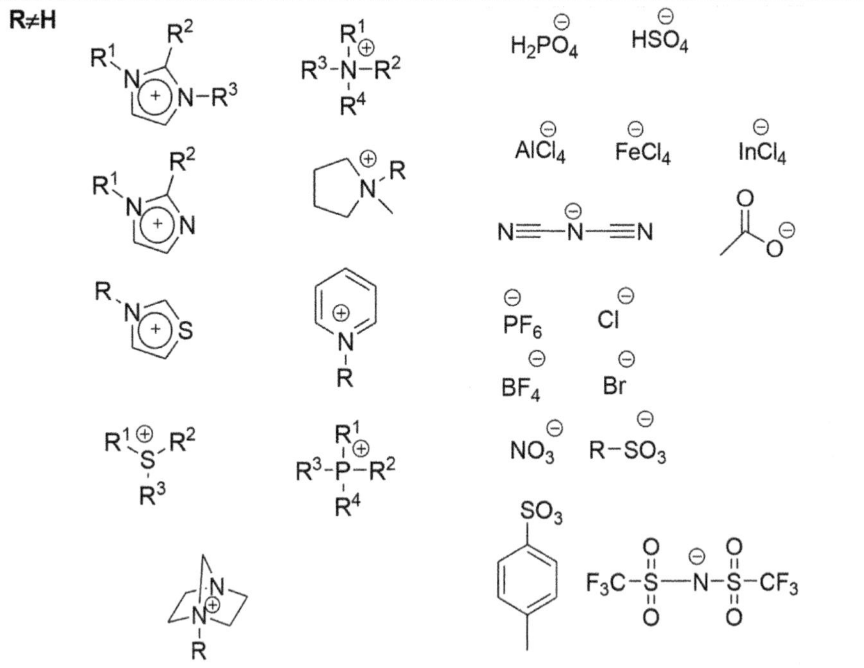 **R≠H**				

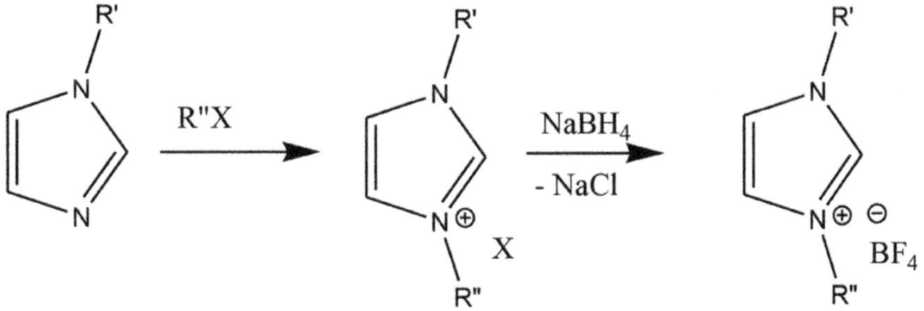

SCHEME 4.1 The most common synthetic route of formation of ILs.[2]

In recent years, the synthesis of metal nanoparticles (NPs) with controlled size and shape is one of the highly important areas of study due to their multiple applications in different segments like material science, organic transformations, and pharmaceutical production. The unique properties of metal NPs can be judged by their composition, structure, shape, size, and their stability. In the area of metal NP synthesis, the major concern of researcher is to create stable, well disperse, and agglomeration free metal NPs with narrow size distribution.[3] The bottom up, the wet chemistry approach is one of the common ways to synthesize metal NPs. Although, in most of the synthesis, the metal NPs are synthesized by the wet chemistry route in aqueous or organic solvent systems, but their properties like density, polarity, solubility, high toxicity, low boiling point, and flammability make them unattractive. At present scenario, researchers are looking ILs as a promising reaction medium for the synthesis of metal NPs,[4–8] due to their unique properties which can overcome the demerits of conventional solvent systems in the synthesis metal NPs.

Various reports have been published, concerning the synthesis of IL mediated NPs.[8] The cationic and anionic species present in ILs are electrostatic in nature, hence they act as electrostatic stabilizers; therefore, in ILs the stable suspension of metal NPs can be synthesized easily.[9] The catalytic rate and selectivity of metal NPs in major organic transformations is higher under IL medium with respect to conventional solvent system.[10] In the ILs, the cations and anions create an unique microenvironment; due to this the desired selective transformation was observed in many reactions.

Hence, we can say that reactions catalyzed by metal NPs are task specific and also the recycling of immobilized NPs is easily possible due to tunable miscibility of ILs. In recent advancement, the IL mediated metal NPs have been employed to various catalytic reactions like, for hydrogenation, oxidation, coupling reactions, etc.[11–13] The good dispersion, high catalytic activity, or selectivity and less agglomeration are the properties of an ideal nano catalytic system. Some of the metal NPs still show negligible dispersion in ILs, poor catalytic activity, and agglomeration in recycling experiment. To overcome these challenges, researchers are working day and night to overcome these problems.

In this chapter, we are going to review IL and its application in metal NPs synthesis. This chapter starts with ILs, their types, and their versatile application focusing NPs synthesis; we also discuss NPs.

4.2 IONIC LIQUIDS

In general, the ILs are defined as salt, which is liquid at room temperature or melts at or below 100°C, having poor coordination of ions, that is, organic cations and inorganic anions. At ambient temperature, ILs are freely flowing liquid. The ILs are unstable crystal lattice because they have one or more than one ion with delocalized charge and having one organic component in its total structure. Among different definitions, four of them are most acceptable.

Definition 1

"The term ionic liquid implies a material that is fluid at (or close to) ambient temperature, is colorless, has a low viscosity, and is easily handled and room temperature ionic liquids [RTILs] are generally salts of organic cations, e.g., tetraalkylammonium, tetraalkylphosphonium, N-alkylpyridinium, 1,3-dialkylimidazolium and trialkylsulfonium cations."[14]

Definition 2

"The most basic definition of a room temperature ionic liquid is a salt that has a melting point at or near room temperature."[15]

Definition 3

"Organic salts with melting points below ambient or reaction temperature."[16]

Definition 4

"Ionic liquid is a salt with a melting temperature below the boiling point of water."[17]

There are many synonyms used like room temperature ionic liquid (RTIL), molten salt, room temperature molten salt, ambient temperature molten salt or IL, task specific ionic liquid (TSIL), liquid organic salt, fused salt, neoteric solvent, etc., to represent IL. In the total structure of ILs, the low symmetry, weak intermolecular interaction, and low charge density make cations bulky, for example, asymmetric ammonium, phosphonium salts, and hetero aromatic. In latest advancement, there are some reports, where inorganic anions have been replaced with organic anions for synthesis of ILs. The nature of cations and anions decides the chemical and physical properties of ILs. The properties of ILs are very unique, which makes ILs to act as solvent for the reactions, electro catalysis, as catalyst, as co-catalyst, as ligand source, and as supporting materials. The properties like relatively low viscosity and, low vapor pressure under ambient condition show its non-volatile nature and immiscibility with numerous organic solvents which makes ILs more promising solvent for chemical transformations.

As compared to conventional solvents, the ILs are quiet viscous (but least viscous at room temperature). The viscosities of ILs are decided by the length of alkyl chain or functionalized alkyl chain. The shorter alkyl chain length makes low viscous ILs, while large alkyl chain increases the viscosity of IL. Also long chain alkyl group provides a low density while increasing the halogen group, it makes high density ILs. The density of ILs is least affected by impurities or temperature. The unsymmetrical cations give the ILs at room temperature, while low melting temperature of ILs was obtained by more weakly coordinating anions. As compared to low boiling point solvents offers much lower toxicity because of negligible volatility. The synthesis and purification of ILs involve organic solvents, which contradicts the greenness of ILs, as the impacts of ILs preparation on the environment are more severe than organic solvent preparation. The recent research study on the greenness of ILs like bioaccumulation, toxicity, and degradability shown that they are not as eco-friendly to the environment as initially predicted. The ILs can be said to be green if the following points can be achieved successfully (1) improvement in synthesis route of ILs, (2) increasing degradability, (3) easy recyclability recovery process, and (4) suitable purification process of ILs.[33]

4.2.1   TYPES OF IONIC LIQUID

The nature of ILs is decided by the type of cations and anions used during its synthesis. Depending upon the group attached to the cations or anions, the IL behaves as an acidic, basic, or neutral. The anionic nature ILs, determines the coordination properties and the acidity of ILs but at some level the cations of ILs also influence the acidity of IL system. The concentrations of cations or anions present in ILs or the combining strength of cations and anions give acidic or basic properties to ILs. The neutral ionic liquid (NILs) can be obtained by matching a wide of anions with the cations components.

The efficiency or productivity of any reaction process substantially influences by the acidity or basicity of the reaction medium, due to this a special attention has been given to acidic or basic ILs as they show a great tendency to increase efficiency of many reaction processes. The basic ionic liquids (BILs) are the combination of ILs and inorganic bases, and they show properties like stability to water and air, high catalytic efficiency and recyclability, and easy separation it makes them a good option to replace conventional catalysts and solvent system. The high selectivity and activity, also recyclability of catalysts, made BILs to replace traditional bases like KOH, NaOH, NaHCO$_3$ NaOAc, trimethylamine, etc. BILs are immiscible with many different organic solvents; also non-volatile and non-corrosive in nature.[23-26] There are many base catalyzed reactions in which BILs are used, like Knoevenagel condensation, Marknovnikov addition, Michael addition, and Mannich reaction. The Bronsted acidic ILs and Lewis acidic ILs are the types of acidic ionic liquid (AILs). In general Bronsted acidic ILs can be prepared by attaching of alkane sulfonic acid or carboxylate acid group to the side chain of the cations, while Lewis acidic ILs are prepared from metal halides and organic halides. The problems like corrosion, residual waste, and environmental issues arise, when using traditional acids and bases, which generated the need of an environmental friendly catalyst. In many organic reactions, the AILs and NILs have been used successfully, while the report related to the BILs shows very rare cases of organic reaction. The NILs [BMIm]BF$_4$ promoted high selective esterification of tertiary alcohols has been achieved by acetic anhydride, while the acidic ILs and basic ILs show less selectivity under similar conditions.[27]

4.2.2 APPLICATIONS OF IONIC LIQUIDS

ILs show great potential for chemical analysis as a solvent. In recent advancements, the ILs have shown various analytical applications in chromatography, extraction, electrochemistry, and sensor technology.[34] There are various applications of ILs as a solvent, catalyst, and as support for several important organic reactions like hydrogenation, hydroformylation, alkoxycarbonylation, cross coupling like Heck, Suzuki, Negishi, Stille reactions, allylic substitution, Friedel–Crafts alkylation, Diels–Alder, Diol or carbonyl protection, Epoxidation/epoxide opening, cyanosilyation of aldehydes, esterification, ring closing metathesis, Knoevenagel condensation, Baylis–Hillman, Robinson annulation, alcohol oxidation Friedlander reaction, nitration of phenols, bromination of alkynes, biocatalysis, and chiral solvent for asymmetric synthesis (Table 4.2). ILs have also been utilized as storage medium for toxic gases, lubricant additives in pigments, and as propellants.

TABLE 4.2 Chemical Application of Ionic Liquid.

Entry	Reaction Name	Ionic Liquid/Solvent/ Catalyst	Substrate 1	Substrate 2	Product	Yield (%)	Reference No.
1.	Diol protection	BF4- (imidazolium H_2N–N–Me)	R_1–C(=O)–R_2	HO–R_3–OH	dioxolane (R_3, R_1, R_2)	100	[18]
2.	Friedlander synthesis	BF4- (imidazolium H_2N–N–Bu)	2-aminoaryl ketone (C(=O)R_3, NH_2)	R_1–C(=O)–CH$_2$–R_2	quinoline (R_3, R_2, R_1, N)	96	[19]
3.	Robinson annulation	PF6- (Me–HN–N–hex) + glycine	NC–CH$_2$–CN	2,6-dichlorobenzaldehyde (Cl, H, C=O, Cl)	Cl-aryl–CH=C(CN)$_2$ ... CN, Cl	86	[20]
4.	Hydrogenation	BF4- (Me–N–N–imidazolium) + [RuCl$_2$- (S)-BINAP]$_2$•NEt$_3$	Ph–CH=...	i–PrOH	Ph–CH(CO$_2$H) (S)	100	[21]

TABLE 4.2 *(Continued)*

Entry	Reaction Name	Ionic Liquid/Solvent/ Catalyst	Substrate 1	Substrate 2	Product	Yield (%)	Reference No.
5.	Diels–Alder reaction	[Me–N⊕N–Bu] PF6- + Sc(OTf)₃				> 99	[22]
6.	Olefin epoxidation	[Me–N⊕N–Bu] PF6- + (R, R)-Jacobsen's Catalyst		–		86	[23]
7.	Heck reaction	[Me–N⊕HN–N–Me] Br-				99	[28]
8.	Wittig reaction	[Me–N⊕N] BF4-	Ph₃P=HC–C(=O)–Me	PhCHO	Ph—C=C—C—Me (H H O)	82	[29]

TABLE 4.2 *(Continued)*

Entry	Reaction Name	Ionic Liquid/Solvent/ Catalyst	Substrate 1	Substrate 2	Product	Yield (%)	Reference No.
9.	Suzuki–Miyaura coupling reaction	[imidazolium BF4- ionic liquid] + Pd(PPh₃)₄	MeO—⟨⟩—Br	⟨⟩—B(OH)₂	MeO—⟨⟩—⟨⟩	81	[30]
10.	Stille reaction	[imidazolium BF4- ionic liquid] + PdCl₂(PhCN)₂	[cyclohexenone structure]	⟍⟍SnBu₃	[product structure]	82	[31]
11.	Friedel–Craft reaction	[imidazolium BF4- ionic liquid] a. + Cu(OTF)₂	OMe ⟨⟩	[benzoyl chloride]	[product with OMe]	p/o ratio 96/4	[25/32]

4.3 NANOPARTICLES

The NPs are composed of identical atoms, molecules, and two or more different species. It is a solid particle with a size ranging from 1 to 100 nm and having specific physical and chemical properties. The NPs have very small size, which means more defects, that is, more edge and corners compared to large particles. The NPs have high catalytic activity in various reactions due to their larger surface area to volume ratio and residual energy. The nano catalyst requires the stabilization with respect to agglomeration to the bulk metal, as they are kinetically quite unstable. The catalyst modulation and catalyst recyclability both are advantages of homogeneous and heterogeneous catalysis, respectively. While combining these two properties the more efficient catalytic system can be achieved. To obtain such catalytic system, the modern nanochemistry is very helpful as it is possible to make soluble analogues of heterogeneous catalyst using this technology.

It is well documented that ILs not only stabilize the metal NPs by providing electrostatic stabilization support, but also give more surface area for the reaction to take place.[35] Therefore, the stability and activity of NPs are controlled by ILs. Thus ILs mediated NPs provide easy recovery of NPs and also improve the total turnover number during the catalytic run. The common transition metals which were studied as NPs are like Ag, Au, Sn, Pd, Pt, Ru, Rh, Ni, and Cu. There are mainly two types of metal NPs such as monometallic and bimetallic.[36] The most commonly used transition metals, which are used for preparation of monometallic NPs catalyst, include mostly Pd, Pt, Cu, Ni, and Rh.[37] Two universal methods are usually applied for the synthesis of metal NPs are as under below.[38]

Method-1: Subdivision of metallic aggregates (physical method).
Method-2: Nucleation and growth of metallic NPs (chemical method).

The chemical process is commonly used to design NPs, as it provides better control over particle size, aggregation, and composition of nano material. Three major routes can be applied for synthesizing NP followed by chemical method:

Route-1 Chemical reduction of metallic salts.
Route-2 Thermal and photochemical decomposition of metal complexes.
Route-3 Electrochemical reduction of metallic salts.

In the chemical reduction process, many reducing agents were used such as molecular hydrogen and organic and inorganic chemicals, but among all the reducing agents, NaBH$_4$, and molecular hydrogen is the most common to obtain agglomeration free, highly stable, well-dispersed, and active NPs, whereas templates for the fabrication of NPs, ILs have attracted a lot of interest in recent years. There are many reports where ILs have been applied as reaction medium as well as template for the synthesis of various metallic NPs such as inorganic oxide materials including zeolite materials, mesoporous titania, and titania NPs. The high nucleation rates in many ILs are due to the low surface tension and in consequence, to small particles. ILs mediated metal NPs can be synthesized through chemical reduction, where metal atoms are in a positive oxidation state in their metal salts.

4.4 APPLICATIONS OF IONIC LIQUIDS IN METAL NANOPARTICLE SYNTHESIS

ILs are mostly applied to obtain metal NPs, metal oxide NPs, and alloy NPs. The physicochemical properties of metal NPs, their surface properties, size, shape, and their molecular arrangement make them attentive for various applications such as medicine, chemicals, and electronics. The surface properties of one metal NPs are mainly different from another because of their different physiochemical properties (as developed after synthesis). In ILs mediated metal NP synthesis, ILs act as in situ stabilizing agent and offers a variety of surface properties and applications. The preparation and stabilization of metal NPs make them more efficient in terms of catalyst loading, catalytic selectivity, and catalyst recycling with respect to IL mediated organometallic or metal salt catalysts.[7] The imidazolium based ILs are most studied and recommended class of IL, which was successfully used to synthesize a series of NPs (Fig. 4.1).

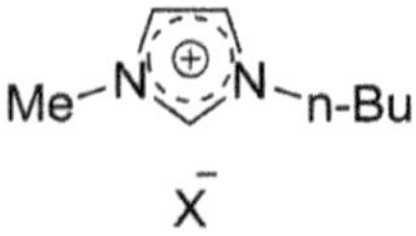

$$Me-N\overset{\oplus}{\underset{}{N}}-n\text{-}Bu$$

$$\overset{-}{X}$$

where X = BF$_4$ or PF$_6$ or N(SO$_2$CF$_3$)$_2$ or CF$_3$SO$_3$.

FIGURE 4.1   Common structure of imidazolium ILs used for preparation of metal NPs.

In IL mediated NPs synthesis, both anions and cations play an important role not only to stabilize the NPs of metal, but also to offer the interactions of anion with metal surface of NPs which is stronger than cations, and hence anions are more active for stabilization of NPs.[9] However, the high catalytic activity can be achieved due to weak interaction between metal NPs and non-functionalized ILs but the problem of poor stability and recyclability also occurs with NPs during the reaction. Such glitches can be resolved by using functionalized ILs, as it is well documented that —OH group functionalized ILs generate more stable NPs.

4.5 IONIC LIQUID MEDIATED METAL NANOPARTICLES OF RU, PD, PT, AND TiO$_2$

4.5.1 SYNTHESIS OF IONIC LIQUID MEDIATED RU-NPS

The reduction of metal salts is the most applicable method to generate metal nanoparticles (MNPs) in ILs. MNPs of metals like Ru, Pd, Pt, Au, Ag, Rh, and Ir with positive oxidation state (M^{n+}) have been prepared in ILs. In general, the reducing agents like $NaBH_4$ and molecular hydrogen (4 bar) were commonly used for synthesis of Ru NPs using a series of metal precursors such as $RuO_2 \cdot 3H_2O$, RuO_2, $[Ru(C_8H_{12})(2\text{-methyallyl})_2]$, and $[Ru(C_8H_{12})$ $(C_8H_{10})]$. The Ru metal precursors found highly dispersed in different ILs like 1-n-butyl-3-methylimidazolium hexafluorophosphate [BMIM][PF$_6$], or 1-n-butyl-3-methylimidazolium trifluoromethanesulphonate [BMIM] [CF$_3$SO$_3$], or 1-butyl-3-methylimidazolium tetrafluoroborate [BMIM][BF$_4$], or in different cations of imidazolium 1-n-decyl-3-methylimidazolium [DMI], and 1-butyronitilre-3-methylimidazolium[(BCN)MI] with anions PF$_6$, NTf$_2$, and BF$_4$.[9]

The chemical reduction method offers very fine, stable, and agglomeration free Ru(0) NPs in IL medium.[39] The H$_2$ reduction is one of the commercially available reducing agents to Ru metal precursor $RuO_2 \cdot 3H_2O$, which easily offers Ru(0) NPs. In addition, this process does not require any organometallic precursor. The Ru(0) NP was successfully applied to benzene hydrogenation and synthesis of hydrogenated benzene derivatives as desired requirements. Biphasic experiments were also utilized to synthesize Ru NPs, where prior to the addition of the substrate, RuO_2 was dispersed in IL (1-n-butyl-3-methylimidazolium hexafluorophosphate [BMIM][PF$_6$], 1-n-butyl-3-methylimidazolium trifluoromethanesulphonate [BMIM][CF$_3$SO$_3$], or 1-butyl-3-methylimidazolium tetrafluoroborate [BMIM][BF$_4$]). Later on, the vacuum was created in a reactor at 75°C under constant stirring. The reactor was connected to the hydrogen gas source and by passing H$_2$ gas (4 atm) the reaction was initiated. The pressure drop of H$_2$ gas confirms the progress of the reaction.[40] The ILs containing *N*-bis(trifluoromethanesulfonyl)imidates give NPs with smaller size with respect to the less coordinated tetrafluoroborate analogs.

[Ru(COD)(COT)] (COD = 1,5-cyclooctadiene, COT = 1,3,5-cyclooctatriene) is zero-valent organometallic compounds, which were used as an significant precursor to obtain Ru(0) metal atom followed by decomposition of COD and COT ligand under hydrogen pressure. The concentration of [Ru(COD)(COT)] probably controls the crystal growth of NPs and hence it limits the size of the non-polar domains. The degree of self-organization as well as the structural morphology of the Imidazolium based IL oversees the size of Ru NPs. The reduction of [Ru(COD)(2-methylallyl)$_2$] (COD = 1,5-cyclooctadiene) associated with the *N*-bis(trifluoromethanesulfonyl)imidates (NTf$_2$) or with the tetrafluoroborates (BF$_4$), offers well-distributed and ultra size Ru NPs at 50°C in hydrogen pressure (4 bar).[41]

The RuCl$_3$ reduction in ILs (1-n-butyl-3-methylimidazolium hexafluorophosphate [BMIM][PF$_6$] and 1-n-butyl-3-methylimidazolium tetrafluoroborate [BMIM][BF$_4$]) with reducing agents NaBH$_4$, is a simple and reproducible method to obtain ruthenium oxide NPs. This well-accepted method offers stable Ru NPs with a narrow size distribution (2–3 nm). Chemists achieved high catalytic activity of ruthenium oxide NPs in the biphasic hydrogenation of olefins and arenes under mild reaction conditions with added advantage of 10 times catalysts recycling for model reaction (hydrogenation of 1-hexene yielding a total turnover number for exposed Ru(0) atoms of 175,000).[49–50] The trinuclear Ru metal carbonyls (Ru$_3$(CO)$_{12}$) in *n*-butylmethylimidazolium tetrafluoroborate, [BMIM][BF$_4$], under an argon atmosphere get deoxygenated. The dispersed metal carbonyl in [BMIM][BF$_4$] IL was thermally decomposed 250°C for several hours. The brown Ru NPs of size 1.5–2.5 nm was obtained, which was stable for several months under argon.[51,52]

4.5.2 SYNTHESIS OF ILS MEDIATED PD-NPS

Imidazolium-based ILs were found highly promising for the synthesis of Pd-NPs from palladium (II) salt without using any additional reducing agent. It is reported that the formation of Pd-NPs takes place after the development of Pd-N-heterocyclic carbine complexes as intermediates (Scheme 4.2).[53]

As a precursor, the imidazolium salts are also counted as mild reducing agents [55,56] and applied for the synthesis of Pd-NPs. Thermal decomposition of Pd(OAc)$_2$ is a well-established protocol and it was well studied in various conventional solvents.[57,58] Simply heating of Pd(OAc)$_2$ in [BMIm]

[Tf$_2$N] at 80°C in the presence of triphenyl phosphine (PPh$_3$). Pd-NPs were obtained with a diameter of nearly 1 nm,[59] while in the absence of an additional reducing agent, Pd-NPs of 5 and 10 nm were obtained by heating the proper mixture of Pd(acac)$_2$ and [HOBMIm][Tf$_2$N].[60]

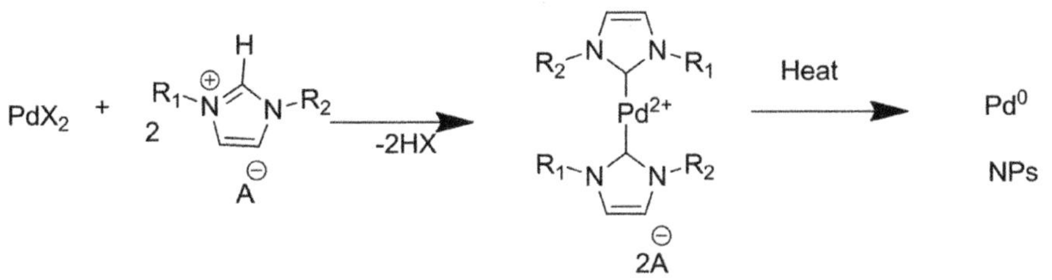

SCHEME 4.2   Reduction of Pd(II) species through intermediate followed by subsequent decomplexation and reduction.[54]

Heating of well-dispersed Pd(OAc)$_2$ in 1-butyronitrile-3-methyl-imidazolium *N*-bis(trifluoromethane sulfonyl)amide [NCBMIm][Tf$_2$N], under reduced pressure, leads to the formation of stable and small-sized Pd-NPs.[61] In a series of hydroxy-functionalized ILs with the [HOEMIm]$^+$ cation, the effect of the anions on the decomposition rate of Pd(OAc)$_2$ was found in the given order [Tf$_2$N]$^-$, [PF$_6$]$^-$ > [BF$_4$]$^-$ > [OTf]$^-$ > [TFA]$^-$, based on a determination of the percentage of Pd(OAc)$_2$ remaining in the sample. Thermal treatment in hydroxyl functionalized ILs with the 1-(20-hydroxyethyl)-3-methyl-imidazolium [HOEMIm]$^+$ cation and non-functionalized control IL, Pd-NPs were synthesized from Pd(OAc)$_2$. The OH-functionalized ILs [HOEMIm][Tf$_2$N] gave smaller Pd-NPs with diameters 4.0 ($\pm$ 0.6 nm)[62] as it is accepted that the alcohol group in the [HOEMIm]$^+$ cation is not working as the reducing agent (this is also confirmed by ^{1}H NMR spectra recorded before and after reduction of Pd(OAc)$_2$ showed no difference).[62]

The Pd(0) stable NPs with a size of 4.9($\pm$0.8) nm were obtained by reduction of Pd(acac)$_2$ (acac = acetylacetonate) under high pressure hydrogen atmosphere (4 atm) at 75°C. Before, going to reduction, the Pd(acac)$_2$ (acac = acetylacetonate) was dissolved in IL like (1-*n*-butyl-3-methylimidazolium hexafluorophosphate [BMIM][PF$_6$] or 1-n-butyl-3-methylimidazolium tetrafluoroborate [BMIM][BF$_4$]). These developed Pd-NPs were further tested for the selective partial hydrogenation 1,3-butadiene.[42] In situ laser radiation process was also trying to break large Pd(0) NPs fragmentation into smaller particles with narrow size distribution in 1-*n*-butyl-3-methylimidazolium hexafluorophosphate IL. In another approach, simple hydrogen reduction of PdCl$_2$ dispersed in 1-*n*-butyl-3-methylimidazolium hexafluorophosphate at 75°C, the formation of Pd(0) NPs indicated by a dark solution which was obtained after 15 min.[43] Guan-ILs (Fig. 4.2) as a mediator for the nucleation and growth process uniform Pd-NPs supported on Vulcan XC-72 carbon were synthesized from H$_2$PdCl$_4$ and NaBH$_4$.[65]

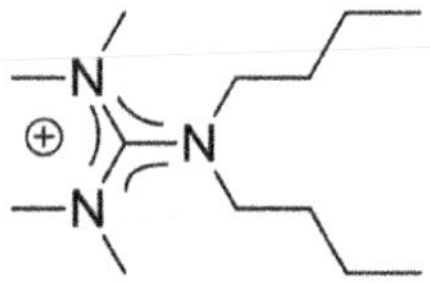

FIGURE 4.2   Guan ILs cation, [Guan]$^+$ = 1,1-dibutyl-2,2,3,3-tetramethyl-guanidinium.

Ultrasonic approach was also applied to irradiate Pd(OAc)$_2$ or PdCl$_2$ salt solution in imidazolium-based ILs such as [BBIm][Br] or [BBIm][BF$_4$] to obtain nearly spherical Pd-NPs (20 nm). The succeeding sonolytic conversion of Pd biscarbene complexes as intermediates to Pd-NPs have been recognized by NMR/MS and TEM analysis, respectively.[53] Electrochemical reduction of palladium foil as the anode and a platinum foil as the cathode was used to prepare stabilized Pd-NPs in morpholinium based IL, [BMMor] [BF$_4$]. At Pd anode, Pd ions migrate toward Pt cathode and NPs forms due to the reduction of Pd ions at the cathode. With an increase in temperature as well as electrolysis duration along with a decrease in the current density, the surprising increase in particle size of Pd was observed. The particles had a crystalline structure was confirmed by the electron diffraction patterns of the resulting NPs.[68]

4.5.3   SYNTHESIS OF IONIC LIQUID MEDIATED PT-NPS

The well-crystalline, distinct, small and longtime stable Pt-nanocrystal synthesis was carried out by thermal, photolytic or microwave assisted decomposition of the air/moisture stable organometallic Pt(IV) precursor (MeCp)PtMe$_3$ in following IL reaction medium [BMIm][BF$_4$] and [BtMA][Tf$_2$N] (full name of IL should be entered). IL mediated Pt NPs (0:0125 mol:%) offered the biphasic hydrosilylation of phenylacetylene with triethylsilane, which leads to give the distal and proximal products triethyl (2- and 1-phenylvinyl) silane with TOF 96,000 h^{-1}.[63] The catalytically sound and active Pt (0) (2.3 nm size) was obtained by controlling thermal decomposition of the organometallic precursor Pt$_2$(dba)$_3$

(dba = dibenzylideneacetone) in reported [BMIM][BF$_4$] and [BMIM][PF$_6$], IL medium.[44,45] Pt NPs also synthesized from H$_2$Pt(OH)$_6$ and formic acid in [BMIm][FEP] (FEP = tris(pentafluoroethyl) trifluorophosphate) which offers chemo selective hydrogenation of 3-nitrostyrene (yields exceeding 90%). Recyclability of Pt NPs found while immobilizing on solid SiO$_2$ or supported carbon nanotube (CNTs).[64–67] Pt sputtering without stabilizing agents onto the IL trimethyl-n-propylammonium bis((trifluoromethyl) sulfonyl)amide [Me$_3$PrN][Tf$_2$N] produce Pt NPs of mean particle diameters of ca. 2.3–2.4 nm but they are independent of sputtering time.[69] Without any pretreatment of SWCNTs or any chemical reagent, the Pt-NP-SWCNT composites was achieved by using the Pt sputtered ILs and the Pt-NP immobilization onto single-walled carbon nanotubes.[70]

4.5.4   SYNTHESIS OF IONIC LIQUID MEDIATED TiO$_2$ NPS

Titanium oxide (TiO$_2$) is one of the most important nanomaterials, which has attracted a great attention due to its unique properties like optical, dielectric, and catalytic properties. Such properties of TiO$_2$ nanomaterial Ti NPs make them nearer to following industrial applications such as pigments, fillers, catalyst supports, and photo-catalysts. It is reported that the overall physiochemical properties of this material may vary with size, morphology, and crystalline phase of Ti NPs. Various significant methods such as liquid process (sol–gel, solvothermal, and hydrothermal), solid state processing routes (mechanical alloying/milling, mechanochemical, and RF thermal plasma), and other routes such as laser ablation have been developed over the last decade. Among a series of ILs, 1-butyl-3-methylimidazolium bis(triflylmethylsulfonyl)imide [BMIM][Tf$_2$N is used several times in the synthesis of titanium oxide NPs. The anatase, brookite, rutile are the polymorph of titanium. The synthesis of nano size rutile phase of titanium is considered as most promising material for the development of photocatalyst as well as electrode material. The under mild condition, IL mediated synthesis of titanium dioxide nanocrystal of 2–3 nm size was done due to their self-assembly toward mesoporous TiO$_2$ spheres. The resulting larger spheres with a considerable high surface area and narrow pore size distribution have shown a great potential for catalysis, solar energy conversion and optics–electronic devices. 1-butyl-3-methylimidazoliumtetrafluoroborate was utilized as solvent for the synthesis of TiO$_2$ NPs.[46] The hydrolysis of titanium tetrachloride in hydrochloric acid with the support of 1-ethyl-3-methyl-imidazolium bromide successfully synthesized the pure rutile and rutile-anatase nanocomposites. Simple optimization of 1-ethyl-3-methyl-imidazolium bromide IL added to the reaction mixture may control the content of rutile in the composites. As per the approach, the reaction can be performed under atmospheric pressure in a glass vessel and with high yield. The composites with arbitrary content of rutile have been rationally prepared and the contents are controllable with high crystallinity. This method can be improved in general way to synthesize other metal oxide NPs on a huge scale.[47] An alkoxide sol–gel method was found applicable for synthesis of nanocrystalline particles of TiO_2 in the presence of water-immiscible ILs as a solvent medium. In addition, later non-ionic surfactant was used as pore-templating material for modified synthesis of TiO_2.[48]

4.6 CONCLUSION

In this chapter, we framed most significant developments in the area of IL mediated important metal NPs. The basic aim of this chapter is to provide comprehensive information in a concise manner to new researchers by taking the support of best scientific reports. Here, not only we are covering the impact of imidazolium based IL, but we also discuss some other important groups of ILs in order to make the chapter most versatile in the area of IL mediated metal nano particle synthesis. The fascinating properties of IL not only control the size and shape of the NP, but also provide a stable environment to make them more reactive toward their applications. As per the reported literatures, IL provides an opportunity to achieve the synthesis of well-dispersed, small size, metallic NPs with different chemical compositions. Anions of IL play a more noteworthy role in stabilizing NPs, as they interact more strongly and effectively with the metal surface other than cationic part of the IL. However, it is worth noting here that NPs are only kinetically stable and neither anion nor cation could prevent the process of agglomeration.

ACKNOWLEDGMENTS

This work is financially supported by DST Fast Track (SB/FT/CS-124/2012). Author is also thankful for the library of NIIT University for providing useful journals and books for the compilation of this review chapter.

KEYWORDS

- ionic liquid
- hydrogenation reaction
- nanoparticle synthesis
- imidazole ionic liquid
- Friedlander reaction

REFERENCES

1. Tarek, A. A. A. Sonochemical Synthesis and Characterization of Metal Oxides Nanoparticles in Ionic Liquids, Ph.D. Dissertation, Ruhr University, Bochum, 2011.
2. Zhao, D. Design, Synthesis and Applications of Functionalized Ionic Liquids, Ph.D. Dissertation, Ecole Polytechnique Federale De Lausanne, 2007.
3. Lu, X.; Rycenga, M.; Skrabalak, S. E.; Wiley, B.; Xia, Y. Chemical Synthesis of Novel Plasmonic Nanoparticles. *Annu. Rev. Phys. Chem.* **2009,** *60,* 167–179.
4. Walden, P. Ueber die Molekulargrösse und elektrische Leitfähigkeit Einiger Geschmolzenen Salze. *Bull. Acad. Imp. Sci.* **1914,** *8,* 405–422.
5. Hurley, F. H.; Wler, T. P. Electrodeposition of Metals from Fused Quaternary Ammonium Salts. *J. Electrochem. Soc.* **1951,** *98,* 203–206.
6. Wilkes, J. S.; Zaworotko, M. Air and Water Stable Lethyl3methylimidazolium Based Ionic Liquids. *J. Chem. Commun.* **1992,** *13,* 965–967.
7. Vipul Bansal; Suresh, Bhargava, K. Ionic Liquids as Designer Solvents for the Synthesis of Metal Nanoparticles. In *Ionic Liquids: Theory, Properties, New Approaches;* Alexander, K., Ed.; InTech: Rijeka, Croatia, 2011; p 367.
8. Dupont, J.; Fonseca, G. S.; Umpierre, A. P.; Fichtner, P. F. P.; Teixeira, S. R. Transition-Metal Nanoparticles in Imidazolium Ionic Liquids: Recyclable Catalysts for Biphasic Hydrogenation Reactions. *J. Am. Chem. Soc.* **2002,** *124,* 4228–4229.
9. Zhang, B.; Yan, N. Towards Rational Design of Nanoparticle Catalysis in Ionic Liquids. *Catalysts.* **2013,** *3* (2), 543–562.
10. Lee, J. W.; Shin, J. Y.; Chun, Y. S.; Jang, H. B.; Song, C. E.; Lee, S. G. Toward Understanding the Origin of Positive Effects of Ionic Liquids on Catalysis: Formation of More Reactive Catalysts and Stabilization of Reactive Intermediates and Transition States in Ionic Liquids. *Acc. Chem. Res.* **2010,** *43,* 985–994.
11. Roucoux, A.; Nowicki, A.; Philippot, K. Rhodium and Ruthenium Nanoparticles in Catalysis, Wiley-VCH Verlag GmbH & Co. KGaA: Weinheim, Germany, 2008; pp 349–388.
12. Seddon, K. R.; Stark, A. Selective Catalytic Oxidation of Benzyl Alcohol and Alkylbenzenes in Ionic Liquids. *Green Chem.* **2002,** *4,* 119–123.
13. Balanta, A.; Godard, C.; Claver, C. Pd Nanoparticles for C–C Coupling Reactions, *Chem. Soc. Rev.* **2011,** *40,* 4973–4985.

14. Sheldon, R. Catalytic Reactions in Ionic Liquids. *Chem. Commun.* **2001,** 2399–2407.
15. Handy, S. T. Room Temperature Ionic Liquids: Different Classes and Physical Properties. *Curr. Org. Chem.* **2005,** *9* (10), 959–988.
16. Miao, W.; Chan, T. H. Ionic-Liquid-Supported Synthesis: A Novel Liquid-Phase Strategy for Organic Synthesis. *Acc. Chem. Res.* 2006, *39* (12), 897–908.
17. Wilkes, J. S. A Short History of Ionic Liquids—from Molten Salts to Neoteric Solvents, *Green Chem.* **2002,** *4,* 73–80.
18. Wu, H. H.; Yang, F.; Cui, P.; Tang, J.; He, M.Y. An Efficient Procedure for Protection of Carbonyls in Brønsted Acidic Ionic Liquid [Hmim]BF$_4$. *Tetrahedron Lett.* **2004,** *45* (25), 4963–4965.
19. Palimkar, S. S.; Siddiqui, S. A.; Daniel, T.; Lahoti, R. J.; Srinivasan, K. V.; Ionic Liquid-Promoted Regiospecific Friedlander Annulation: Novel Synthesis of Quinolines and Fused Polycyclic Quinolones. *J. Org. Chem.* **2003,** *68,* 9371–9378.
20. Forbes, D. C.; Law A. M.; Morrison, D. W. The Knoevenagel Reaction: Analysis and Recycling of the Ionic Liquid Medium. *Tetrahedron Lett.* **2006,** *47,* 1699–1703.
21. Monteiro, A. L.; Zinn, F. K.; de Souza, R. F.; Dupont, J. Asymmetric Hydrogenation of 2-Arylacrylic Acids Catalyzed by Immobilized Ru-BINAP Complex in 1-n-butyl-3-Methylimidazolium Tetrafluoroborate Molten Salt. *Tetrahedron. Asymmetry.* **1997,** *8,* 177–179.
22. Song, C. E.; Shim, W. H.; Roh, E. J.; Lee, S.; Choi, J. H. Ionic Liquids as Powerful Media in Scandium Triflate Catalysed Diets-Alderreactions: Significant Rate Acceleration, Selectivity Improvement and Easyrecycling of Catalyst. *Chem. Commun.* **2001,** *12,* 1122–1123.
23. Song, C. E.; Roh, E. J. Practical Method to Recycle a Chiral (salen)Mn Epoxidation Catalyst by Using an Ionic Liquid. *Chem. Commun.* **2000,** *10,* 837–838.
24. Xu, L.; Chen, W.; Ross, J.; Xiao, Palladium-Catalyzed Regioselective Arylation of an Electron-Rich Olefin by Aryl Halides in Ionic Liquids. *J. Org. Lett.* **2001,** *3,* 295–297.
25. Jin-Hong, L.; Cheng-Pan, Z.; Zhi-Qiang, Z.; Qing-Yun, C.; Ji-Chang, X. A Novel Pyrrolidinium Ionic Liquid with 1,1,2,2-tetrafluoro-2-(1,1,2,2-tetrafluoroethoxy) Ethanesulfonate Anion as a Recyclable Reaction Medium and Efficient Catalyst for Friedel–Crafts Alkylations of Indoles with Nitroalkenes, *J. Fluor.Chem.* **2009,** *130* (4), 394–398.
26. Hajipour, A. R.; Rafiee, F. Basic Ionic Liquids. A Short Review. *J. Iran. Chem. Soc.* **2009,** *6* (4), 647–678.
27. Zhiying, D.; Yanlong, G.; Youquan, D. Neutral Ionic Liquid [BMIm]BF$_4$ Promoted Highly Selective Esterification of Tertiary Alcohols by Acetic Anhydride. *J. Mol. Catal A. Chem.* **2006,** *246,* 70–75.
28. Xu, L.; Chen, W.; Xiao, J. Heck Reaction in Ionic Liquids and the in Situ Identification of N-Heterocyclic Carbene Complexes of Palladium. *Organometallics.* **2000,** *19* (6), 1123–1127.
29. Boulaire, V. L.; Gree, R. Witting Reactions in the Ionic Solvents [bmim]BF$_4$. *Chem. Commun.* **2000,** *21,* 2195–2196.
30. Mathews, C. J.; Smith, P. J.; Welton, T. Palladium Catalysed Suzuki Cross-Coupling Reactions in Ambient Temperature Ionic Liquids. *Chem. Commun.* **2000,** *14,* 1249–1250.
31. Handy, S.T.; Zhang, X. Organic Synthesis in Ionic Liquids: The Stille Coupling, *Org. Lett.* **2001,** *3* (2), 233–236.
32. Ross, J. Xiao, J. Friedel-Crafts Acylation Reactions Using Metal Triflates in Ionic Liquid. *Green Chem.* **2002,** *4,* 129–133.

33. Zhu, S.; Chen, R.; Wu, Y.; Chen, Q.; Zeng, X. Yu, Z. A Mini Review on Greenness of Ionic Liquids. *Chem. Biochem. Eng. Q.* **2009,** *23* (2), 207–211.

34. Pandey, S. Analytical Application of Room Temperature Ionic Liquids: A Review of Recent Efforts. *Anal. Chim. Acta.* **2005,** *556* (1), 38–45.

35. Astruc, D.; Lu, F.; Aranzaes, J. R. Nanoparticles as Recyclable Catalysts: The Frontier between Homogeneous and Heterogeneous Catalysis. *Angew. Chem. Int. Ed.* **2005,** *44,* 7852–7872.

36. Ahmed, M. A.; Katori, E. E. E.; Gharni, H. Z. Photocatalytic Degradation of Methylene Blue Dye Using Fe_2O_3/TiO_2 Nanoparticles Prepared by Sol–Gel Method. *J. Alloys. Compd.* **2013,** *553,* 19–29.

37. Choi, H.; Kim, Y. J.; Varma, R. S.; Dionysiou, D. D. Thermally Stable Nanocrystalline TiO2 Photocatalysts Synthesized via Sol–Gel Methods Modified with Ionic Liquid and Surfactant Molecules. *Chem. Mater.* **2006,** *18,* 5377–5384.

38. Didier, A.; Feng, L.; Jaime, R. A. Nanoparticles as Recyclable Catalysts: The Frontier between Homogeneous and Heterogeneous Catalysis. *Angew. Chem. Int. Ed.* **2005,** *44,* 7852–7872.

39. Fonseca, G. S.; Umpierre, A. P.; Fichtner, P.; Teixera, F. P.; Dupont, S. R. The Use of Imidazolium Ionic Liquids for the Formation and Stabilization of Ir(0) and Rh(0) Nanoparticles: Efficient Catalysts for the Hydrogenation of Arenes. *Chem A. Eur. J.* **2003,** *9* (14) 3263–3269.

40. Rossi, L. M.; Machado, G. Ruthenium Nanoparticles Prepared from Rruthenium Dioxide Precursor: Highly Active Catalyst for Hydrogenation of Arenes under Mild Conditions. *J. Mol. Catal A. Chem.* **2009,** *298* (1–2), 69–73.

41. Prechtl, M. H. G.; Scariot, M.; Scholten, J. D.; Machado, G.; Teixeira, S. R.; Dupont, J. Nanoscale Ru(0) Particles: Arene Hydrogenation Catalysts in Imidazolium Ionic Liquids. *Inorg. Chem.* **2008,** *47* (19), 8995–9001.

42. Umpierre, A. P.; Machado, G.; Fecher, G. H.; Morais, J.; Dupont, J. Selective Hydrogenation of 1,3-Butadiene to 1-Butene by Pd(0) Nanoparticles Embedded in Imidazolium Ionic Liquids. *Adv. Synth. Catal.* **2005,** *347* (10), 1404–1412.

43. Gelesky, M. A.; Umpierre, A. P.; Machado, G.; Correia, R. R. B.; Magno, W. C.; Morais, J.; Ebeling, G.; Dupont, J. Laser-Induced Fragmentation of Transition Metal Nanoparticles in Ionic Liquids. *J. Am. Chem. Soc.* **2005,** *127* (13), 4588–4589.

44. Scheeren, C. W.; Machado, G.; Dupont, J.; Fichtner, P. F. P.; Texeira, S. R. Nanoscale Pt(0) Particles Prepared in Imidazolium Room Temperature Ionic Liquids: Synthesis from an Organometallic Precursor, Characterization, and Catalytic Properties in Hydrogenation Reactions. *Inorg. Chem.* **2003,** *42* (15), 4738–4742.

45. Scheeren, C. W.; Machado, G.; Teixeira, S. R.; Morais, J.; Domingos, J. B.; Dupont, J. Synthesis and Characterization of Pt(0) Nanoparticles in Imidazolium Ionic Liquids. *J. Phys. Chem B.* **2006,** *110* (26), 13011–13020.

46. Zhou, Y.; Antonietti, M. Synthesis of Very Small TiO_2 Nanocrystals in a Room Temperature Ionic Liquid and their Self-Assembly toward Mesoporous Spherical Aggregates. *J. Am. Chem. Soc.* **2003,** *125* (49) 14960–14961.

47. Zheng, W.; Liu, X.; Yan, Z.; Zhu, L. Ionic Liquid-Assisted Synthesis of Large-Scale TiO_2 Nanoparticles with Controllable Phase by Hydrolysis of $TiCl_4$. *ACS Nano.* **2009,** *3* (1), 115–122.

48. Choi, H.; Kim, Y. J.; Varma, R. S.; Dionysiou, D. D. Thermally Stable Nanocrystalline TiO_2 Photocatalysts Synthesized via Sol–Gel Methods Modified with Ionic Liquid and Surfactant Molecules. *Chem. Mater.* **2006,** *18* (22), 5377–5384.

49. Rossi, L. M.; Dupont, J.; Machado, G.; Fichtnerb, P. F. P.; Radtke, C.; Baumvol I. J. R.; Teixeira, S. R. Ruthenium Dioxide Nanoparticles in Ionic Liquids: Synthesis, Characterization and Catalytic Properties in Hydrogenation of Olefins and Arenes. *J. Braz. Chem. Soc.* **2004**, *15* (6), 904–910.

50. Alexandrina, N.; Jürgen, L. Ionic Liquids as Advantageous Solvents for Preparation of Nanostructures. In *Applications of Ionic Liquids in Science and Technology*; Prof. Scott Handy Ed.; InTech: Rijeka, Croatia, 2011; ISBN: 978-953-307-605-8.

51. Krämer, J.; Redel, E.; Thomann, R.; Janiak, C. Use of Ionic Liquids for the Synthesis of Iron, Ruthenium, and Osmium Nanoparticles from Their Metal Carbonyl Precursors. *Organometallics.* **2008**, *27,* 1976–1978.

52. Kim, K.S.; Demberelnyamba, D.; Lee, H. Size-Selective Synthesis of Gold and Platinum Nanoparticles Using Novel Thiol-Functionalized Ionic Liquids. *Langmuir.* 2004, *20,* 556–560.

53. Deshmukh, R. R.; Rajagopal, R.; Srinivasan, K. V. Ultrasound Promoted C–C Bond Formation: Heck Reaction at Ambient Conditions in Room Temperature Ionic Liquids. *Chem. Commun.* **2001**, 1544–1545.

54. Prechtl, M. H. G.; Scholten, J. D.; Dupont, J. Carbon-Carbon Cross Coupling Reactions in Ionic Liquids Catalysed by Palladium Metal Nanoparticles. *Molecules.* **2010**, *15,* 3441–3461.

55. Scholten, J. D.; Ebeling, G.; Dupont, J. On the Involvement of NHC Carbenes in Catalytic Reactions by Iridium Complexes, Nanoparticle and Bulk Metal Dispersed in Imidazolium Ionic Liquids. *Dalton Trans.* **2007**, *47,* 5554–5560.

56. Zhao, L.; Zhang, C.; Zhuo, L.; Zhang, Y.; Ying, J. Y. Imidazolium Salts: A Mild Reducing and Antioxidative Reagent. *J. Am. Chem. Soc.* **2008**, *130,* 12586–12587.

57. Tano, T.; Esumi, K.; Meguro, K. Preparation of Organopalladium Sols by Thermal Decomposition of Palladium Acetate. *J. Colloid Interface. Sci.* **1989**, *133,* 530–533.

58. Bradley, J. S.; Hill, E. W.; Klein, C.; Chaudret, B.; Duteil, A. Synthesis of Monodispersed Bimetallic Palladium Copper Nanoscale Colloids. *Chem. Mater.* **1993**, *5* (3), 254–256.

59. Anderson, K.; Fern´andez, S. C.; Hardacre, C.; Marr, P. C. Preparation of Nanoparticulate Metal Catalysts in Porous Supports Using an Ionic Liquid Route; Hydrogenation and C-C Coupling. *Inorg. Chem. Commun.* **2004**, *7,* 73–76.

60. Ruta, M.; Laurenczy, G.; Dyson, P. J.; Kiwi-Minsker, L. Pd Nanoparticles in a Supported Ionic Liquid Phase: Highly Stable Catalysts for Selective Acetylene Hydrogenation under Continuous-Flow Conditions. *J. Phys. Chem. C.* **2008**, *112,* 17814–17819.

61. Venkatesan, R.; Prechtl, M. H. G.; Scholten, J. D.; Pezzi, R. P. Machado, G. Dupont, J. Palladium Nanoparticle Catalysts in Ionic Liquids: Synthesis, Characterisation and Selective Partial Hydrogenation of Alkynes to Z-Alkenes. *J. Mater. Chem.* **2011**, *21,* 3030–3036.

62. Yuan, X.; Yan, N.; Katsyuba, S. A.; Zvereva, E.; Kou, Y.; Dyson, P. J. A Remarkable Anion Effect on Palladium Nanoparticle Formation and Stabilization in Hydroxyl-Functionalized Ionic Liquids. *Phys. Chem. Chem. Phys.* **2012**, *14,* 6026–6033.

63. Marquardt, D.; Barthel, J.; Braun, M.; Ganter, C.; Janiak, C. Weakly-coordinated Stable Platinum Nanocrystals. *Cryst. Eng. Comm.* **2012**, *14,* 7607–7615.

64. Beier, M. J.; Andanson, J.M.; Baiker, A. Tuning the Chemoselective Hydrogenation of Nitrostyrenes Catalyzed by Ionic Liquid-Supported Platinum Nanoparticles. *ACS Catal.* **2012**, *2,* 2587–2595.

65. Zhao, X.; Hua, Y.; Liang, L.; Liu, C.; Liao, J.; Xing, W. Ionic Liquid-Mediated Synthesis of 'Clean' Palladium Nanoparticles for Formic Acid Electrooxidation. *Int. J. Hydrogen Energy.* **2012,** *37,* 51–58.
66. Marquardt, D.; Beckert, F.; Pennetreau, F.; Tolle, F.; M"ulhaupt, R.; Riant, O.; Hermans, S.; Barthel, J.; Janiak, C. Hybrid Materials of Platinum Nanoparticles and Thiol-Functionalized Graphene Derivatives. *Carbon.* **2013,** *66,* 285–294. DOI: 10.1016/*j. carbon*.2013.09.002.
67. Marquardt, D.; Janiak. C. Getrennt und Geschützt Mit Flüssigen Salzen. *Nachr. Chemie.* **2013,** *61,* 754–757.
68. Cha, J. H.; Kim, K. S.; Choi, S.; Yeon, S. H.; Lee, H.; Lee, C. S.; Shim, J. J. Size-controlled Electrochemical Synthesis of Palladium Nanoparticles Using Morpholinium Ionic Liquid. *Korean J. Chem. Eng.* **2007,** *24,* 1089–1094.
69. Tsuda, T.; Yoshii, K.; Torimoto, T.; Kuwabata, S. Oxygen Reduction Catalytic Ability of Platinum Nanoparticles Prepared by Room-Temperature Ionic Liquid Sputtering Method. *J. Power Sour.* **2010,** *195,* 5980–985.
70. Yoshii, K.; Tsuda, T.; Arimura T. Imanishi A.; Torimoto T.; Kuwabata S. Platinum Nanoparticle Immobilization onto Carbon Nanotubes Using Pt-Sputtered Room-Temperature Ionic Liquid. *RSC Adv.* **2012,** *2,* 8262–8264.

CHAPTER 5

POLY METHYL METHACRYLATE COATED HEMATITE/ALUMINA NANOPARTICLES: ULTRASOUND ASSISTED SYNTHESIS AND ITS CHARACTERIZATION

SHABANA SHAIK and SHIRISH H. SONAWANE*

Department of Chemical Engineering, National Institute of Technology, Warangal 506004, Telangana, India

Corresponding author. E-mail: shirishsonawane09@gmail.com

CONTENTS

ABSTRACT

This chapter reports the simple method for in situ synthesis of Fe_2O_3 and Al_2O_3 with poly methyl methacrylate (PMMA) and its characterization. The hematite (Fe_2O_3) and aluminum oxide (Al_2O_3) nanoparticles were prepared by ultrasonication method using iron III nitrate and sodium hydroxide as precursors and for alumina the synthesized aluminum nanoparticles from aluminum isopropoxide and lithium aluminum hydride were exposed to open air to undergo oxidation and to form aluminum oxide or alumina. By using methyl methacrylate (MMA) as a monomer and sodium dodecyl sulfate (SDS) as a surfactant a combined ultrasound effect and proper initiation of potassium persulfate (KPS) initiator the PMMA is synthesized by emulsion polymerization technique. The conversion of MMA to PMMA is approximately 90%. The average size of the PMMA is of 3 nm, Fe_2O_3 loaded PMMA is around 1 nm and Al_2O_3 loaded PMMA is around 6 nm. The novelty of the work is seen in obtaining nanosize range (less than 10 nm) of the particles by sonicating it for 45 min.

5.1 INTRODUCTION

The work on polymer materials has grown continuously in the last two decades. Poly methyl methacrylate (PMMA) is a transparent thermoplastic material and has a number of applications. It is also known as acrylic glass, acrylite, Plexiglas, optix, oroglas, altuglas, etc. It is used in a number of applications due to its versatile physical and chemical properties, ease of handling, and ease of processing. Non-modified PMMA behaves brittle when loaded. PMMA is generally produced using different techniques such as emulsion, solution, bulk, and anionic polymerization.[1] PMMA acquired major attention of material scientist due to its optical properties[2] and it also exhibits excellent material properties such as hardness, high rigidity, transparency, mechanical strength, and good insulation properties.[3] Being transparent and durable, it has many applications like an acrylic glass, signs and displays, sanitary ware (bath tubs), furniture's, biomedical and electronics industrial appliances.[4] It is also used as a construction material in microfluidic devices, micro electro mechanical systems and a material for artificial organ in transplantology.[5] Inorganic nanoparticles/metal/metal oxide loaded polymers now undergo more attention due to their combined effect of polymer and inorganic particles such as elasticity, transparency, dielectric properties, specific absorption of light, magnetic properties, chemical activity, and catalyst characteristics.[6]

Metal and metal oxide nanoparticles like hematite and alumina loaded polymers exhibit various unique properties with respect to sorption behavior, chemical reduction, ligand sequestration,[7] magnetic properties (super paramagnetism, low curie temperature, and high magnetic susceptibility), non toxic, biocompatibility, and low cost drags them into many applications. Magnetic nanoparticles loaded polymers are also used in biomedical applications. These particles have the ability to interact with the various biological molecules due to its special magnetic properties.[8]

In this work, it is focused on the in situ batch emulsion polymerization technique for the preparation of PMMA and metal oxides loaded polymer by cavitation method. Where the chemical effects of the ultrasound are best explained by the localized hot spots created during the collapse of cavitations bubbles. The collapse of bubbles produces intense change in temperatures and creates hot spots.[9] The different ways in which ultrasonication can affect a chemical reaction have been used for carrying out emulsion polymerization[10] of methyl metacrylate monomer.[11,12] Emulsion polymerization is an important industrial approach used for the production of polymers which are used in many applications. Emulsion polymerization offers significant advantages like controlled reaction and narrow size distribution. PMMA has its properties rich in transparency and high modulus and mechanical properties such as abrasion are relatively low.[13]

5.2 MATERIALS AND METHOD

5.2.1 MATERIALS USED

Iron nitrate (III) non anhydrate purified (ferric nitrate—$Fe(NO_3)_3.9H_2O$) from Merck, oleic acid (LR) ($C_{18}H_{34}O_2$) from S D Fine Chem. Ltd., sodium hydroxide pellets (GR) (NAOH), and potassium persulfate (KPS—$K_2S_2O_8$) from Molychem, methyl methacrylate ($C_5H_8O_2$) from Loba Chemie, sodium dodecyl sulfate (SDS—$C_{12}H_{25}SO_4$) from SRL, Mumbai, ethanol (C_2H_5OH)—analytical CS reagent and distilled water.

5.2.2 EXPERIMENTAL PROCEDURE

5.2.2.1 SYNTHESIS OF IN SITU BATCH EMULSION POLYMERIZATION OF PMMA

Synthesis of PMMA using in situ batch emulsion polymerization is carried out in a glass reactor of 250 mL by maintaining a constant temperature throughout the reaction time which equipped under 2 cm tip diameter stainless steel sonic probe sonicator connected to an ultrasonic generator operating at a frequency of 22 KHz (Dakshin Ultrasound 22 KHz, 120 W). The process involves the making of SDS solution (0.282 g of SDS is added in 60 mL of distilled water) later adding the monomer (10 mL of MMA) and then initiator (0.38 g of KPS) and sonicating for reaction to be happened for 45 min . The individual SDS solution is sonicated for 10 min and the required amount of MMA monomer is added and sonicated for about 10 min and then the initiator is added and sonicated for about 45 min. For every 3 min of sonication the samples were collected along with the initial sample collected before adding initiator, the polymerization conversion of MMA to PMMA is calculated.

5.2.2.2 SYNTHESIS OF FE$_2$O$_3$ LOADED PMMA

Oleic acid encapsulated hematite nanoparticles are synthesized using ferric nitrate and sodium hydroxide as a precursors. Overall, 0.1 N solution of ferric nitrate solution is prepared and 1 mM of oleic acid is added to it and sonicated under ultrasonic bath sonicator maintaining water at 20°C (Bath Ultrasonication—Make of PCI Analytics, operating frequency of 50 KHz, 170 V) for 30 min. Addition of oleic acid acts as a capping agents on hematite particles. Then, 0.1 N of NaOH solution is prepared and sonicated for 30 min. Later NaOH solution is slowly added to ferric solution under sonication with a time span of 10 min. A reddish black colored precipitate solution is formed. The obtained solution is filtered, washed with ethanol and distilled water, reddish-black colored particles are obtained, dried for atmospheric air for a day and the particles are collected and used in the preparation of metal oxide loaded polymer. Out of the hematite nanoparticles, 0.47 g is taken and 0.50 µL of oleic acid is added to the particles and then 10 mL of MMA solution is added to it. This whole solution is added to 60 mL of SDS solution where it is sonicated for 10 min. On 10 min sonication of both the solutions together, the initiator KPS of 0.38 g dissolved in 5 mL water is added to the above solution and sonicated for about 45 min. The thus-prepared samples are characterized for further analysis and applications.

5.2.2.3 SYNTHESIS OF AL$_2$O$_3$ LOADED PMMA

Aluminum isopropoxide of 0.5 g is taken in 50 mL toluene and 0.2–0.3 µL of oleic acid is added to it and sonicated on ultrasonic probe sonicator of frequency 22 KHz continuously for 15 min. Lithium aluminum hydride of 0.2 g in 50 mL toluene was sonicated separately for 15 min. Lithium aluminum hydride of 0.2 g in 50 mL toluene was sonicated separately for 15 min. The temperature while sonicating reaches to a range of 50–90°C, to abolish this overheating and maintain a constant temperature a cooling bath is maintained at a temperature of 20–30°C and then sonicated. Both the solutions are mixed and then sonicated for 15 min; the solution retains the gray colored product settling at the bottom of the beaker. The above conditions generate a blackish gray color solution at inert atmosphere that gradually precipitates, yielding a grayish black powder. These solid particles are separated by vacuum filtration through a glass frit with grade 3 porous size filter paper and kept open to the atmospheric air for drying and there-fore the oxidation takes place and the formed aluminum is converted into alumina. These alumina nanoparticles are used for the synthesis of Al$_2$O$_3$ loaded PMMA by following the same procedure as above.

5.2.3 CHARACTERIZATION SECTION

The Fourier transform infrared (FTIR) analysis was carried out on Perkin–Elmer 100S spectrophotometer IR by preparing the sample pellet by adding well-heated KBr in presence of inert gas N$_2$. The spectra analysis recorded are used to characterize the chemical structure of the synthesized polymer and loaded metal oxides in polymer. The particle size distribution (PSD) analysis was carried out on Malvern Zetasizer analyzer—Nano S90 version 7.02. The overall particle analyzing processes were recorded with the dynamic light scattering method. The sample minimum of 3 mL is taken in Cuvette cell of glass or fiber depending on the solvent properties. The plot is drawn between the size and light intensity and shows a clear average size of the particle. The sample preparation is the major task carried out during analysis. The X-ray diffraction (XRD) was recorded on Bruker D8 Advanced X-ray diffractom-eter. The X-rays were produced using copper $K\alpha$ (Cu) having wavelength of 0.154 nm and are recorded on the 2θ scale range from $10°$ to $120°$ with a step size of $0.019°$ and at a fixed step counting time of 46 min 5 s. The X-rays were detected using a slow counting detector based on silicon strip technology. FTIR, PSD, and XRD analyses were carried out in National Institute of Technology, Warangal.

5.3   RESULTS AND DISCUSSION

5.3.1   CONVERSION ANALYSIS

A polymerization reaction assembly consisting of a beaker of 250 mL capacity placed in a 1000 mL capacity beaker with its continuous input and output of the utility. The sampling vowel is taken and the weight of each of these labeled vowels is measured by sensitive balance, and noted. A sample after reaction mixture (about 1 mL) is withdrawn from the reactor at an equal interval of 1–3 min each and poured in the sample bottles. The collected samples in the sample bottles are weighed again and noted. The sample bottles are now dried (with their lids removed) in the hot air oven at a temperature of about 100°C for 6 h so that all the monomer and water gets evaporates and only the dry polymer remains. Thus dried sample bottles are weight is measured again and noted. The gravimetric analysis is used for calculating the conversion of monomer to polymer. The percentage conversion of polymer is calculated using the mass of monomer and polymer as shown in equation. The above equation can be used only when it is sure that all the left over monomer and water has been evaporated in drying.

$$X_A = \frac{R_1 - R_2}{M} \tag{5.1}$$

where X_A = Percentage conversion of the monomer;
  R_1 = Weight of sampling bottle after drying (which is weight of polymer + weight of sampling bottle);
  R_2 = Weight of empty sampling bottle;
  M = Mass of monomer which was initially present in 1 mL of solution.

Where to calculate M:

  10 mL of MMA is used in the reaction (10 mL of MMA = 10.605 g) (density of MMA = 0.943 g/mL at 20°C);
  Volume of water = 60 mL;

KPS (initiator) dissolved in 5 mL water.
Total volume of the solution = 60+ 10+5=75 mL.

Amount of monomer present in each 1 mL of solution initially (as the contents of the reactor are well mixed, uniform concentration can be assumed) = 10.605g/75 mL = 0.1414 g/mL. An aliquot of 1 mL contains 0.1414 g of monomer.

Hence M = 0.1414 g.

5.3.2 CHARACTERIZATION ANALYSIS

From the XRD pattern, the amorphous natures of the PMMA are clearly shown. All the peaks match very well with the XRD pattern reported in earlier reports.[2,3,14] The PMMA shows the predominate maximum peak at a 2Θ value of 14.54°, and the minor peaks at 24° and 28° appeared in Figure 5.1. From the XRD patterns of Fe_2O_3 (Fig. 5.2) the major peak was observed at 35.66° and the other minor peaks were appeared at 30°, 43°, 57°, and 62° similar to the patterns shown in references, whereas in hematite loaded PMMA-the major peak is absorbed at 13.54° and minor peaks like 29° and 43°is visible in Figure 5.4. The major peak of the alumina loaded PMMA is absorbed at 13.96° and the other peaks at 28°, 40°, 72°, and 88° are also seen in Figure 5.5. The pure alumina nanoparticles in Figure 5.3 shows their diffraction peaks at 11.99°, 23°, 35°, 40°, 47°, 63°, and 64°

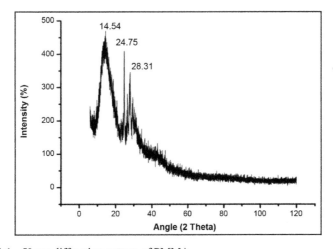

FIGURE 5.1 X-ray diffraction pattern of PMMA.

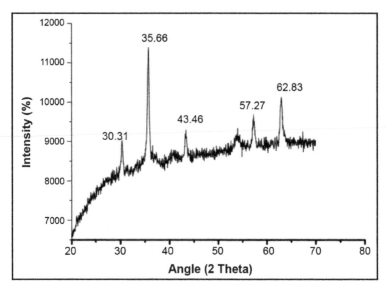

FIGURE 5.2   X-ray diffraction pattern of Fe₂O₃ nanoparticles.

and are similar to JCPDS 461212 and JCPDS 800956. The major peak of hematite from Figure 5.4 is absorbed at 35.66° and the other peaks are observed at 30°, 43°, 57°, and 62°. According to the FTIR values, metals lie in the range of 500–900 cm⁻¹. From IR spectra's (Figs. 5.6–5.8) it is found that the C—H stretches remain same as reported for the spectrum of oleic acid, and further the carbonyl peaks are disappeared and OH band visible at 3200–3600 cm⁻¹ and alkane CH bands are stretched at a range of 2900 cm⁻¹. The stretch at approximately 1700 cm⁻¹ is of C=O of ester carboxyl bonds and peak at 1400 cm⁻¹ are of aromatic C=C multiple bands. Ether C—O bonds are absorbed at 1182 and at 1154 cm⁻¹ —C—H alkene strong bending spectra's are observed. At 900 cm⁻¹ alkane C—H bonds occurs, at 818 cm⁻¹ strong alkyl halide C—Cl bonds appears and in the range of 900–500 cm⁻¹ metals and metal oxides are formed. The average particle size of the polymer and the metal oxide loaded polymer (Figs. 5.9–5.13) are of 3.349, 278, 1.009, 59.89, and 6.876 nm for PMMA, Fe₂O₃, PMMA loaded Fe₂O₃, Al₂O₃ nanoparticles and Al₂O₃ loaded PMMA, respectively. The TEM images of liquid PMMA, Al₂O₃, Fe₂O₃, and Al₂O₃ loaded PMMA and Fe₂O₃ loaded PMMA are shown in Figures 5.14–5.18. As PMMA is a highly conducting material the reflection can be seen in the image. From Figure 5.16, Fe₂O₃ nanoparticles are shown as needle-shaped flowers where as Fe₂O₃ loaded PMMA are seem to be like dots. The TEM images of Al₂O₃ nanoparticles

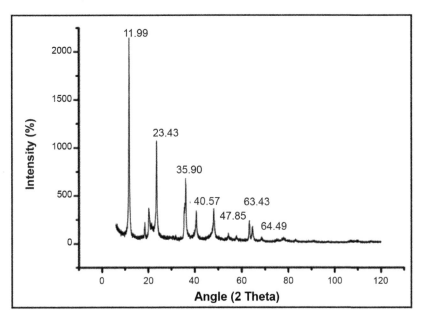

FIGURE 5.3 X-ray diffraction pattern of Al_2O_3 nanoparticles.

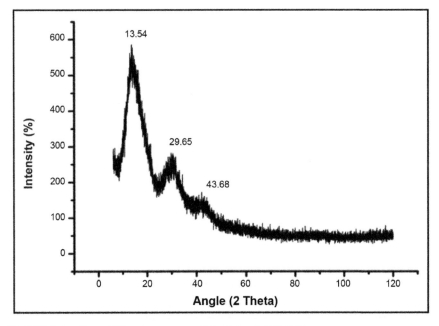

FIGURE 5.4 X-ray diffraction pattern of Fe_2O_3 loaded PMMA.

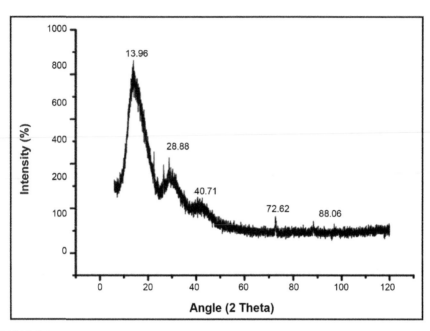

FIGURE 5.5   X-ray diffraction pattern of Al$_2$O$_3$ loaded PMMA.

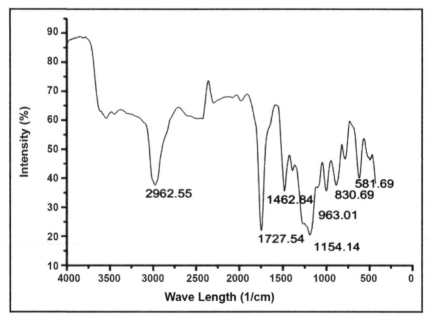

FIGURE 5.6   FTIR spectroscopic plot of PMMA.

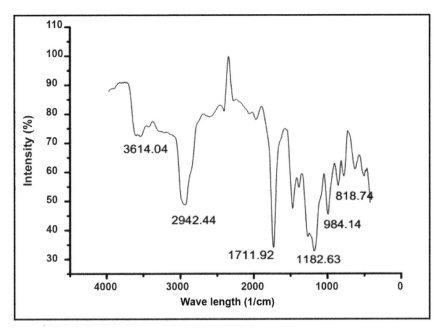

FIGURE 5.7   FTIR spectroscopic plot of Fe_2O_3 loaded PMMA.

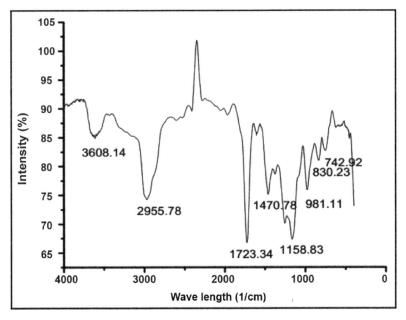

FIGURE 5.8   FTIR spectroscopic plot of Al_2O_3 loaded PMMA.

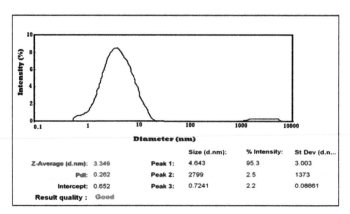

FIGURE 5.9 Particle size distribution curve of PMMA synthesized by ultrasound assisted in situ batch emulsion polymerization.

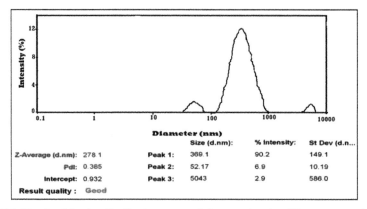

FIGURE 5.10 Particle size distribution curve of ultrasound assisted Fe₂O₃ nanoparticles.

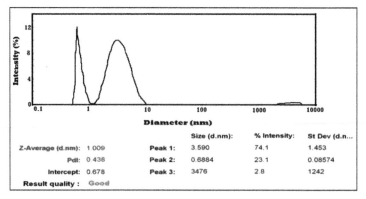

FIGURE 5.11 Particle size distribution curve of ultrasound assisted Fe₂O₃ loaded PMMA.

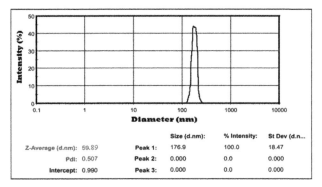

FIGURE 5.12 Particle size distribution curve of ultrasound assisted Al_2O_3 nanoparticles.

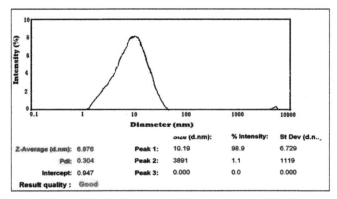

FIGURE 5.13 Particle size distribution curve of ultrasound assisted Al_2O_3 loaded PMMA.

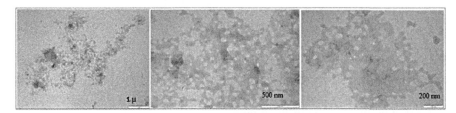

FIGURE 5.14 TEM images of liquid PMMA.

FIGURE 5.15 TEM images of Al_2O_3 nanoparticles.

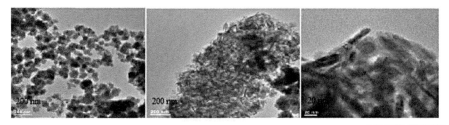

FIGURE 5.16   TEM images of Fe_2O_3 nanoparticles.

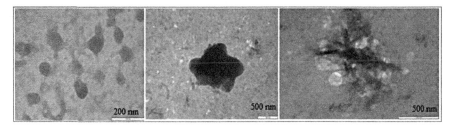

FIGURE 5.17   TEM images of Al_2O_3 nanoparticles loaded PMMA.

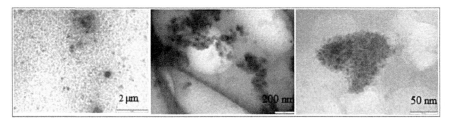

FIGURE 5.18   TEM images of Fe_2O_3 nanoparticles loaded PMMA.

are shown in Figure 5.15 which seems to have a needle-like structure and alumina loaded PMMA shown as clusters spread over the solution and well dispersed in the solution. The strong interaction between the nanoparticles and polymer can be seen in the images.

5.4   CONCLUSION

The in situ batch emulsion polymerization of polymethyl methacrylate using ultrasonication has been successfully carried out and the metal oxides like hematite and alumina loaded polymer has also been synthesized. The uniform functionalization of PMMA and metal oxide loaded PMMA with oleic acid was achieved in the solution. The above-synthesized solutions have many applications. Polymer is strictly used for LCD screens, electronic devices, dental purposes, and furniture, and the iron oxide loaded polymer can exhibit magnetic and biocompatibility properties which can be used in the hyptherimia, LPG sensing and as interpenetrating polymer networks sensors. Al_2O_3–PMMA composite had a small amount of PMMA, the addition of alumina increases the mechanical properties of the polymer and strengthens the products prepared by it. The average particle size of the polymer was 3 nm and metal oxide loaded polymer achieved below 10 nm by sonicating it for 45 min.

KEYWORDS

- **poly methyl methacrylate**
- **iron oxide**
- **emulsion polymerization**
- **in situ**
- **ultrasonication**

REFERENCES

1. Joe Schwarcz. *The Right Chemistry;* Doubleday Canada: Toronto, 2006, eISBN: 978–0–385–67160–6.
2. Shobhana. E. X-Ray Diffraction and UV–Visible Studies of PMMA Thin Films. *Int. J. Modern Eng. Res. (IJMER).* **2012,** *2* (3), 1092–1095.
3. Vijay Kumar, T.; Danny, V.; Samy, A. M.; Michael. R. K. Bio-Inspired Green Surface Functionalization of PMMA for Multifunctional Capacitors. *Royal Soc. Chem.* **2014,** *4,* 6677–6684.
4. Lapshin, R. V.; Alekhin, A. G.; Kirilenko, A. G.; Odintsov, S. L.; Krotkov, V. A. Vacuum Ultraviolet Smooting of Nanameter – Scale Asperities of Poly Methyl Methacrylate Surface. *J. Sur. Inves.* **2010,** *4* (1), 1–11.
5. Lapshin, R. V.; Alekhin, A. P.; Kirilenko, A. G.; Odinstov, S. L.; Krotkov, V. A. Vacuum Ultraviolet Smoothing of Nanometer-Scale Asperities of Poly (methyl methacrylate) Surface. *J. Sur. Invest.* **2010,** *4* (1), 5–16.
6. Wang, Z. G.; Zu, X. T.; Xiang, X. Preparation and Characterization of Polymer/Inorganic Nanoparticles Composites through Electron Irradiation. *J. Ma. Sci.* **2006,** *41,* 1973–1978.
7. Sudipta, S.; Guibal, E.; Quignard, F.; SenGupta, A. K. Polymer-Supported Metal and Metal Oxide Nanoparticles: Synthesis, Characterization, and Applications. *J. Nanopart. Res.* **2012,** *14,* 715.

8. Businova, P.; Chomoucka, J.; Prasek, J.; Hrdy, R.; Drbohlavova, J.; Sedlacek, P.; Hubalek, J. Polymer-Coated Iron Oxide Magnetic Nanoparticles – Preparation and Characterization. *Brno, Czech Republic, EU*, **2011,** *9,* 21–23.

9. Krishnamurthy, P.; Shirish, S.; Meifang, Z.; Ashokkumar, M. Ultrasound Assisted Synthesis and Characterization of Poly (mMthyl Methacrylate)/CaCO$_3$ Nanocomposites. *Chem. Eng. J.* **2013,** *219,* 254–261.

10. Price, G. J. *New Methods of Polymer Synthesis;* Blackie: Glasgow, 1995.

11. Teo, B. M.; Prescott, S. W.; Ashokkuma, M.; Grieser, F. Ultrasound Initiated Miniemulsion Polymerization of Methacrylate Monomer. *Ultrason. Sonochem.* **2008,** *15,* 89–94.

12. Albano, C.; Gonzalez, G.; Parra, C. Sonochemical Synthesis of Polymethyl Methacrylate to be Used as a Biomaterial. *Polym. Bull.* **2010,** *65,* 89–94.

13. Bharat, A. B.; Shirish, H. S.; Dipak, V. P.; Parag, R. G.; Aniruddha, B. P. Kinetic Studies of Semi Batch Emulsion Copolymerization of Methyl Methacrylate and Styrene in the Presence of High Intensity Ultrasound and Initiator. *Chem. Eng. Process.* **2014,** *85,* 168–177.

14. Devikala, V.; Kamaraj, P.; Artanareeswari, M. Conductivity and Dielectric Studies of PMMA Composites. *Chem. Sci. Trans.* **2013,** *2* (S1), 129–134.

CHAPTER 6

A TECHNICAL COMPARISON ON DIFFERENT NANO-INCORPORATIONS ON CEMENT COMPOSITES

MAINAK GHOSAL[1*] and ARUM KUMAR CHAKRABORTY[2]

[1]Indian Institute of Engineering Science and Technology, Shibpur, Howrah, West Bengal, India

[2]Department of Civil Engineering, Indian Institute of Engineering Science and Technology, Shibpur, Howrah, West Bengal, India

*Corresponding author. E-mail: mainakghosal2010@gmail.com

CONTENTS

ABSTRACT

Cement is the only standard material in concrete composition. The present results simply corroborated the fact that nano-additions on ordinary Portland cement (OPC) influences its hardened properties appreciably in both short term (1 day, 3 day) and medium term (28 days). As 90–95% cement hardens within 28 days from its casting, it is a standard protocol to test the cement properties at 28 days. The experimental work was on the hardened properties of basic cement mortar of a suitable w/c ratio of 0.4 (based on cement's consistency) and sand (river):cement = 3:1 which were initially mixed and then modified with nano-silica (nS) (in proportions ranging 0, 0.5, 0.75, 1, 1.25, and 1.5% @ cement wt.), multi-walled Industrial Grade carbon nanotubes (CNT) (0, 0.02, 0.05, and 0.1% @ cement wt. dispersed in a suitable admixture-superplasticizer-1% poly-carboxylate ether (PCE) @ cement wt.) and nano-titanium oxide (n-TiO$_2$) (0, 1, and 2.5% @ cement wt.) also dispersed in a suitable admixture-superplasticizer-1% PCE @ cement wt. as the latter two were found unsuitable in dissolving in water in its original powder form.

6.1 INTRODUCTION

Concrete is a macro-material strongly influenced by its micro (or nano) properties. Concrete is the most consumptible material on earth after water and Portland cement is the most common and widely used construction material. But concrete made with Portland cement has many disadvantages such as lack of sustainability, durability, carbon dioxide permissibility apart from the financial restraints. These constraints, however, could be avoided by reducing the particle size of cement or by using nanotechnology. Today, we wear nanotechnology application based clothes, we talk using nano-technology application based cell phones, we drive to office in cars that have nanotechnology based paints (not the world's cheapest TATA NANO car), and last, but not the least, we eat food that have nanotechnology based applications. Nanomaterials are increasingly coming into use in healthcare, aviation, automotive, telecommunications and networking, textiles, energy, electronics, cosmetics, and numerous other useful areas having high potential for applications. Their physical and chemical properties often differ from those of bulk materials, so they often call for specialized treatment both among themselves and for those who uses them for material benefit. There have been many successful nanotechnology based applications which could have been almost impossible without the utility of nano-sized particles.

For example, anti-scratch paints, anti-bacterial paints, anti-fouling concrete, dirt repellant textiles, clothes that need no ironing, non-reflective glasses, wonder drugs, etc., are only the tip of the iceberg. Materials with structure at the nanoscale often have unique optical, electronic, or mechanical properties. Nanomaterials are not new, they were known to us from the pages of history.[1]

The emergence of this technology was caused in 1980s due to the convergence of experimental advances like invention of scanning tunneling microscope (1981) and discovery of fullerenes (1985). Nano-silica (nS), the first nanomaterial to be used in construction, are produced from the vaporization of silica or by feeding worms with rice husk or by precipitation method, while techniques have been developed to produce carbon nanotube (CNT) in sizeable quantities, including arc discharge, laser ablation, high-pressure carbon monoxide disproportionation (HiPco), and chemical vapordeposition (CVD). Nano-titanium oxide (n-TiO_2), a nano-additive known for its self-cleansing properties is the most widely used nanomaterial today with an entire church in Rome sprinkled with it. Nano-titanium oxide (n-TiO_2) is readily mined in its purest form of beach sand from Sankaramangalam, Kerala, India.

The rheology of cement paste, mortar, and concrete is influenced with nano-additions. It also improves the microstructure of the whole concrete system matrix. Work presented by many authors reveals that the nano-additions improve the performance of concrete resulting in high performance concrete (HPC) or ultra-high performance concrete (UHPC) and in turn making concrete durable. The contribution on SiO_2 and silica fume (SF) to various micro structural properties of cement paste, mortar and concrete[13] was reported. Also the possibility of modifying the matrix of a cement mortar mixes with and without SF by incorporation of nS solids during the mixing stage[15] was studied.

6.2 LITERATURE REVIEW

6.2.1 ON EFFECT OF NANO-SILICA

Belkowitz and Armentrout[3] developed relationships to distinguish the benefits when using different sizes of nS in cement hydration paste through experimenting and measuring the heat of hydration of multiple mix designs and showed that as silica particles decreased in size with increased size distribution, the C—S—H became more rigid and in turn increasing the compressive strength.

Quercia and Brouwers[4] aimed to present in their chapter the nS production process from olivine dissolution, their addition and their application in concrete.

Valipour et al.[5] studied the influence of nS addition on properties of concrete when compared with SF through measurement of compressive strength, electrical resistivity, and gas permeability. The results show that the replacement of a portion of SF with nS is more active in early age due to larger specific surface and fineness, and will also improve the durability aspects of HPC.

Jemimath and Arulraj[6] reported that cement replaced with nanocement (NC), nanoflyash (NFA), and nano-silica fume (NSF) showed an unaffected consistency, but it was found that the addition of NC decreases the initial and final setting times while the addition of NFA and NSF increases the initial and final setting times.

Maheswaran et al.[8] attempted to highlight the influence of Ns toward pore filling effect and its pozzolonic activity with cement for the improvement of mechanical properties and durability aspects. The paper also says that there is a scope for development of crack free concrete.

Yang.[9] presented the laboratory investigations that when nano-silicon powder mixing content is 0.5, 0.75, and 1.0% compared with ordinary concrete ,the bending tensile strength at 28 days were increased by 3.2, 7.5, and 4.0% and the shrinkage rate at the same age reported increase by 75.5, 127.1, and 163.0%.

Yuvraj[10] described when nS is added, it makes the concrete less alkaline as C–H in concrete is reduced, which reduces the corrosion of steel bars. He also added that more C–S–H is produced at the Nanoscale, thus increasing the compressive strength.

Abyaneh et al.[11] investigated the compressive strength, electrical resistivity and water absorption of the concrete containing nS and micro-silica at 7, 14,and 28 daysreported that concrete with micro-silica and nS have high compressive strength than those with only micro-silica. He further deliberated that specimens with 2% nS and 10% micro-silica have less water absorption and more electrical resistance in comparison with others.

Rajmane et al.[15] showed that nS cannot be used as an admixture to improve the microstructure of the cement composites (with and without 5% SF) at the w/c ratio of 0.5.

6.2.2 ON EFFECT OF CARBON NANOTUBES

Kumar et al.[7] discussed the effect of multiwalled CNTs (MWCNT) on strength characteristics of hydrated Portland cement paste by mixing various proportions of MWCNT and found an increase in compressive and tensile strength of 15 and 36% at 28 days.

Madhavi et al.[14] found an increase in compressive and split-tensile strengths of samples with increasing MWCNT. 0.045% of MWCNT has improved the 28 days compressive strength by 27% while the split tensile strength increased by 45%. Crack propagation was reduced and water absorption decreased by 17% at 28 days curing.

6.2.3   ON EFFECT OF NANO-TITANIUM DIOXIDE

Jayapalan et al.[2] threw light on the effect of nano-sized titanium dioxide on early age hydration of Portland cement. He said that the addition of TiO_2 to cement modifies the hydration rate primarily due to dilution, modification of particle size distribution and heterogeneous nucleation. But when the dosage is increased, cement dilution is resulted with an increase in w/c ratio.

Lucas et al.[12] concluded that photo catalytic activity of TiO_2 increases with increase in dosage but its mechanical strength decreases for addition of more than 1% wt of TiO_2.

6.3   SCOPE OF PRESENT WORK

The nS, CNT (multi-walled, Industrial grade), and n-TiO_2 are products of nanotechnogy are commercially available. The overall aim of our work is to find the optimum dosages of nanomaterials that would produce the maximum mechanical strength at 28 days when 90–95% of cement hydration gets completed. These optimum dosages as found out in this paper could throw a alight on concrete behavior when repeated the same for it.

6.4   EXPERIMENTAL PROGRAM

The experimental program is as follows:

6.4.1   MATERIALS USED

1. Cement: Ordinary Portland cement (OPC) Type
2. Reinforcement Bar: NA

3. Stone Aggregate:　　　NA
4. Fine Aggregate:　　　Natural (River) Sand
5. Water:　　　　　　　Drinking (Tap) Water
6. Chemical Admixture:　Superplasticizer type: Melamine Formaldehyde (MF) and Poly Carboxylate Ether (PCE)
7. Nanomaterials:　　　Nano-Silica (SiO_2), Multiwalled Carbon Nanotubes (MWNTs) and Nano-Titanium Oxide

6.4.2 LABORATORY EXPERIMENTS

The laboratory test procedure involved testing nS in variable quantities (0, 0.5, 0.75, 1.0, 1.25, and 1.5% by cement wt), Carbon-Nanotubes dispersed in Superplastcizer (PCE and MF) at 1% by wt of cement) at various dosages (0.02, 0.05, and 0.1% by wt of cement as per literature review) and n-TiO_2 in variable quantities (0, 1.0, and 2.5 by cement wt) dispersed in PCE at 1% by wt of cement in Portland cement. Mortar cubes of $70.7 \times 70.7 \times 70.7$ mm^3 dimensions are filled with 1 part of cement + 3 parts of river sand with water added, according to the standard formula $P = (P/4 + 3)$ (1 part cement + 3 parts sand). Here P = quantity of water and P = consistency of cement used, that is, the amount of water used to make 300 g cement paste to support a penetration of 5–7 mm in a standard Vicat mould with a Vicat needle. Now we would be testing the Compressive Strength of both composite and ordinary cement mortar after 1, 3, 7, 28, 90, and 180 days ordinary curing in a compression cube crushing/testing machine (Figs. 6.1–6.3).

FIGURE 6.1　Lycurgus cup turns into different colors indicating the presence of poison.

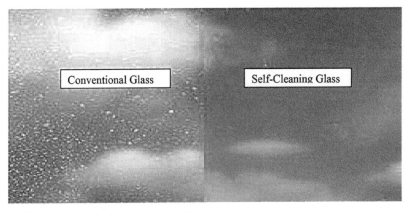

FIGURE 6.2   Use of nanomaterials in self-cleaning glasses.

6.4.3   DISPERSION OF CNTs

It is to be noted here that CNTs were insoluble in water so it could not be dissolved in aqueous medium and sonication method was used to disperse CNTs in MF admixture [ultrasonication for about 30 min in an external ultra-sonicator bath (250 W Piezo-U-Sonic Ultrasonic Cleaner)] (Fig. 6.4), whereas in PCE, CNTs got well dispersed without sonication (Figs. 6.3–6.13).

Figures 6.1–6.3 show the X-ray diffraction (XRD) images of nS, CNT, and titanium di-oxide used.

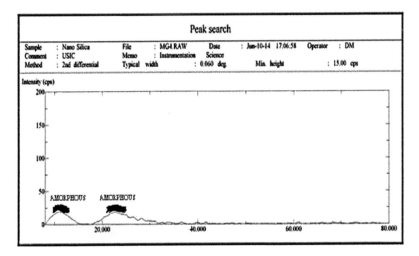

FIGURE 6.3   XRD image of nano-silica used.

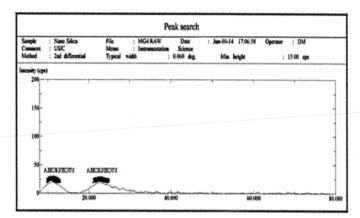

FIGURE 6.3   XRD image of nano-silica used.

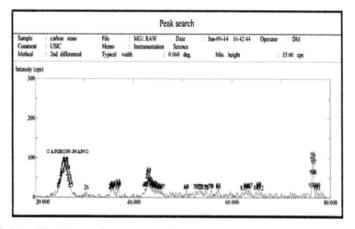

FIGURE 6.4   XRD image of carbon nanotubes used.

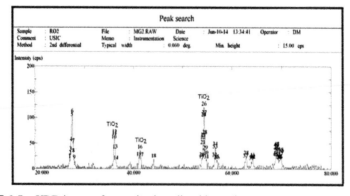

FIGURE 6.5   XRD image of nano-titanium di-oxide used.

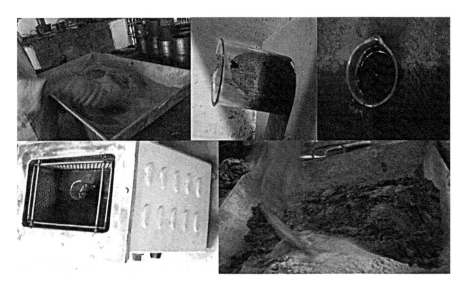

FIGURE 6.6 Some apparatus used for mortar castings.

FIGURE 6.7 Mortar stirrer.

FIGURE 6.8   Ext. vibrator for mortar cubes.

FIGURE 6.9   Compression testing machine.

FIGURE 6.10   (a) Cup and cone failure (b) eccentric failure.

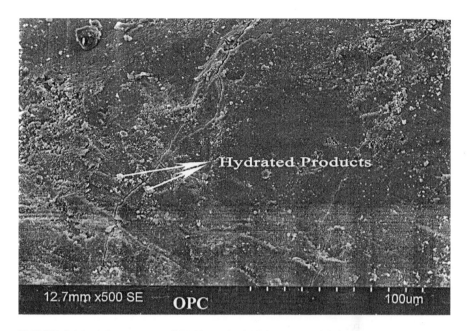

FIGURE 6.11   Microstructure (SEM image) of OPC mortar at 28 days.

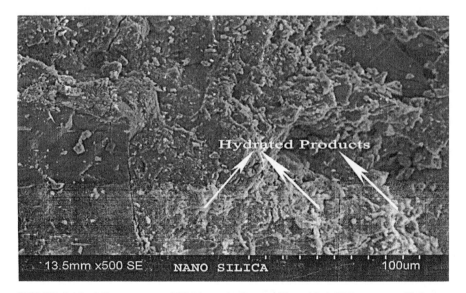

FIGURE 6.12   Microstructure (SEM image) of nano-silica added OPC mortar at 28 days.

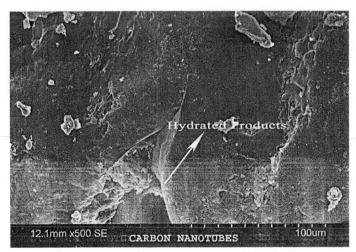

FIGURE 6.13   Microstructure (SEM image) of CNT added OPC mortar at 28 days.

Tables 6.1–6.6 show the specific properties of nS, CNT, and titanium di-oxide used here.

TABLE 6.1   The Specific Properties of Nano-Silica (SiO$_2$) Used.

Sample	% Content (Lit.)	Specific Gravity (Lab.)	% Content (Lab.)	Specific Gravity (Lit.)
XLP	14–16	1.12	21.4	1.08–1.11
XTX	30–32	1.16	40.74	1.20–1.22
XFXL	40–43	1.24	41.935	1.30–1.32

TABLE 6.2   The Specific Properties of Multi-Walled CNTs (Industrial Grade) Used.

Item	Description
Diameter	20–40 nm
Length	25–45 nm
Purity	80–85% (a/c Raman Spectrometer and SEM analysis)
Amorphous carbon	5–8%
Residue (calcination in air)	5–6% by wt.
Average interlayer distance	0.34 nm
Specific surface area	90–220 m^2/g
Bulk density	0.07–0.32 g/cc
Real density	1–8 g/cc
Volume resistivity	0.1–0.15 ohm.cm (measured at pressure in powder)

TABLE 6.3 Specific Properties of TiO$_2$ Used.

Nano-titanium oxide%	97
Rutile content%	98
pH	7
Average particle size (TEM)	30–40 nm
Treatment	Nil
Moisture%	1.75–2
Bulk density	0.31 g/cc
Water solubility	In-soluble

6.5 RESULTS

6.5.1 MECHANICAL STRENGTH TEST

The mechanical or compressive strength is done as per the standard conditions of IS: 516 by applying uniaxial loading (say P) and the compressive strength is measured by $S = P_{max}/A$, where A is the c/s area of cube. The compression-testing machine (Fig. 6.11) has a capacity of 200 Tons (max.) and its rate of loading is 140 kg/cm^2/min. The standardized end conditions were censured by the machine's plane surface to eliminate the local stress concentrations which has a tendency to induce reduction in the test results.

Tables 6.4–6.6 show the strength development at various ages in various types of cement mortar.

TABLE 6.4 Strength (N/mm^2) for Various Proportions/Ages of nS Added OPC Mortar (% Increase with Respect to Ordinary Control Cement Cubes).

Cement	1 Day (% incr.)	3 Days (% incr.)	7 Days (% incr.)	28 Days (% incr.)
OPC	19.59	23.72	21.08	31.89
(0% nS	(–)	(–)	(–)	(–)
OPC	16.28	27.16	23.85	35.51
(0.5% nS	(−16.89%)	(14.50%)	(13.14%)	(11.35%)
OPC	16.83	30.10	27.73	42.27
(0.75% nS	(−14.08%)	(26.89%)	(31.54%)	(32.55%)
OPC	18.36	19.38	25.07	37.36
(1.0% nS	(−6.27%)	(−18.29%)	(18.93%)	(17.15%)
OPC	20.16	27.54	23.17	30.85
(1.25% nS	(2.91%)	(16.10%)	(9.91%)	(−3.26%)
OPC	20.69	23.35	23.81	37.79
(1.5% nS	(5.62%)	(−1.56%)	(12.95%)	(18.50%)

TABLE 6.5 Strength (N/mm²) for Various Proportions/Ages of CNT added OPC Mortar (% Increase with Respect to Ordinary Control Cement Cubes).

Cement	1 Day (% incr.)	3 Days (% incr.)	7 Days (% incr.)	28 Days (% incr.)
OPC	19.59	23.72	21.08|	31.89
	(–)	(–)	(–)	(–)
OPC	NA	NA	18.88	28.73
(1% PCE)			(–10.44%)	(–9.91%)
OPC	NA	11.03	17.69	43.75
(1% PCE + 0.02% CNT)		(–53.5%)	(–16.08%)	(37.19%)
OPC	NA	7.98	27.19	34.88
(1% PCE + 0.05% CNT)		(–66.36)	(28.98%)	(9.37%)
OPC	NA	14.21	21.69	24.83
(1% PCE + 0.1% CNT)		(–40.09%)	(2.89%)	(–22.14%)
OPC	NA	13.68	28.59	41.15
(1% MF)		(–42.33%)	(35.63%)	(29.04%)
OPC	NA	15.86	30.85	31.84
(1% MF+ 0.02% CNT)		(–33.13%)	(46.35%)	(–0.15%)

TABLE 6.6 Strength (N/mm²) for Various Proportions/Ages of TiO_2 Added OPC Mortar (% Increase with Respect to Ordinary Control Cement Cubes).

Cement Type	1 Day (% incr.)	3 Days (% incr.)	7 Days (% incr.)	28 Days (% incr.)
OPC	16.08	18.25	16.58	32.60
(0% PCE + 0% TiO_2)	(–)	(–)	(–)	(–)
OPC	NA	19.27	21.27	37.01
(1% PCE + 0% TiO_2)		(5.59%)	(28.28%)	(13.53%)
OPC	NA	27.81	25.24	36.71
(1% PCE + 1% TiO_2)		(52.38%)	(52.23%)	(12.59%)
OPC	NA	21.67	20.34	34.97
(1% PCE + 2.5% TiO_2)		(18.74%)	(22.68%)	(7.27%)

6.6 DISCUSSION OF TEST RESULTS

The results of mortar compressive strength determined as per IS:4031 obtained when compared with a controlled cement-mortar sample are: (1)

For nS, there was no gain in early strength (1 day) but with increase in age, the gain in strength was observed to be more than 30% at 7 and 28 days with the maximum strength optimized at 0.75% nS for all ages. (2) For CNTs, there was no gain in early strength (3 days) but with increase in age, strength gain was more than 35% at 28 days with the maximum strength optimized at 1% PCE + 0.02% CNT. (3) For TiO_2 there was early high gain in strength of more than 52% (for 3 and 7 days), 12% for 28 days and the maximum strength was found to be optimized at 1% TiO_2. As CNTs was insoluble in an aqueous medium another superplasticizer, namely MF @ 1% by cement wt. was tried with a same optimization of 0.02% CNTs. A decrease of 33% at 3 days and an increase of 30% and 41% at 7 and 28 days of compressive strength were observed.

6.6.1 FRACTURE MECHANISM

There are two types of failure of mortar specimens as shown in Figure 6.12a—cup and cone failure and Figure 6.12b—eccentric failure. Under uniaxial compressive loading, the pores in the paste tend to grow parallel to the direction of loading, resulting in an Eccentric fracture plane (Eccentric failure). However, as concrete is weak in tension, pores also tend to grow in perpendicular directions under poisson effect, resulting in a cup and cone fracture plane due to combined effects of both (cup and cone failure).

6.7 CONCLUSIONS

In this chapter, an attempt was made to find out the optimized proportions of the nanomaterials namely, nS, CNT, titanium di-oxide, etc., to be added in the cement mortar matrix for maximum strength gain, keeping w/c ratio fixed at 0.4.

1. The results of the investigation showed that for OPC, addition of 0.75% nS and 0.02% CNT with respect to cement wt. shows maximum compressive strength gain at all ages, while addition of more than 1% nano-titanium di-oxide decreases its compressive strength.
2. Further research on micro structural studies is necessary for the characterization of nanomaterials in cement mortar.

KEYWORDS

- **cement**
- **nano**
- **term**
- **properties**
- **test**

REFERENCES

1. http://en.wikipedia.org/wiki/Lycurgus_Cup
2. Jayapalan, A. R., et al. Effect of Nano-sized Titanium Dioxide on Early Age Hydration of Portland Cement. In *Nano Technology in Construction 3*; Springer Publications: New York, 2009; pp 267–273.
3. Belkowitz, S. J.; Armentrout, D. In *An Investigation of Nano Silica in the Cement Hydration Process,* Concrete Sustainability Conference, National Ready Mixed Concrete Association, Europe, 2003.
4. Quercia, G.; Brouwers, H. J. H. In *Application of Nano-Silica (nS) in Concrete Mixtures,* 8th fib Ph.D. Symposium in Kgs, Lyngby, Denmark, 2010.
5. Valipour, M., et al. Comparative Study of Nano-SiO$_2$ and Silica Fume on Gas Permeability of High Performance Concrete (HPC), Korea Concrete Institute: Seoul, South Korea, 2010.
6. Jemimath, C. M.; Arulraj, G. Effect of Nano-Flyash on Strength of Concrete. *Inter. J. Civil Struct. Eng.* **2011,** *2* (2), 475.
7. Kumar, S., et al. Effect of Multi-walled Carbon Nanotubes on Mechanical Strength of Cement Paste. *J. Mater. Civil. Eng.* **2012,** *24* (1), 84–915.
8. Maheswaran, S., et al. An Overview on the Influence of Nano Silica in Concrete a Research Initiative. *Res. J. Recent. Sci.* **2012,** *2* (2012), 17–24.
9. Yang, H. In *Strength and Shrinkage Property of Nano Silica Powder Concrete,* 2nd International Conference on Electronic & Mechanical Engineering & Information Technology, China, 2012.
10. Yuvraj, S. Experimental Research on Improvement of Concrete Strength and Enhancing the Resisting Property of Corrosion and Permeability by the Use of Nano Silica Fly-ashed Concrete. *Inter. J. Emerg. Technol. Advan. Eng.* **2012,** *2* (6), 105–110.
11. Abyaneh, M. R. J., et al. Effects of Nano-Silica on Permeability of Concrete and Steel Bars Reinforcement Corrosion. *Aus. J. Basic. App. Sci.* **2013,** *9*, 464–467.
12. Lucas, S. S., et al. Incorporation of Titanium Dioxide Nanoparticles in Mortars-Influence of Microstructure in the Hardened State Properties and Photo Catalytic Activity. *Cem. Concr. Res.* **2013,** *43*, 112–120.
13. Ghosh, A., et al. Effect of Nano-Silica on Strength and Microstructure of Cement Silica Fume Paste, Mortar and Concrete. *Ind. Conc. J.* **2013,** *14* (1), 11–25.

14. Madhavi, T. Ch., et al. Effect of Multiwalled Carbon Nanotubes on Mechanical Properties of Concrete. *Inter. J. Sci. Res.* **2013,** *2* (6), 166–168.
15. Rajmane, N. P., et al. Application of Nano-Technology for Cement Concrete, Proceedings of National Seminar, Karnataka, Institution of Engineers (I), Karnataka centre, Dec 15–18, 2013.

CHAPTER 7

STUDIES IN PREPARATION OF REDISPERSIBLE POWDER PUTTY AND POWDER PAINT

R. H. DONGRE and P. A. MAHANWAR[*]

Department of Polymer and Surface Engineering, Institute of Chemical Technology, Mumbai, India

[*]*Corresponding author. E-mail: pa.mahanwar@ictmumbai.edu.in*

CONTENTS

ABSTRACT

The requirements of coatings are increasing day by day for protecting different substrates from various types of stresses. Today, the decorative paints, specif-ically water-based paints are in liquid form resulting in increased handling cost. The purpose of this chapter is to develop a single component powder putty and powder paint using easily available nonconventional materials at affordable cost. This chapter deals with studies in effect of concentration and mode of addition of various components, on the properties of paint for decorative applications. These powder putty and powder paint formulations have been evaluated and compared with conventional decorative paints. The powder paint formulations have also been evaluated for exterior applica-tions, namely mechanical, thermally insulating, and chemical performance properties. The effects of concentration of glass microspheres on properties of powder paints have also been studied.

It has been seen that powder putty and powder paint show satisfactory performance as compared to conventional putty paste and liquid paint. The new paint formulations, added with glass microspheres are found to give very good heat insulation where temperature difference is 15–20°C with respect to surrounding temperature and the coating. These products are based on simple and easily available raw materials. The existing equip-ment/machinery can be used for preparation of these products. Also this new coating is applicator friendly and easy to apply.

7.1   INTRODUCTION

Prior to painting interiors of house, surface preparation includes three main stages, that is, applying the primer to the surface, filling of putty, and applying finish coat (one or two coats). The material usually consists of a primer in liquid stage (either water base or hydrocarbon solvent based), while putty is a dry powder (blend of different raw materials), which is mixed with water to make suitable thick paste which can be applied by available means to desired thickness. This putty coat after drying is sanded and leveled before coating with the finish coat.

The primer or topcoat in case of a latex paint is usually an emulsion of acrylics, styrene-acrylic, vinyl, etc. The average molecular weight of these materials ranges from 100,000 to 1,000,000.[1] Acrylic polymers are not biodegradable and may accumulate in the soil of surface waters if not properly treated from the effluent treatment plant.[2]

The alkyd-based paints and varnishes are found to produce 250–350 g/L of VOC from their formulations.[3] The alkyds are prepared from condensation reaction of polybasic acids and polyhydric alcohol along with fatty acids. The materials used apart from oil/fatty acids in both forms of paints are not environmental friendly and requires high temperature processing.[4]

The putty used is most conventionally based on mixed material extenders like calcium carbonate or other mineral extenders. Although calcite is nonhazardous but has poor performance in terms of flaking, cracking, poor alkali and acid resistance, poor dry scrub, wet scrub resistance, etc.

In the 19th century, several workers experimented with alkaline silicates. Such paints were two-component systems based on potassium silicate solutions and were required to mature before application. Such systems are rarely used now and are mostly limited to professional specialist applicators.[5] Sodium silicate also has been used in present formulation as one of the ingredient having molecular formula Na_2SiO_3. The material is alkaline and hence is best suitable for alkaline surfaces such as cement, plaster of Paris, etc. This material has good water solubility, and is found to be excellent in water resistance, heat resistance, etc. It can be reacted with a variety of materials such as CO_2 gas, $Na_2(CO_3)_2$, and NaH_2PO_4. The solution of this, when applied as a thin layer on surfaces of other materials, dries to form a tough, tightly adhering inorganic bond, or film.

The setting agents such as finely divided sodium silicofluoride dissolve slowly in water and are used up to 7% of weight of liquid silicate. The coatings prepared with this, exhibit longer working lives.[6] The setting agents for sodium silicate can be classified as: inorganic salts, mineral acid, organic acids, and inorganic oxides. The sodium silicate reacts with calcium salts $[CaCl_2, Ca(OH)_2]$ to forms calcium silicate hydrate. Mineral acids such as H_2SO_4, HCl, and organic acids such as citric acid, acetic acid, and stearic acid are found to act as setting agents for sodium silicate. The inorganic oxides such as magnesium oxide, zinc oxide, calcium oxide, etc., are also found to work as setting agents for sodium silicate.[7] However, use of sodium silicate along with careful selection of other raw materials in definite proportion is found to deliver promising results.

Historical monuments found in medieval Indian architecture made use of jaggery, lime, and calcium oxide for the construction of buildings. These coatings had superior alkali, moisture resistance, and color stability. In old Indian mural paintings also, the use of sugar/jaggery (an unrefined sugar) has been reported. One of the old palm leaf manuscripts found in the Padmanabhapuram palace contains the reference for the use of a variety of herbs, fruits, and a particular species of cactus, blended with palm jaggery and was left to ferment for 15 days prior to use it as plaster along with lime.[8]

In Central India, the famous Charminar (the mosque of four minarets) is now an iconic and historical monument built in1591–1593 by Quali Qutub Shah, the fifth ruler of the Qutub Shahi. In the construction, use of granite and lime mortar was made along with use of shell, jaggery, and white bulk of eggs to enhance the properties of mortars and plasters.[9] Similarly the historic St. Lourdes Church in Tiruchirappalli, India, was rehabilitated and reno-vated. Here also, the lime mortar used for the preparation contained lime and sand in the ratio of 1:2 along with jaggery and gall nut.[10] Taking the concept of use of jaggery in these historical monuments, we have used jaggery along with calcium oxide in a few of our formulations.

Vinyl acetate-ethylene copolymer has been in use in coating industries for a long time.[11] Vinyl acetate/ethylene copolymer powders (redispersible powder termed as RDP) are obtained by spray-drying of corresponding aqueous polymer dispersions, while simultaneously adding fine inorganic materials such as anti-caking agents. The polymer powders thus obtained have excellent storage stability. They are easily redispersible in water and give plastics mortars of high water resistance. The plaster made from this system is free from cracks.[12]

The RDP powders are free flowing with very good storage life even at elevated temperature. They can be dispersed easily in water within a few seconds, forming creamy aqueous dispersions having a constant viscosity and good shelf life. The emulsion obtained from the polymer powders dispersion has good alkali stability. They are also used for the preparation of glue, adhesives, and paints. Plasters prepared with mortars containing these powders are, surprisingly, free from cracks and are water resistant. This RDP powder has been used as binder in our powder putty and powder paint formulations.

There are a number of manufacturers of vinyl acetate/ethylene redispers-ible polymer powder, with specific features in every product. Some of these types of products offer specialized properties such as vapor permeability, self-leveling, high decorative value, etc.[13]

The use of hollow microspheres in coatings has been the subject of study in the past decade. The microspheres can be classified as organic hollow micro-spheres (e.g., polystyrene, polyacrylonitrile, or phenolic materials) and inor-ganic hollow microspheres (e.g., glass, ceramic, or fly ash from thermal power plants).[14] Thermally insulating latex paint (polyacrylic emulsion) for exterior purpose was prepared with difibered sepiolite and hollow glass microspheres as the main functional additives. It has been found that with the use of 8% difibered sepiolite fibers and 6% hollow glass microspheres in the coating, effective reflection of light and heat insulation effect can be achieved.[15]

The objective of this study is to develop finely ground dry blend of powder putty and powder paint based on principle ingredients such as sodium silicate, aluminum oxide powder, china clay, and RDP powder along with different setting agents and compare the properties of these products against the standard established putty and standard emulsion paint from the market. The formulation has been tried with jaggery and calcium oxide and checked for mechanical and chemical resistance properties. 1% and 2% glass microspheres have been added to this paint to check heat insulation properties. The results have been studied in a specially prepared electric oven as shown in schematic diagram no.1 as reported by P.A. Mahanwar and P.K. Sahu up to 70°C.[16]

7.2   EXPERIMENTAL

7.2.1   MATERIALS

a.   Sodium silicate procured from Otto Chemicals Company of 99% purity was used as such without further testing of its purity.
b.   Aluminum oxide powder purchased from S.D. Fine Chemicals Ltd. with 99 % purity has been used in the formulation.
c.   China clay from local supplier has been used with testing of oil absorption value of 37 (g of linseed oil/100 g of pigment).
d.   Stearic acid (L.R. Grade) used was supplied by PCL Company, Pune.
e.   Glass microspheres sample supplied by 3M of K 1 Grade having typical density of 0.125 g/cc and average particle size of 30–120 μ has been used as such without any further testing.
f.   Jaggery from local grocery shop has been used.
g.   Calcium oxide (L.R. Grade) supplied by PCL Company, Pune.
h. TiO_2 (A) from KMLL has been used in the formulations.
i.   RDP powder (vinyl acetate-ethylene copolymer powder) from Indofill Chemicals, Mumbai has been used in the formulation.

7.2.2   METHOD FOR THE PREPARATION OF PUTTY AND PAINT

All the raw materials used in the formulations (putty/paint) are in dry form and were added in the lab scale jar mill. The mill was allowed to run overnight. The finely ground material was added with required quantity of water in the ratio from 4:1 to 6:1 to prepare putty paste. The product prepared this way was exhaustively tested for various physiochemical properties.

The paint powder was prepared in the similar manner with addition 5% of TiO_2. After grinding, the powder was diluted with water in the ratio of 1:1 (paint powder:water) to prepare paint and again detailed testing of this type of product for various types of formulations was carried out. The asbestos panels were used for testing purpose.

7.2.3 METHOD FOR THE PREPARATION AND EVALUATION OF POWDER PUTTY + POWDER PAINT SAMPLES (MONOCOAT SYSTEM)

The developed powder putty coated with powder paint samples (monocoat system) were tested extensively against established putty sample from market, coated with established standard emulsion paint sample. The putty samples prepared according to formulations, are applied as per standard application process for putty using putty blade/spatula over asbestos panels of dimensions $100 \times 150 \times 5$ mm^3. The putty samples were allowed to dry and become tack free and were immediately coated with similar types of paint samples, which are prepared as per standard formulations. The putty PY-2 Sample was coated with paint PT-2 Sample, whereas putty PY-3 was coated with PT-3 type of top coat, and sample PY-4 was coated with Paint samples of PT-4(2). These composite systems are referred to as Monocoat System and are named as MCS-2, MCS-3, and MCS-4D2, respectively. The composite system is also evaluated for its physiochemical (performance) properties.

7.2.4 METHOD OF PREPARATION OF HEAT INSULATION COATING WITH GLASS MICROSPHERES

The prepared paint samples, from powder paint were added with 1–2% glass microspheres under specially developed glass stirrer with stirring rate of 40–60 rpm. The prepared paint samples have been applied on metal panels of dimensions $150 \times 96 \times 0.8$ mm^3 and were tested in specially developed electric oven up to temperatures of 50–70°C.

7.2.5 FORMULATIONS

(A) Putty: The different sets of formulations prepared, were as follows:–
All the raw materials were weighed accurately in accordance with the

formulations. Based on previous experimental work, four formulations are finalized. The sodium silicate selected as the main ingredient has been used up to 20% (max) in the set of formulations. As per our previous experience, aluminum powder which imparts property of hardness in the film is used maximum up to the level of 10% of total formulation weight. RDP is used to basic minimum level to act as binder in putty and is kept 10% in the formulation. China clay as an extender has been added to the level of 50–60% in the formulation. Several reactive ingredients/setting agents for sodium silicate such as calcium chloride, calcium hydroxide, sodium carbonate, sulfuric acid, hydrochloric acid, etc., have been tried in the formulations. Organic acids such as citric acid, acetic acid and inorganic oxides such as magnesium oxide, zinc oxide were also used. However, stearic acid and calcium oxide along with jaggery are found to act as excellent setting agents for sodium silicate.

Based on number of experiments, four set of formulations PY-1, PY-2, PY-3, and PY-4 were finalized for preparation of powder putty.

(B) Powder Paint: The selected putty formulations were extended for powder paint formulations with increase in content of RDP powder in the formulations and addition of TiO_2 for hiding purpose. Thus, the PY-1 formulation was further developed as PT-1, PY-2 formulation as PT-2, PY-3 as PT-3 and PY-4 were extended as PT-4. This PT-4 formulation was further developed as PT-4D1 and PT-4D2 with marginal increase of RDP powder in the formulation. Thus having kept all other ingredients at constant level, in line with putty formulations, RDP% was increased from 10 to 20% in powder paint formulations. The TiO_2 was added up to 5% for the hiding purpose. Looking at the poor performance of primary properties of PT-1 powder paint formulation, further tests were not carried out. So, for powder paint formulations, a total of four powder paint formulations namely PT-2, PT-3, PT-4, and its extension to PT-4D1 and PT-4D2 were developed. The samples were exhaustively tested for their physiochemical properties and the results are shown in Figures 7.1–7.14.

7.2.6 CHARACTERIZATION OF PUTTY, PAINT, AND MONOCOAT SYSTEM

The coated panels for putty, paints as well as for monocoat systems were tested for their mechanical and chemical properties according to the following standard methods:

Weight/Liter: Standard ASTM D-1475 method was followed to determine weight per liter content of putty paste samples and paint samples.

Percentage Solids: The ASTM D2697-03 standard procedure was used to determine the percentage solids of the powder putty and powder paint.

Viscosity by Brookfield Viscometer: The viscosity of putty paste was measured in accordance to ASTM D2196-99 standard using Brookfield Viscometer (RV Model) for testing consistency of prepared putty materials. Spindle number 7 was used for measurement of viscosity. The viscosity was tested at variable spindle speeds such as 5, 10, 20, and 50; speed at 10 RPM was found to be more suitable for the purpose. The viscosity is calculated using the following formula:

$$V = fs$$

where V is the viscosity of sample in centipoises, mPa·s,

f is the scale factor furnished with instrument, and

s is the scale reading of viscometer

Viscosity by Ford Cup B-4: The viscosity of powder paint after thinning to 1:1 ratio (powder: H_2O on w/w basis) was determined by following the ASTM D1200–94) standard procedure.

Hiding: Paint samples tested for their respective hiding power using ASTM D 5150–92 standard.

Drying Time: The drying time of applied coating was tested according to ASTM D 5895–03 standards. The time was recorded in three stages as set to touch time (touch dry time), tack free time and dry-hard (hard dry) time.

Pencil Hardness: In the pencil hardness test. ASTM D3363 method was followed to measure pencil hardness of coatings.

Dry Scrub Resistance: Applied coating specimens of putty as well as paint were tightly rub against finger pressure and the results were checked for any traces of removal of paint, powder from the coated panel to finger.

Wet Scrub Resistance: Wet scrub resistance was carried out according to ASTM D2486, which covers a procedure for determining the resistance of wall paints to erosion caused by scrubbing. Wet scrub resistance was carried out by using wet scrub resistance apparatus by Sheen.

Humidity Resistance: All the panels of putty powder and paint powder were tested for their humidity resistance in accordance to the ASTM 2247–02 standard procedure.

Chemical Resistance Properties: The exposure of paint film to various chemicals could reduce its gloss, change the color, or produce a swelling or softening of paint film. Resistance to acid and alkali was determined by using Indian Standard 101-1985. For this test, the coating was applied on asbestos panels; these panels were half immersed in 2% aqueous solution of HCl (acid) and 2% aqueous solution of NaOH (alkali). The immersed panels were maintained at constant room temperature. The panels were removed for examination after 6, 12, 18, and 24 h from the start of the test and observed for loss of adhesion, blistering, popping, or any other deterioration of the film.

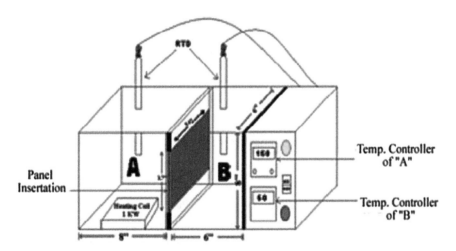

FIGURE 7.1 Schematic diagram of oven used for testing of heat insulation properties of redispersible powder paint system.

7.3 RESULTS AND DISCUSSION

7.3.1 EVALUATION OF POWDER PUTTY AND POWDER PAINT SAMPLES

7.3.1.1 REDUCIBILITY IN WATER

The prepared putty samples and paint samples were diluted with water to bring them to suitable application consistency. The samples showed very good water reducibility and were readily thinned down with water. The putty samples were diluted in the ratio of 4:1 to 6:1 (powder: water) to make them into stiff consistency mass like conventional putty. The paint samples are diluted with water in 1:1 ratio (wt/wt basis). The results of wet putty and paint material have been shown in Tables 7.1–7.6.

7.3.3 PHYSIOCHEMICAL TESTING OF PAINT

TABLE 7.1   Putty Testing.

Sr. No.	Batch No.→ Property↓	Putty 1 Batch [PY-1]	Putty 2 Batch [PY-2]	Putty 3 Batch [PY-3]	Putty 4 Batch [PY-4]	Std. Putty from Market-SPY
1	Mixing ratio putty:water	4:1	4:1	3:1	6:1	3:1
2	Specific gravity	1.56	1.58	1.40	1.52	1.80
3	% solids	80%	81%	67%	65%	78%
4	Viscosity by Brookfield [spindle no.7]	432 poise	436 poise	1145 poise	400 poise	452 poise
5	Consistency	Smooth and uniform	Smooth and uniform	Smooth and uniform	Smooth and uniform	Smooth and uniform
6	Spreading tendency	Excellent	Excellent	Excellent	Excellent	Excellent

TABLE 7.2   Mechanical Barrier Properties of Applied Putty Coatings.

Sr. No.	Testing	PY-1	PY-2	PY-3	PY-4	Standard Putty from Market-SPY
1	Pencil hardness	5 H Passes	6 H passes	HB, H pass, 2H fail	HB, H pass, 2H fail	HB, H pass, 2H fail
2	Dry scrub resistance	Excellent	Excellent	Excellent	Excellent	Moderate
3	Wet scrub resistance	160 double rub test pass	175 double rub test pass	800 double rub test pass	1100 double rub test pass	440 double rub test pass
4	Water vapor transmission (after 7 days)	Test could not completed because of fragile nature of putty sample	25%	30%	29%	Cracks observed so test could not completed

TABLE 7.3 Wet Paint Testing.

Sr. No.	Batch No.→ Property ↓	Paint Batch [PT-2]	Paint Batch [PT-3]	Paint Batch [PT-4]	Paint Batch [PT-4D1]	Paint Batch [PT-4D2]	Premium Quality Emulsion Paint from Market (PQP)
1	Mixing ratio putty:water	1:1	1:1	1:1	1:1	1:1	1:0.65
2	Specific gravity	1.08 g/cc	0.956 g/cc	1.31 g/cc	1.33 g/cc	1.36 g/cc	1.22 g/cc
3	% solids	51%	50%	44%	43%	44%	35%
4	Viscosity by Brookfield [spindle 7]	31 s @ 27°C	37 s @ 27°C	40 s @ 27°C	25 s @ 27°C	46 s @ 27°C	41 s @ 27°C
5	Hiding power	5.74 m²/L	6.37 m²/L.	5.04 m²/L	8.40 m²/L	8.60 m²/L	4.54 m²/L
6	Consistency	Smooth and uniform	Smooth and uniform	Smooth and uniform	Smooth and uniform	Smooth and uniform	Smooth and uniform
7	Spreading tendency	Excellent	Excellent	excellent	excellent	Excellent	Excellent

TABLE 7.4   Mechanical and Barrier Properties of Paint.

Sr. No.	Testing	PT-2	PT-3	PT-4	PT-4D1	PT-4D2	Premium Quality Emulsion Paint from Market (PQP)
1	Pencil hardness	2 H passes	3H pass, 4H fail	HB pass, H pass	H pass, 2H passes	3H passes	4 H passes
2	Dry scrub resistance	Excellent	Excellent	Excellent	Excellent	Excellent	Excellent
3	Wet scrub resistance	55 Double rub test pass	25 Double rub test pass	330 Double rub test pass	560 Double rub test pass	745 Double rub test pass	600 Double rub test pass
4	Water vapor transmission (after 7 days)	7.66%	7.05%	6.9%	5.35%	3.33%	1.83%

TABLE 7.5 Chemical Resistance Properties for Putty Film.

Sr. No	Testing	PY-1	PY-2	PY-3	PY-4	Standard Putty from Market-SPY
1	2% acid resistance after 24 h	Failed after 2 min	Failed after 5 min	Color changes and cracking/flaking observed	Passes ; color changes to dark	Failed after 3 min; film washed out
2	2% alkali resistance after 24 h	Failed after 2.45 hH	Failed after 3 h	Passes, slight puffing observed	Slight color change and fine cracks observed	Failed after 2 h and 35 min; film washed out
3	Humidity resistance	Puffing Observed after 448 h	550 H pass	Puffing observed after 240 h	Excellent passes 650 h	Puffing observed after 268 h.

TABLE 7.6 Chemical Resistance Properties for Paint Film.w

Sr. No.	Testing	PT-2	PT-3	PT-4	PT-4(D1)	PT-4(D2)	Premium Quality Emulsion Paint from Market (PQP)
1	2% acid resistance after 24 h	Failed after 2 min	Color change to yellowish, film is intact	Film intact but color change occurs	Film intact but blackens	Film intact: color changes to dark black	Color change to yellowish, film is intact
2	2% alkali resistance after 24 h	Slight puffing observed after 3 h	Pass, flaking, chalking observed	Excellent no change in color, film is intact	Excellent no change in color, film is intact	Excellent no change in color, film is intact	Excellent no change in color, film is intact
3	Humidity resistance	Excellent 500 H pass	Excellent 500 H pass	Excellent 500 H pass	Excellent 524 H pass	Excellent 542 H pass	Only 224 h pass; puffing observed

7.3.2 PHYSIOCHEMICAL PROPERTIES OF THE PUTTY

7.3.1.2 FILM FORMATION

The prepared putty samples as well as paint samples were found to have excellent spreading tendency with continuity in film, without any sign of breaking or delamination. The applied putty and paint samples were found to have good adhesion on prepared asbestos sheets of dimension $150 \times 100 \times 5$ mm^3. The samples were found to have better spreading rate and formation of smooth and uniform film compared to those of conventional putty and emulsion paint samples from market.

7.3.1.3 DRYING RATE

The putty samples and paint samples are tested for drying rate. The results show that putty samples have touch dry time from 10 min to maximum 30 min and hard dry time from 98 min to overnight. In general, all putty samples were found to surface dry within 30 min. Similarly, paint powder samples when diluted with water in the 1:1 ratio, the drying rate of paint samples were found to touch dry from 7 to 30 min and hard dry time from 108 min to overnight. This test showed that the developed powder putty and powder paint formulation had shorter drying time compared to conventional products. Thus, these innovative formulations can be classified as fast drying powder putty and powder paint. The results are shown in Figure 7.2.

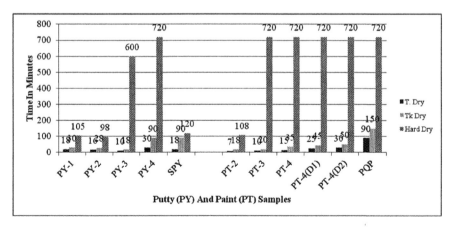

FIGURE 7.2 Drying rate of putty (PY) and paint (PT).

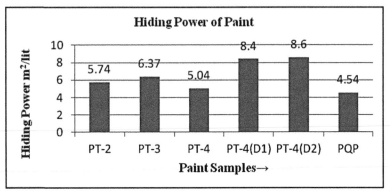

FIGURE 7.3   Hiding power of paint.

7.3.4   FILM TESTING OF POWDER PUTTY AND POWDER PAINT

7.3.4.1   PENCIL HARDNESS TEST

This test is significant, as it indicates the hardness of film against external scratches. The putty samples and paint samples were tested for pencil hardness test after 7 days of hard drying. The results for putty and paint samples against conventional putty and emulsion paint samples are depicted in Figures 7.4 and 7.5. The following table illustrates the gradation for pencil hardness test. From the result, we can conclude that dried film of powder putty and paint samples were at par with existing standard commercial products in terms of surface hardness.

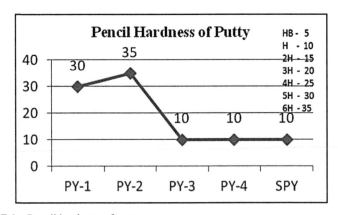

FIGURE 7.4   Pencil hardness of putty.

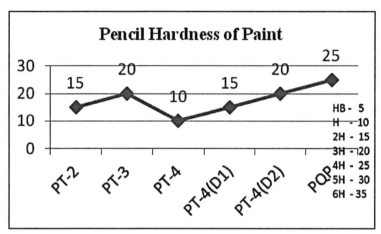

FIGURE 7.5   Pencil hardness of paint.

7.3.4.2   DRY RUB RESISTANCE

Water based coating systems meant for architectural coatings are primarily checked for their dry scrub resistance after film dried. The samples containing stearic acid and jaggery along with calcium oxide were found to have excellent dry scrub resistance compared to available standard samples.

7.3.4.3   WET SCRUB RESISTANCE

This important and indicative test of durability for any water based architectural coating is carried out here. The samples were tested for wet scrub resistance. In the putty sample PY-3, where stearic acid is used, found to pass 800 double rub test and for paint PT-3, film passed 330 double wet scrub test. The samples containing jaggery and calcium oxide PY-4, is found to pass 1100 double wet scrub test for putty, and for paint PT-4, the film found to pass from 330 to 745 double scrub test. Once again, the results were found to be impressive when compared with wet scrub resistance of standard putty and standard emulsion paint samples. The comparative results are shown in Figures 7.6 and 7.7 and also with photo/figures in Figure 7.8 for PT-4, PT-4D1, and PT-4D2 samples.

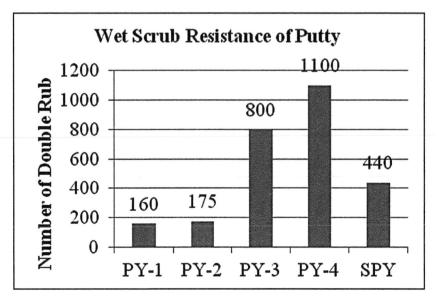

FIGURE 7.6   Wet scrub resistance of putty.

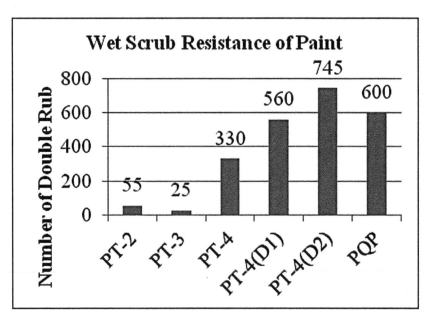

FIGURE 7.7   Wet scrub resistance of paint.

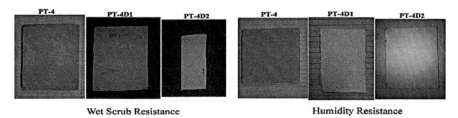

Wet Scrub Resistance   **Humidity Resistance**

FIGURE 7.8 Test panels of wet scrub resistance and humidity resistance test of powder paint.

7.3.4.4 HUMIDITY RESISTANCE

The coated putty samples as well as paint samples were tested for humidity resistance for a minimum of 500 h. duration. The putty sample PY-1 showed film puffing after 448 h while, sample PY-2 passed 550 h test. Sample PY-3 shows puffing after 240 h while sample PY-4 passes 650 h humidity resistance tests. Surprisingly standard putty powder sample (SPY) could not last more than 268 h. Similar is the case with developed powder paint samples, where the premium quality paint sample (PQP) could withstand only up to 224 h humidity resistance test, as against samples PT-4(D1), PT-4(D2), which are found to pass 524 and 542 h test duration, respectively. The results are shown in the form of comparative Figures 7.9 and 7.10 and also in Figure 7.8 of panels. Thus we can say that, our developed powder putty and powder paint samples have excellent humidity resistance property.

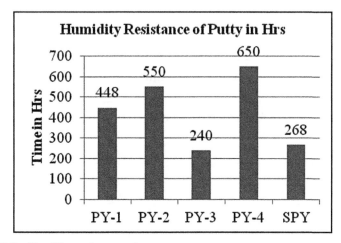

FIGURE 7.9 Humidity resistance of putty.

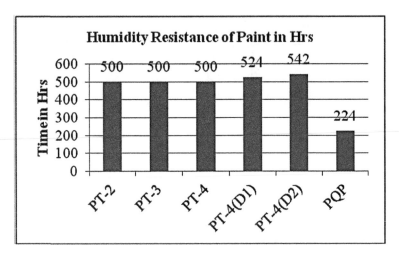

FIGURE 7.10   Humidity resistance of paint.

7.3.4.5   WATER VAPOR TRANSMISSION RATE (WVTR)

Water vapor transmission rate for every putty sample could not be tested due to the fragile nature of few samples particularly for PY-1 and standard putty samples (SPYs). However, for every paint samples (top coat), WVTR has been tested for 168 h. (7 days). These results are shown in Figure 7.11. The test results are significant in view of breathing nature of cement/concrete substrate and requirement of similar nature from the coating applied on it. The developed powder putty and powder paint samples have been found to perform better than standard samples.

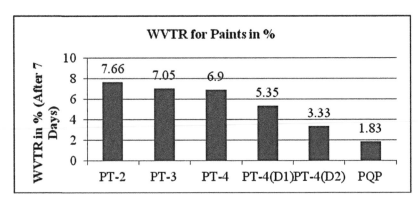

FIGURE 7.11   Water vapor transmission rate for paint.

7.3.4.6   ACID RESISTANCE (2% HCL FOR 24 H)

The coated panels of putty and paint samples after 7 days of hard dry were immersed in the 2% HCl Solution for 24 h. The putty PY-1 panel failed just after 2 min after immersion in acid bath, PY-2 panel after 5 min duration, and for the panel PY-3, slight degree of cracking and flaking was observed with change in color. In case of film of PY-4 panel, it was found to be intact with a dark black color change. The SPY also failed after 3 min. For paint samples except PT-2, all other samples including market sample (PQP) have shown no visible damage to the film except change in color, that is, change from yellowish to dark black or from off white to yellowish. The results can be summarized as good to excellent. Sample photo/figures are shown in Figure 7.12.

FIGURE 7.12   Test panels of acid and alkali resistance test of powder putty and powder paint.

7.3.4.7   ALKALI RESISTANCE (2% NAOH FOR 24 H)

Since the products have been developed for cement/concrete surfaces, their resistances for alkali have been checked by this test method. The coated panels of putty and paint samples after 7 days of hard drying were immersed in the 2% NaOH solution for 24 h. Here, putty panel PY-1 was found to be failed after 2:45 h. and putty panel PY-2 failed after 3:00 h, respectively. In the panel PY-3, slight puffing observed with slight black color change. While for PY-4 (sample containing jaggery and calcium oxide), no effect of alkali was observed, even after 24 h, except change of color of film (off white to reddish). The same putty formulation extended for paint formulation with the addition of 5% TiO_2, that is, paint samples PT-4, PT-4(D1) and PT-4(D2) are found to be excellent for alkali resistance; no effect of alkali was observed even after 24 h only with color of film changed to reddish.

Other paint samples such as PT-2 failed immediately after 3 h, while sample PT-3 passes the test but, the dried film shows slight degree of chalking and cracking.

7.3.4.8   CHALKING, CRACKING, AND FLAKING

The fresh applied coating panels of powder putty and powder paint were allowed to expose at ambient air temperature for 500 h and were thoroughly checked for any signs of chalking, cracking, and flaking tendency. Except in putty sample PY-1, all other putty samples including all applied powder paint coating samples were found to be absolutely free from any chalking, cracking tendency.

7.3.5   PHYSIOCHEMICAL BARRIER PROPERTIES OF MONOCOAT SYSTEM (PUTTY+PAINT)

The physiochemical properties of monocoat systems of MCS-2, MCS-3, and MCS-4D2 have been checked in accordance to evaluation methods explained in Section 7.3.4. The significant results of pencil hardness, wet scrub resistance have been indicated in Figures 7.13 and 7.14.

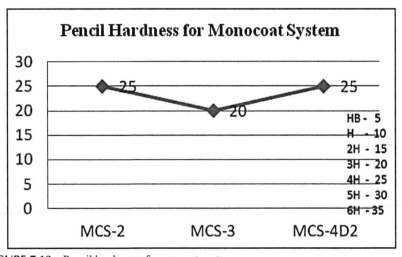

FIGURE 7.13   Pencil hardness of monocoat system.

a. **Pencil Hardness for Monocoat System**: The pencil hardness for MCS-2 panels is found to be of degree 4H, MCS-3 panel is found to withstand up to 3H degree of pencil hardness. For MCS-4D2, pencil hardness is found to withstand up to 4H degree. The results are shown in Figure 7.13. Here again, the results of composite system is at par with the existing standard system, used conventionally.

b. **Dry Scrub Resistance of Monocoat System**: All the samples from MCS-2 to MCS-4D2 show excellent degree of dry scrub resistance without even a slight degree of chalking/powdering of coating.

c. **Wet Scrub Resistance of Monocoat System**: The wet scrub resistance for MCS-2 panel was found to pass 300 double rub scrubbing test. Sample MCS-3 passes 868 double rub scrub resistance test. The panel MCS-4D2 shows the highest level of 1650 double rub scrub resistance test. The results are shown in Figure 7.14. The results are definitely impressive and at par with conventional paint systems being used in architectural painting.

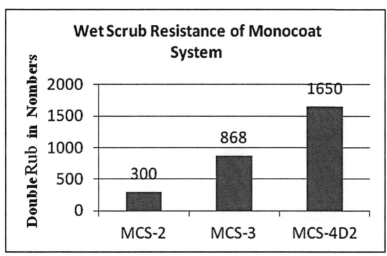

FIGURE 7.14   Wet scrub resistance of monocoat system.

d. **Humidity Resistance of Monocoat System**: All the painted panels of monocoat system passes the minimum 500 h test. The sample MCS-2 has no effect on film till 600 h. Sample MCS-3 shows no adverse sign, even after 750 h. MCS-4D2 panel surpasses 800 h of humidity resistance without having any adverse effect on the film.

e. *Acid Resistance (2% HCl for 24 h):* Panel of MCS-2 failed after 5 min of immersion in acid solution. Panel of MCS-3 shows no visible damage on the film after 24 h, but dark color change occurs. Also, the panel of MCS-4D2 showed no visible damage to the film, except the occurrence of dark color change of the film.

f. *Alkali Resistance (2% NaOH for 24 h):* Panel of MCS-2 withstands up to 6 h of immersion in alkali solution. Panel of MCS-3 shows no visible damage on the film after 24 h of alkali immersion, but slight puffing of the film was observed. Film of Panel of MCS-4D2 was found to be intact with very little color change.

7.3.6   EFFECT OF ADDITION OF GLASS MICROSPHERES (GM) ON HEAT INSULATION PROPERTIES OF REDISPERSIBLE POWDER PAINT

The thermal insulating sodium silicate paint system has been developed in two parts as under:

a. Addition of glass microspheres in 1 and 2% level of total formulation weight of powder paint is carried out under very slow stirring rate (at 40–60 rpm) in the above prepared and optimized best paint formulation. The specialty type of glass stirrer has been used for the dispersion purpose.

b. The prepared glass microsphere dispersion into paint is tested for its heat insulation properties.

The heat insulation properties were tested on metal panels and in specially developed laboratory electric oven as shown in Figure 7.1. The panels on which coatings were applied were of the dimensions $150 \times 96 \times 0.8$ mm^3 (L×B×T). The temperature limit over which the heat insulation properties were tested, was up to the range of 50°C, since in the Indian subcontinent, 50°C temperature is very common in summer season.

The results obtained for heat insulation effect with glass microspheres are as under

a) For PT-2 formulation: After addition of 1% glass microspheres in the formulation PT-2, and after application of coating, the very next day, metal panels were found to be heavily corroded (severe pitting corrosion). The reason may be due to alkalinity of sodium silicate used in the formulation without any setting/curing agent. So, the heat insulation test could not be performed for this sample.

b) For PT-3 formulation: In PT-3 formulation along with sodium silicate, stearic acid was used. Here, the panels applied for heat insulation properties were found to be in good condition. So, the effect of heat insulation with this formulation has been tested with 1 and 2% glass microspheres. The results have been shown in Graphs 7.1 and 7.2.

c) For PT-4D2 formulation: The formulation PT-4(D2) (powder paint with use of jaggery and calcium oxide) is found to deliver very good results. The results have been indicated in Graphs 7.3 and 7.4.

7.3.6.1   HEAT INSULATION FOR PT-3 SAMPLES WITH 1 AND 2% GLASS MICROSPHERES (GM) LOADING

In the paint formulation PT-3, 1 and 2% glass microspheres are added in varying degrees of film thickness. It is observed that for 1% glass loading, heat insulation properties is found to increase at 4–5% at average DFT of 75 μ. It is also observed that with the increase in film thickness for the same degree of loading of glass microspheres (1%), heat insulation effect is found to increase to the tune of 4–6%. The results are shown in Graph 7.1.

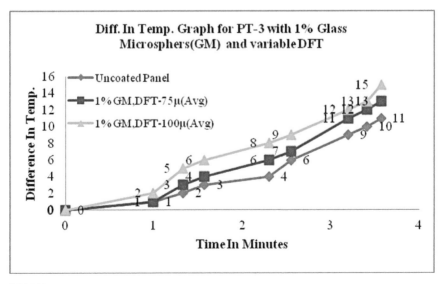

GRAPH 7.1

With 2% loading of glass microspheres in PT-3 formulation, heat insulation is found to increase to the tune of 4–5% at average DFT of 75 μ. However, with the same degree of loading (2%) and with increase in film thickness (average 100 micron DFT), heat insulation is found to increase at the rate of 6–8%. The results are shown in Graph 7.2.

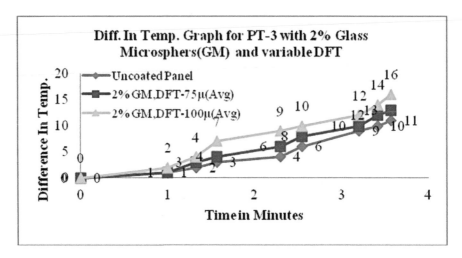

GRAPH 7.2

7.3.6.2 HEAT INSULATION FOR PT-4(D2) SAMPLES WITH 1 AND 2% GLASS MICROSPHERES LOADING

In the paint formulation PT-4(D2), 1 and 2% glass microspheres are added for varying degrees of film thickness. It is observed that for 1% glass loading, heat insulation properties is found to increase at 2.5–3.5% at average DFT of 75–78 μ. It is also observed that, with increase in film thickness of 110–115 μ, for the same degree of loading of glass microspheres (1%), heat insulation effect is found to increase to the tune of 7–8%. The results are shown in Graph 7.3.

With 2% loading of glass microspheres in PT-4(D2) formulation, heat insulation is found to increase to the tune of 4–5% at average DFT of 90–100 μ. However, with the same degree of loading (2%) and with increase in film thickness (140–160 μ), heat Insulation is found to increase at the rate of 6–8%. The results are depicted in Graph 7.4.

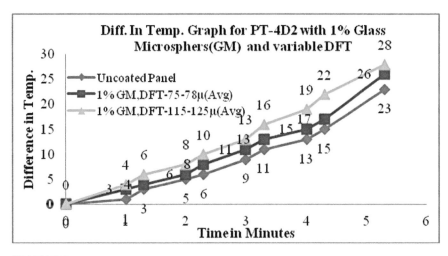

GRAPH 7.3

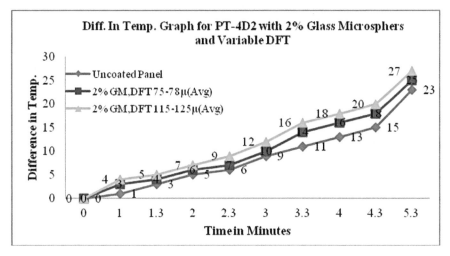

GRAPH 7.4

Thus the promising samples such as PT-3 and PT-4(D2) have been tested for their heat insulation properties in the specially developed oven used for testing of heat insulation properties of powder paint samples loaded with 1–2% glass microspheres. With developed powder paint samples, containing glass microspheres, total heat insulation effect to the tune of 15–20°C was observed with respect to the surrounding temperature and coating.

7.3.7   COST OF DEVELOPED POWDER PUTTY AND POWDER PAINT

The cost of developed powder putty and powder paint has been compared with tested standard samples and available other brands in the market. The cost of developed putty is just above/higher than tested SPY. However, the cost of developed powder paint (excluding glass microspheres) is well below half the price of the tested paint sample and other water based paint samples available in the market. The developed composite system (putty + paint) which is referred to as Monocoat system can definitely be offered at 40–45% lower cost than the tested/ established standard putty and water based paint products.

7.4   CONCLUSIONS

The Developed Technology Provides Ease of Manufacturing, Packing, and Transportation: The unique formulation of dry powder putty and powder paint based on sodium silicate, aluminum oxide, RDP powder, and China clay is successfully developed. These formulations are based on very simple natural raw materials and have comparable properties (mechanical and chemical resistance), in line with the current days' formulations based on modern emulsion. The present simple manufacturing machines and equipments (like Pebble Mill) can be used for processing. The product is convenient for manufacturing, packing, and transportation (no extra addi-tion of H_2O as solvent). The applicator has much convenience at work site to prepare the material as putty for application. The same type of formulation with very little addition of TiO_2 and slight increase in the PVA (15–20%) can work as topcoat over this putty.

Water Reducible and Environment Friendly Coatings: The developed powder putty and powder paint products are readily reducible in water, contains no harmful chemicals, are environment friendly and no health hazards are associated with it. The typical drying rate concern found with water reducible architectural coating is not a matter of concern here. All the putty samples are found to touch dry with 10–30 min while all the paint samples are found to touch dry from 7 to 30 min. The hard dry time in both cases is almost overnight.

Economy Associated with Developed Products: The developed powder putty and powder paint formulation contains easily available raw mate-rials at much lower prices, thus the formulation is very cost effective. The formulations (dry blend of powder) do not contain water, so a greater volume can be accommodated in packing and transportation. This has direct effect on price reduction of the product.

Suitability on Cement/Concrete Surface: Since both powder putty and powder paint contains sodium silicate as one of the ingredient, this raw material is alkaline in nature and is best suitable to strong alkaline surfaces like cement/concrete. Moreover, the formulated products after application to cement surface dries to form strong crystals, having good to excellent adhesion on the cement surfaces. The results have been verified by different physical tests in this experimental work.

Excellent Breathing Tendency and Humidity Resistance Exhibited by Powder Putty and Powder Paint: As proven by water vapor transmission test and humidity resistance test, the developed powder putty and powder paint samples have been found to outperform the existing standard putty and emulsion paint samples in the market.

Effect of Hallow Glass Microspheres on Heat Insulation Property: The powder paint formulations PT-3 and PT-4(D2), which have exhibited the best physiochemical properties were selected to test the heat insulation properties by incorporating 1 and 2% glass microspheres. Comparing the heat insulation properties among PT-3 and PT-4 (D2) samples for 1% glass microsphere loading, PT-3 sample (containing stearic acid as setting agent) is found to have marginal higher heat insulation property (to the tune of 1%) at the DFT of 75 μ. However, with the same degree of loading of glass microspheres (1% glass microspheres) at the DFT of 100 μ, PT-4(D2) formulation (containing jaggery + calcium oxide) is found to have excellent degree of heat insulation property as compared to PT-3. The heat insulation found to increase to the tune of 7–8%.

The PT-3 and PT-4(D2) formulations were loaded with 2% glass microspheres and again their heat insulation effect was studied. However, in both the cases, the average heat insulation effect was found to be in the tune of 4–6%.

Thus, the formulation based on sodium silicate, jaggery and calcium oxide along with RDP powder can be successfully formulated to work as heat insulation coatings. With 1% glass microsphere loaded in the formulation, heat insulation effect to the tune of 15–20°C was observed with respect to the surrounding temperature of the coating. In the Indian continent, in summer season, 50° C temperature is very common. Formulating coating system based on easily available natural raw materials will open new gates to the development of cost effective, durable, and environment friendly coatings.

KEYWORDS

- putty powder
- paint powder
- monocoat system
- mechanical and chemical properties
- glass microspheres
- sodium silicate
- Rdp powder
- heat insulation coating

REFERENCES

1. Breedveld, L.W.; W. M. G. M. van Loon. Watergedragen Polymeren: Een Problem Voor Het Aquatisch Milieu? H_2O, **1997,** *30* (21), 649–651.
2. Sande. *Water Pollution Due to the Changeover to Water-Borne Paints*; Short Study within the Framework of the Decopaint Project: University of Amsterdam: Amsterdam, 2000.
3. European Commission, Final Report– Study on the Potential for Reducing Emissions of Volatile Organic Compounds (VOC) Due to the Use of Decorative Paints and Varnishes for Professional and Non-professional Use, June 2000, pp 1–317.
4. Taylor, C. J. A. *"Alkyd Resins," Oil and Colour Chemists Association, "Convertible Coatings,"* Part 3, Champan and Hall: London, 1966, pp 87–106.
5. Gettwert, G.; Rieber, W.; Bonarius, J. One-Component Silicate Binder systems for Coatings. *Surf. Coat. Int.* **1998,** *81* (12), 596–603.
6. PQ Corporation. Bonding and Coating Applications of PQ Reducible Silicates, Industrial Chemical Division, Bulletin 12–31/107, 2006.
7. Mike McDonald, Janice, H. Recent Developments in Reducible Silicate Based Binders and Coating; PQ Corporation: Valley Forge, PA. http://pqcorp.com/Portals/1/lit/Recent%20Silicate%20Binder%20Developments.pdf
8. Thirumalini, P.; Ravi, R.; Sekar, S. K.; Nambirajan, M. Study on the Performance Enhancement of Lime Mortar Used in Ancient Temples and Monuments in India. *Ind. J. Sci. Technol.* **2011,** *4* (11), 1484–1487.
9. Dorn, C.; John, H.; Alick, L.; Craig, K. A Short History of the Use of Lime as a Building Material beyond Europe and North America. *Int. J. Archit. Herit.* **2011,** *6* (2), 117–146.
10. Natarajan, C.; Chen, S.; Syed, M. Rehabilitation and Preservation of the St. Lourdes Church, Tiruchirappalli. *J. Perform. Constr. Facil.* **2010,** *24* (3), 281–288.
11. Herbert, P.; Beardsley, Wilmington, Del. Vinyl Acetate-Ethylene Copolymer Aqueous Paint Compositions. U. S. Patent 3,578,618, May 11, 1971.

12. Klaus, M.; Karl Josef, R.; Detlev, S. Wolfgang Zimmermann. Process for the Preparation of a Dispersible Vinyl Acetate/ethylene Polymer Powder. U. S. Patent, US3883489 A, May 13, 1975.

13. Special Chem, New Vinyl Acetate and Ethylene-based Dispersions and More, "Wacker at ABRAFATI 2013," Sep 17, 2013.

14. Gade, S. Application of Hollow Microspheres in Coatings, 2011. Available at: www.pcimag.com/articals/93074-application-Of-hollow-microspheres-in-coatings (accessed Jan 5, 2012).

15. Fei, W.; Jin Sheng, L. Preparation and Properties of Thermal Insulation Latex Paint for Exterior Wall Based on Difibred Sepiolite and Hollow Glass Microspheres. *Adv. Mater. Res.* **2008,** *58,* 103–108.

16. Sahu, P. K.; Dr. Mahanwar, P. A. Effect of Hollow Glass Microspheres and Cenospheres on Insulation Properties of Coatings. *J. Pigm. Resin Technol.* **2013,** *42* (4), 223–230.

PART II
Nanomaterials as Catalysts

CHAPTER 8

SONOCHEMICAL SYNTHESIS OF GO– ZnO NANOPHOTOCATALYST FOR THE DEGRADATION OF MALACHITE GREEN USING A HYBRID ADVANCED OXIDATION PROCESS

BHASKAR BETHI[1], SHIRISH H. SONAWANE[1*], BHARAT A. BHANVASE[2], VIKAS MITTAL[3], and MUTHUPANDIAN ASHOKKUMAR[4]

[1]Department of Chemical Engineering, National Institute of Technology, Warangal 506004, Telangana, India

[2]Department of Chemical Engineering Laxminarayan Institute of Technology, Rashtrasant Tukadoji Maharaj Nagpur University, Nagpur 440033, Maharashtra, India

[3]Department of Chemical Engineering, The Petroleum Institute, Abu Dhabi, UAE

[4]Department of Chemistry University of Melbourne, VIC 3010, Australia

*Corresponding author. E-mail: shirish@nitw.ac.in

CONTENTS

ABSTRACT

In this article, a novel approach for the synthesis of graphene oxide–zinc oxide (GO–ZnO) nanophotocatalyst by acoustic cavitation technique is reported. Initially, graphene oxide (GO) was prepared from graphite powder by using modified Hammer's method in the presence of ultrasonic irradiation. The prepared GO and GO–ZnO nanophotocatalysts were characterized by XRD, FESEM, and TEM analysis. Further prepared GO–ZnO nanophotocatalyst was used for the degradation of malachite green (MG) dye (500 mg/L) in aqueous solution using a hybrid advanced oxidation process (AOPs) consisted of hydrodynamic cavitation (HC), GO–ZnO photocatalysis, and hydrogen peroxide (H_2O_2). Hydrodynamic cavitation setup consists of a centrifugal pump having the power rating of 0.5 HP, 2 mm diameter of orifice hole as a cavitation device and 9 W UV–visible light for a typical hybrid AOPs. It has been observed that MG was effectively decolorized up to 99% at pH 8.5, and at inlet pressure of 2 bars in 90 min using the hybrid process.

8.1 INTRODUCTION

Material scientists are interested in graphene for a number of applications due to its remarkable and versatile properties, more specifically its improved electrical, thermal, and mechanical properties.[1,2] Graphene oxide (GO) contains high reactive oxygen functional groups: Due to this property, GO is considered as one of the good candidates for chemical functionalization.[3] Different inorganic nanoparticles such as Ag, Au, Fe_3O_4, CdS, TiO_2, SnO_2, and ZnO are used for doping GO to obtain a number of functional materials such as antibacterial materials, targeted drug carriers, optoelectronic materials, electrode materials, and photocatalytic materials.[4–12] In recent years, researchers have proposed several methods for the synthesis of graphene oxide–zinc oxide (GO–ZnO) nanocomposites, for example, microwave-assisted synthesis[13] and hydrothermal route.[14] These methods have shown advantage for the synthesis of GO–ZnO nanocomposite, such as uniform heating and high purity GO. However, these methods did not generate uniform distribution of ZnO on to graphene oxide support.

Malachite green is a basic dye and is traditionally used in textile and leather industries.[15] Due to heavy industrial use, the dye is considered as a pollutant in aqueous environment. Different traditional treatment techniques such as coagulation/flocculation, membrane separation, adsorption process were used to treat dye-contaminated industrial wastewater. However, these methods generate secondary pollutants, while biological treatment has limitation to implement, as some dyes show biological resistance.[16] Advanced oxidation processes (AOPs) have been used for the degradation of organic pollutants in aqueous environment. Some AOPs have shown considerable promising effect for wastewater treatment applications, which include cavitation process, Fenton reagent, and photocatalytic oxidation.[17,18] Usually a combination of two or more AOPs has been found to be more efficient for wastewater treatment,[19-24] as compared to individual oxidation process. Cavitation-based hybrid AOPs have been widely investigated for wastewater treatment applications, though a number of articles reported the use of ultrasonically induced cavitational process, which have the limitation to scale up.[25,26] The recent trend is the development of efficient hybrid systems for wastewater treatment using cavitation, coupled with various AOPs such as UV photocatalysis, H_2O_2, cavitation coupled with photo Fenton, etc.

Recent studies show that ZnO doped graphene oxide can be used as photocatalyst for wastewater treatment. Ameen et al.[27] have synthesized the ZnO–GO hybrid nanophotocatalyst for the photodegradation of crystal violet (CV) dye. Ahmad et al.[28] have synthesized visible light-responsive ZnO/graphene composites. Chen et al.[29] have synthesized the ZnO/GO nanocomposites and evaluated the photocatalytic activity using methyl orange (MO) as a pollutant under UV irradiation. They found that the equilibrium adsorption capacity (D_e%) of MO on ZnO/GO composite material is much higher than that observed with pure ZnO and GO. Benxia et al.[30] have synthesized the ZnO/GO nanocomposite via a facile chemical deposition route and used it for the photodegradation of methylene blue from water under visible light. Wang et al.[31] have studied the decolorization of an azo dye, C.I. reactive red 2 (RR2) using TiO_2 photocatalysis coupled with water jet cavitation. Bagal et al.[32] have studied the degradation of diclofenac, a widely detected pharmaceutical drug in wastewater samples, using a combined approach of hydrodynamic cavitation, and heterogeneous photocatalysis. They found that the degradation of diclofenac is about 95% with 76% reduction in TOC using hydrodynamic cavitation in conjunction with $UV/TiO_2/H_2O_2$ under the optimized operating conditions. Behnajady et al.[33] have studied the effect of acoustic cavitation (US) on the degradation of MG by direct photolysis with ultraviolet (UV) radiation and UV/H_2O_2 processes. They concluded that $US/UV/H_2O_2$ was the most effective process for degradation of MG from aqueous solution. The degradation of malachite green using hydrodynamic cavitation and hydrodynamic cavitation coupled with photocatalysis (GO–ZnO) and other AOP's have not been reported. Hydrodynamic cavitation could be one of the commercially viable processes for the degradation of dyes in wastewater released by textile industries.

In this work, we report the ultrasound assisted synthesis of GO–ZnO nanophotocatalyst. Ultrasonically prepared GO–ZnO nanophotocatalyst has shown a good deposition of ZnO on GO sheets. Most of the literature related to GO–ZnO synthesis via conventional methods has shown flower-like structure and agglomerated crystalline phases. In the case of ultrasonic assisted synthesis of GO–ZnO, agglomerated crystalline phase has not appeared. Degradation of malachite green using hydrodynamic cavitation and hydrodynamic cavitation coupled with photocatalysis (GO–ZnO) and other AOP's have been reported.

8.2 EXPERIMENTAL

8.2.1 MATERIALS

For the synthesis of graphene oxide from graphite powder, the following chemicals were procured: 98 wt% H_2SO_4 (SDFCL, Mumbai), 30 wt% H_2O_2 (Finar Chemicals, Ahmadabad), 98% $NaNO_3$ (Qualigens Fine Chemicals) and $KMNO_4$ (Finar Chemicals, Ahmadabad). For the synthesis of GO–ZnO nanocomposite, precursor $Zn(NO_3)_2.6H_2O$ (zinc nitrate) was procured from Himedia Lab, Mumbai, India. Malachite green dye (96% pure) was procured from Universal Laboratories, Mumbai.

8.2.2 SONOCHEMICAL SYNTHESIS OF GRAPHENE OXIDE (GO)

The GO was prepared from graphite powder by using modified Hammer's method[34] in the presence of ultrasonic irradiation using an ultrasonic probe (Dakshin ultrasonic probe sonicator, 25 kHz frequency). In this method, graphite powder (3 g) and sodium nitrate (3 g) were added to 138 mL of H_2SO_4 solution (98 wt%) at 0°C by keeping the solution in an ice bath. The mixture was sonicated for 10 min and 15 g of $KMNO_4$ was gradually added to the above mixture under sonication for 30 min. During the addition of $KMNO_4$, a small amount of fumes was released, and the reaction progressed until the solution turned in to a paste. The reaction was stopped by the gradual addition of 300 mL DI (Deionized water) to the above reaction mixture. The solution was sonicated for another 5 min. Then, 24 mL of H_2O_2 (30% w/v) was added drop wise to the above reaction mixture. The mixture was filtered and washed with 450 mL HCl solution (10 wt%) to remove metal ions and then washed with the 300 mL DI water to remove the acid. Graphene oxide from the solution was separated by filtration. Exfoliated graphene oxide (GO) was obtained under sonication for 30 min. The solution was then filtered and the powder was dried in an oven at 40–50°C for 2 h.

8.2.3 SONOCHEMICAL SYNTHESIS OF GO–ZnO NANOCOMPOSITE

In this method, 6 g of zinc nitrate was dissolved in 50 mL of DI water, and the resultant mixture was sonicated for 5 min and 0.05 g of GO powder was dispersed in 5 mL of DI water, and transferred to the above-sonicated mixture and sonicated for another 30 min. During sonication, the pH was maintained between 6 and 7 by the addition of small amount of NaOH solution. The solution was then filtered and the powder was dried in an oven at 50°C for 1 h.

8.2.4 CHARACTERIZATION

GO–ZnO nanocomposite was characterized by using powder X-ray diffractometer (Phillips PW 1800, range is 6–80 2θ) and the structure and morphology were studied by field emission scanning electron microscopy (FESEM). The morphology and particle size of GO–ZnO composites were characterized by transmission electron microscopy (TEM, H-7650 accelerating voltage of 120 kv).

8.3 HYDRODYNAMIC CAVITATION COUPLED WITH PHOTOCATALYTIC REACTOR

The experimental setup of hydrodynamic cavitation coupled with photocatalytic reactor is shown in Figure 8.1. This set up includes a 5 L holding tank along with UV–visible light assembly inside the tank. Then, 9 W UV–visible light was used for the photocatalytic degradation studies. This light was fixed vertically in the centre of the tank with the support of a flat lid. A centrifugal pump of power rating 0.5 HP was used for pumping the dye solution. The suction line of the pump was dipped into the tank through a hole on the flat lid. Discharge from the pump consisted of a flange, which houses the cavitation device (orifice plate) and the flow rate in this line was adjusted by placing a control valve near to the entrance of the orifice plate. The end of this bypass line was inserted into the tank through the hole provided on the flat lid of the tank. The end of the discharge line after orifice device has been connected to the tank through the lid. Diameter of the main pipeline was 12.7 mm and orifice hole diameter was 2 mm. Schematic diagram for geometrical dimensions of an orifice is shown in Figure 8.2.

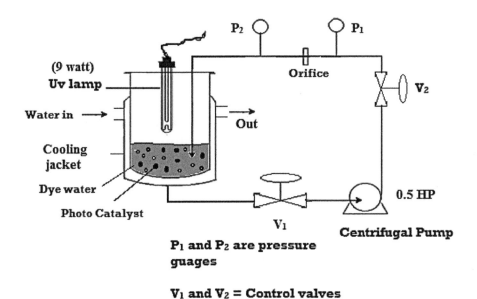

FIGURE 8.1   Hydrodynamic cavitation setup with photocatalysis.

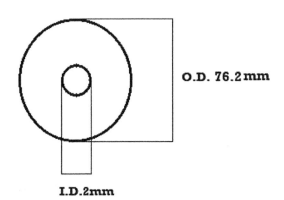

FIGURE 8.2   Schematic diagram for geometrical dimensions of an orifice.

Hydrodynamic characteristics of the cavitation device (orifice) were studied by measuring the main line flow rate and by using a dimensionless parameter called as cavitation number (Cv). The cavitation number is a dimensionless number used to characterize the condition of cavitation in hydraulic devices.[19] Cavitation number for orifice device was calculated by considering the main line flow rate and cross sectional area of the orifice device. The calculated Cv for an orifice at an inlet pressure of 2 kg/cm² is 0.11 ($Cv = 0.11$). The cavitation number is defined as

$$Cv = \frac{P_1 - P_v}{\frac{1}{2} p v^2} \qquad (7.1)$$

where
  P_1 = inlet pressure at the orifice in kg/cm²,
  P_v = vapor pressure at the outlet of the orifice in kg/cm²,
  p = density of the dye solution in kg/m³, and
  v = velocity of the dye solution in m/s.

Hydrodynamic cavitation-based degradation of malachite green was carried out at different conditions using fixed solution volume of 5 L and for a constant circulation time of 90 min. The initial concentration of malachite green was kept constant at 539 µM (500 ppm). The temperature of the solution during experiments was kept constant in all cases at about 35°C. Experiments were conducted at constant inlet fluid pressure of 2 kg/cm² and various pH values of 6, 7, 8, 8.5, 9, and 10. Also experiments were conducted for molar ratio 1:10 of dye to H_2O_2. The samples were withdrawn at a regular interval of 15 min and the total time fixed for decolorization was 90 min. The degradation of malachite green (MG) through hydrodynamic cavitation was analyzed using UV-spectroscopy.

8.4   RESULTS AND DISCUSSION

8.4.1   MECHANISM OF FORMATION OF GO/ZnO NANOCOMPOSITE PHOTOCATALYST USING SONOCHEMICAL APPROACH

The formation mechanism of GO/ZnO nanocomposite in sonochemical environment has been shown in Figure 8.3. GO was generated by oxidizing natural graphite powder in the presence of ultrasonic irradiation and strong oxidizing agent $KMnO_4$ along with H_2SO_4. Zinc oxide nanoparticles was formed by the addition of sodium hydroxide in to the mixture of GO and Zinc nitrate solution. In the reaction mixture, Zn^{2+} ions were hydrolyzed to form ZnO nanaoparticles at 100°C and the cavitational effect generated by the ultrasonic irradiation results in the formation of uniform loading of ZnO nanoparticles on GO.

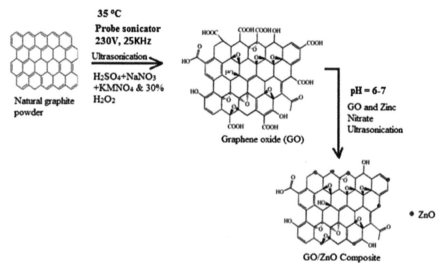

FIGURE 8.3 Diagram of GO–ZnO nanocomposite formation mechanism.

XRD analysis of ultrasonically prepared GO was carried out and the results are shown in Figure 8.4. A diffraction peak is observed at 2θ~6.45° with a d spacing value of 8.84A°.

The XRD results are in good agreement with the previously published data for ZnO and GO–ZnO nanocomposites.[35] The XRD pattern of ultrasonically prepared ZnO nanoparticles is shown in Figure 8.5. Seven peaks were observed at $2\theta = 31.7°, 34.8°, 36.7°, 47.5°, 56.8°, 63°$, and 68.5° which correspond to the (1 0 0), (0 0 2), (1 0 1), (1 0 2), (1 1 0), (1 0 3), and (1 1 2) crystalline planes of ZnO, respectively. The XRD pattern of ultrasonically prepared GO–ZnO nanocomposite is shown in Figure 8.6. There are nine peaks at $2\theta = 31.5°, 33.5°, 35.5°, 45.5°, 55.2°, 56.5°, 65.3°$, and 68°. The XRD pattern of the GO–ZnO nanocomposite reveals the formation of ZnO with wurtzite crystal structure.[36] There is a small peak in the XRD pattern at 2θ~26° with a much lower intensity, because of incomplete exfoliation in the presence of ultrasound. This peak indicates a very small amount of

graphene presence in the GO/ZnO nanocomposite. The broad XRD peak at
$2\theta \sim 31.5°$ of GO–ZnO composite sample indicates the crystalline form with
pure phase.

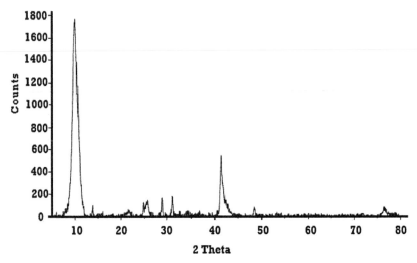

FIGURE 8.4   X-ray diffraction pattern of GO.

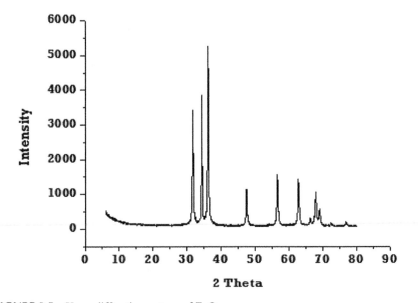

FIGURE 8.5   X-ray diffraction pattern of ZnO.

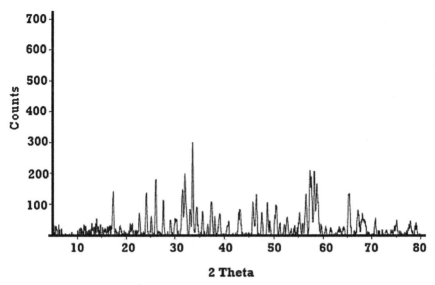

FIGURE 8.6 X-ray diffraction pattern of GO–ZnO nanocomposite.

The average crystal size of ZnO and ZnO loaded on GO has been calculated using Scherrer's equation.[35] The estimated crystal sizes of ZnO and GO–ZnO has been shown in Table 8.1.

TABLE 8.1 Estimated Particle Sizes of Photocatalysts Prepared by Sonochemical Method.

Type of Photocatalyst	2θ	Peak Width (β)	Crystallite Size (nm)
Bare ZnO	31.7	0.1	86
GO–ZnO	31.5	0.3	28

$$d = \frac{k\lambda}{\beta \cos\theta}$$

where d is average size of crystal, k is the Scherrer constant that is 0.89, $\lambda = 1.540$ nm that is the wavelength of X-ray, and β is the width of the XRD peak in radians at half maximum intensity. This quantity is also sometimes denoted as $\Delta (2\theta)$ and θ is the Bragg diffraction angle that is 15.5° for both ZnO and GO–ZnO.

The morphology of the GO was studied by FESEM analysis. Many exfoliated layers of sheets can be seen in the FESEM image of GO (see Fig. 8.7). FESEM image of conventionally prepared GO shows the agglomerated crystalline phase.[37] However, in the case of ultrasonically prepared samples, agglomerated crystalline phase has not been observed.

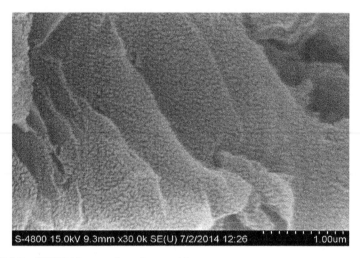

S-4800 15.0kV 9.3mm x30.0k SE(U) 7/2/2014 12:26   1.00um

FIGURE 8.7   FESEM image of graphene oxide.

The FESEM image of GO–ZnO nanocomposite is shown in Figure 8.8. It can be seen that ZnO nanoparticles anchored onto the edges and planes of curled GO sheets. TEM image of GO sheets has been shown in Figure 8.9. It was observed from the TEM image of GO that the GO sheets present a crumpled surface, which is consistent with the observation from FESEM image. It can be clearly seen in Figure 8.10 that the exfoliated GO sheet was decorated with 20 nm ZnO particles.

S-4800 15.0kV 9.6mm x40.0k SE(U) 7/2/2014 12:29   1.00um

FIGURE 8.8   FESEM image of exfoliated GO–ZnO nanocomposite.

FIGURE 8.9   TEM image of exfoliated GO sheets using sonochemical approach.

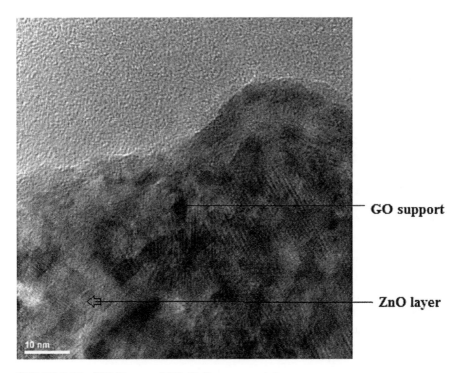

FIGURE 8.10   TEM image of GO–ZnO nanocomposite.

8.4.2 MG DEGRADATION BY HYDRODYNAMIC CAVITATION

8.4.2.1 EFFECT OF pH

A fixed solution volume of 5 L having the concentration of 500 mg/L MG dye was taken in HC reactor. The temperature of the solution during experiments was kept constant in all cases at about 35°C. Experiments were conducted at constant inlet fluid pressure of 2 kg/cm². To observe the effect of solution pH on the degradation rate, experiments were performed at different pH values of 6, 8, 8.5 (natural pH), 9, and 10. The absorption spectra recorded of MG de-colorization during hydrodynamic cavitation treatment are shown in Figure 8.11.

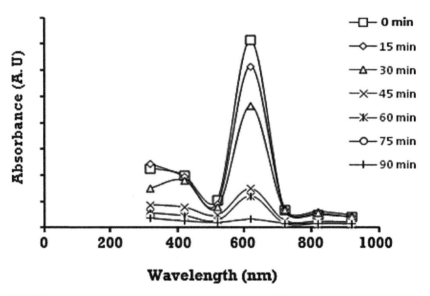

FIGURE 8.11   UV spectra profile of MG absorbance at different time intervals.

As the pH increases from 6 to 10, the degradation time decreases. It can be seen that the degradation of MG is favored in basic media. From Figure 8.12, it can be observed that percentage of the MG decolorization increased by increasing the pH of dye solution. However, the optimum pH has been taken as 8.5, because beyond pH = 9, malachite green has the pH transition stage.

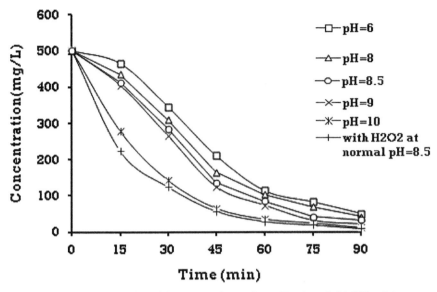

FIGURE 8.12 Concentration of dye removal at various pH values (with HC only).

8.4.3 EFFECT OF H_2O_2+HC

Hydrogen peroxide is a commonly used oxidizing agent which can be used in wastewater treatment due to its high oxidation potential. Experiments were carried out using the hybrid process of HC and H_2O_2 at 2 kg/cm^2 inlet pressure, solution pH of 8.5, and 1:10 molar ratio of MG to H_2O_2. The addition of H_2O_2 resulted in a very slight increase in the decolorization of MG. Percentage of decolorization of HC and HC+H_2O_2 are 93 and 98, respectively.

8.4.4 HC+UV PHOTOCATALYSIS

The degradation of MG in the presence of GO–ZnO nanophotocatalyst coupled with hydrodynamic cavitation was studied. A 9 W Phillips UV–visible light was used for the degradation studies of MG in the presence of 2 g of GO–ZnO nanophotocatalyst and hydrodynamic cavitation. The degradation efficiency was 95% when employing UV and hydrodynamic cavitations simultaneously.

8.4.5 DEGRADATION THROUGH HC+UV PHOTOCATALYSIS+H$_2$O$_2$

The degradation effect on MG using hybrid process of HC+ UV Photocatalysis+H$_2$O$_2$ was studied. In this case, initially 2 mL of H$_2$O$_2$ was added in to the hydrodynamic cavitation reactor and all the remaining parameters are kept contant for comparing the results that are obtained in the previous system. It was oberved that the degradation of MG using HC+UV photocatalysis showed of 95% degradation. Whereas in the case of degradation of MG using using HC+UV, photocatalysis+H$_2$O$_2$ showed 99.9% degradation. Degradation percentages of MG with respect to time using HC+photocatalysis and HC+photocatalysis+ H$_2$O$_2$ has been shown in Figure 8.13.

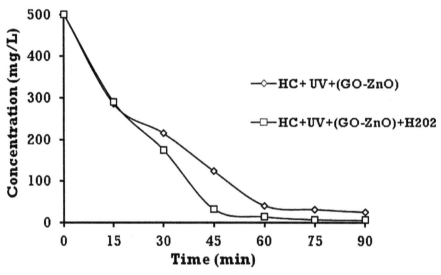

FIGURE 8.13 Concentration of MG removal in various systems (at pH = 8.5).

8.4.6 COMPARISON OF PHOTOCATALYSTS

To know the photocatalytic performance of ZnO and GO–ZnO, experiments were carried out with 1 L dye solution, which consists of 0.4 g of ZnO and GO–ZnO individually. For every 15 min time interval, samples were collected over 90 min of study. UV absorbance analysis was carried out to know the concentration of MG dye at each interval over 90 min. The obtained results show that the GO–ZnO photocatalysts have better performance than the ZnO photocatalysts. Reason for attaining such performance is graphene oxide (GO) which contains high reactive oxygen functional groups in GO–ZnO nanocomposite and also that the synthesized GO–ZnO nanoparticles are of very less size (It indicates that more surface area consists of more active sites for adsorption of dye) compared with ZnO nanoparticles size. The removal of dye concentration at every 15 min over 90 min using different photocatalysts alone is shown in Figure 8.14.

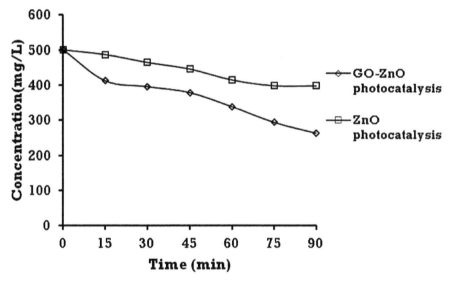

FIGURE 8.14 Concentration of MG removal using different photocatalysts alone at pH = 8.5.

8.4.7 RATE KINETICS OF MG DEGRADATION USING HC+PHOTOCATALYSIS

The degradation of organic pollutants in hydrodynamic cavitation will occur by the generation of ·OH radicals. These ·OH radicals rapidly react with MG dye which is present at the interface of the generared cavities and wastewater. This mechanisim leads to the degradation of dye pollutants in the wastewater. Most of the investigators have found that the degradation reactions of many organic pollutants in AOPs followed the first order reaction rate.[38–41] According to their procedure, it was assumed that the degradation reaction of MG followed the pseudo first order reaction rate in the presence of hydrodynamic cavitation, coupled with UV photocatalysis. The rate constants of degradation reaction of MG dye has been calculated using the following equation.

$$\ln\left(\frac{C_o}{C}\right) = k \times t \qquad (8.2)$$

where C is the concentration of dye molecules present in mol/L, k is the rate constant in min^{-1}, and t is the time in minutes.

The reaction kinetic mechanism for degration of MG dye using hydrodynamic cavitation coupled with photocatlysis followed the first order reaction, and has been confirmed by plotting the graph for $\ln\left(\frac{C_o}{C}\right)$ vs time (t). The plot of $\ln\left(\frac{C_o}{C}\right)$ vs time (t) gave a straight line and it is passed through the origin. If the line is passed through the origin then we can say that the reaction of that system followed the first order reaction kinetics. The plot of $\ln\left(\frac{C_o}{C}\right)$ vs time (t) for investigating the rate kinetics of MG removal at different pH values using HC alone, at normal pH 8.5 using HC+H$_2$O$_2$, using HC+(GO–ZnO)+H$_2$O$_2$, and using photocatalysis is shown in Figures 8.15–8.18, respectively. Reaction rate constants (k) and R^2 values for MG decolourization at normal solution pH 8.5 in various systems are given in Table 8.2.

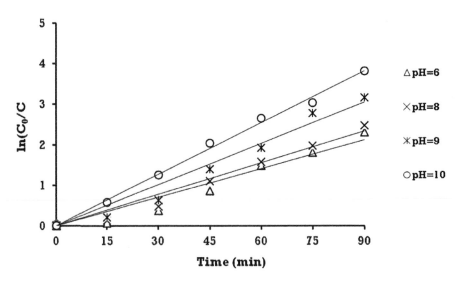

FIGURE 8.15 Rate kinetics of MG removal at different pH values using HC alone.

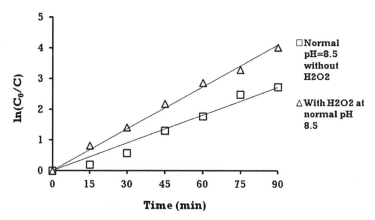

FIGURE 8.16   Rate kinetics of MG removal using HC+H$_2$O$_2$ at normal pH 8.5.

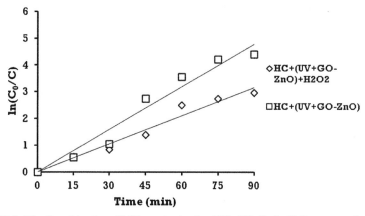

FIGURE 8.17   Rate kinetics of MG removal using HC+GO–ZnO+H$_2$O$_2$ at normal pH 8.5.

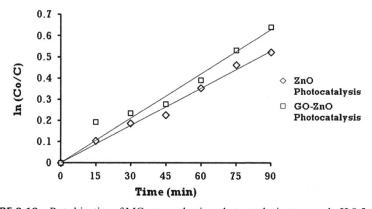

FIGURE 8.18   Rate kinetics of MG removal using photocatalysis at normal pH 8.5.

TABLE 8.2　Reaction Rate Constants (k) of Methylene Blue Decolourization in Different Systems (at pH = 8.5, Inlet Pressure = 2 bar, Temperature = 35°C, Initial Dye Concentration 500 mg/L).

Method	K Value (min⁻¹)	R₂ Value
HC	0.03	0.966
HC + H₂O₂	0.0454	0.994
GO–ZnO photocatalysis	0.007	0.962
ZnO photocatalysis	0.0059	0.989
HC + (UV+GO–ZnO)	0.0351	0.965
HC + (UV+GO–ZnO)+H₂O₂	0.0531	0.958

8.4.8　ADSORPTION KINETICS AND ADSORPTION ISOTHERM MODELS

The mechanism of adsorption and adsorption kinetics for ZnO nanoparticles was determined using a pseudo-first and second order equation. It begins with the consideration of 0.4 g of ZnO nanoparticles in 1 L MG dye solution. Magnetic stirrer was used to mix the solution. For every 15 min time interval, samples were collected and carried out for UV absorbance analysis. Concentrations at every time interval were obtained from the calibration curve prepared for the MG. The pseudo-first-order equation was used to plot a graph of log (q_e-q) against time, as shown in Figure 8.19. K_1 and q_e values were found from the slope and intercept of this equation, respectively. The pseudo first order rate constant and R^2 has been reported in Table 8.3. Further, if the adsorption rate is pseudo-second-order, then plotting t/q against t straight lines were obtained, as shown in Figure 8.20. The values of K_2 and q_e were calculated and are reported in Table 8.3. It has been observed from the R^2 values of both, that the pseudo-second-order kinetics is better than the pseudo-first-order model.

The linear plot of C_e/q_e versus C_e had shown Langmuir isotherm model and it is shown in Figure 8.21. It was observed that $R^2 = 0.99$ was obtained for the monolayer Langmuir adsorption isotherm. Linear plot of log q_e versus log C_e represents the Freundlich isotherm model and it is shown in Figure 8.22. $R^2 = 0.808$ was obtained in the freundlich isotherm model. It was observed that the adsorption of MG dye on the ZnO surface follows the Langmuir adsorption isotherm model.

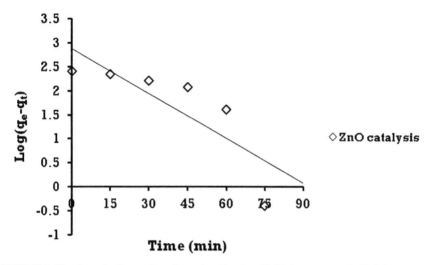

FIGURE 8.19   Pseudo-first order adsorption kinetic of MG dye at normal pH 8.5.

TABLE 8.3   Constants for the First-order and Second-order Kinetics of MG Dye Adsorption on ZnO Catalyst.

Type of Catalyst	Pseudo First-order Kinetics		Pseudo Second-order Kinetics	
	K_1 (min^{-1})	R^2	K_1 (min^{-1})	R^2
ZnO	0.0215	0.876	0.0099	0.99

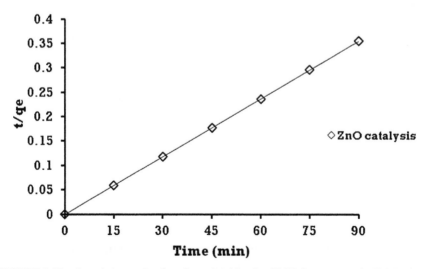

FIGURE 8.20   Pseudo-second order adsorption kinetic of MG dye at normal pH 8.5.

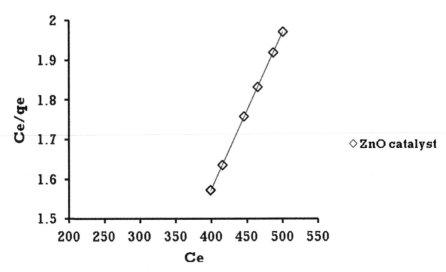

FIGURE 8.21   Langmuir isotherm model for MG dye on ZnO at normal pH 8.5.

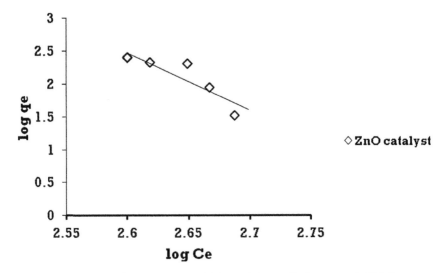

FIGURE 8.22   Freundlich Isotherm Model for MG dye on ZnO at normal pH 8.5.

8.5 CONCLUSION

In summary, GO–ZnO nanophotocatalyst has been synthesized successfully using effective ultrasound assisted solution-based synthesis route and the same applied for the degradation of malachite green in aqueous solution using hydrodynamic cavitation coupled with UV photocatalysis and the addition of H_2O_2 Malachite green in aqueous solution having basic pH is more favorable for degradation. Hybridizing of UV photocatlysis with hydrodynamic cavitation did not show much enhancement in rate of degradation. A percentage of degradation of MG of about 99% has been attained by the addition of H_2O_2 in to the above hybrid process.

ACKNOWLEDGMENTS

Shirish H. Sonawane acknowledges the Ministry of Environment and Forest (MoEF, Govt of India) for providing the financial support thorough Grant No: 10-1/2010-CT (WM).

KEYWORDS

- **GO–ZnO nanophotocatalyst**
- **hydrodynamic cavitation**
- **hybrid advanced oxidation process**
- **wastewater treatment**
- **aqueous environment**

REFERENCES

1. Lee, C.; Wei, X.; Kysar, J. W.; Hone, J. Measurement of the Elastic Properties and Intrinsic Strength of Monolayer Graphene. *Science* **2008,** *321,* 385–388.
2. Ghosh, S.; Calizo, I.; Teweldebrhan, D.; Pokatilov, E. P.; Nika, D. L.; Balandin, A. A.; Bao, W.; Miao, F.; Lau, C. N. Extremely High Thermal Conductivity of Graphene: Prospects for Thermal Management Applications in Nanoelectronic Circuits. *Appl. Phys. Lett.* **2008,** *92,* 151911–151913.
3. Dreyer, D. R.; Park, S.; Bielawski, C. W.; Ruoff, R. S. The Chemistry of Graphene Oxide. *Chem. Soc. Rev.* **2010,** *39,* 228–240.
4. Pham, T. A.; Choi, B. C.; Lim, K. T.; Jeong, Y. T. A Simple Approach for Immobilization of Gold Nanoparticles on Graphene Oxide Sheets by Covalent Bonding. *Appl. Surf. Sci.* **2011,** *257,* 3350–3357.
5. Williams, G.; Seger, B.; Kamat, P. V. TiO_2-Graphene Nanocomposites. UV-Assisted Photocatalytic Reduction of Graphene Oxide. *ACS Nano.* **2008,** *2,* 1487–1491.

6. Kamat, P. V. Graphene- based Nanoarchitectures: Anchoring Semiconductor and Metal Nanoparticles on a Two-Dimensional Carbon Support. *J. Phys. Chem. Lett.* **2010**, *1*, 520–527.

7. Yang, X. Y.; Zhang, X. Y.; Ma, Y. F.; Huang, Y.; Wang, Y. S.; Chen, Y. S. Superparamagnetic Graphene Oxide–Fe$_3$O$_4$ Nanoparticles Hybrid for Controlled Targeted Drug Carriers. *J. Mater. Chem.* 2009, *19*, 2710–2714.

8. Paek, S. M.; Yoo, E.; Honma, I. Enhanced Cyclic Performance and Lithium Storage Capacity of SnO2/Graphene Nanoporous Electrodes with Three-Dimensionally Delaminuteated Flexible Structure. *Nano Lett.* 2009, *9*, 72–75.

9. Lightcap, I. V.; Kosel, T. H.; Kamat, P. V. Anchoring Semiconductor and Metal Nanoparticles on a Two-Dimensional Catalyst Mat. Storing and Shuttling Electrons with Reduced Graphene Oxide. *Nano Lett.* 2010, *10*, 577–583.

10. Shen, J. F.; Shi, M.; Li, N.; Yan, B.; Ma, H. W.; Hu, Y. Z.; Ye, M. X. Facile Synthesis and Application of Ag-Chemically Converted Graphene Nanocomposite. *Nano Res.* **2010**, *3*, 339.

11. Li, J.; Liu, C. Y. Ag/Graphene Heterostructures: Synthesis, Characterization and Optical Properties. *Eur. J. Inorg. Chem.* **2010**, *2010*, 1244–1248.

12. Williams, G.; Kamat, P. V. Graphene-Semiconductor Nanocomposites: Excited-State Interactions Between ZnO Nanoparticles and Graphene Oxide. *Langmuir* 2009, *25*, 13869.

13. Akhavan, O. Photocatalytic Reduction of Graphene Oxides Hybridized by ZnO Nanoparticles in Ethanol. *Carbon* **2010**, *49*, 11.

14. Lu, T.; Pan, L.; Li, H.; Zhu, G.; Lv, T.; Liu, X.; Sun, Z.; Chen, T.; Chua, D. H. C. Microwave-Assisted Synthesis of Graphene–ZnO Nanocomposite for Electrochemical Supercapacitors. *J. Alloys Compd.* **2011**, *509*, 5488–5492.

15. Saravanakumar, B.; Mohan, R.; Kim, S. J. Facile Synthesis of Graphene/ZnO Nanocomposites by Low Temperature Hydrothermal Method. *Mater. Res. Bull.* **2013**, *48*, 878–883.

16. Dai, S.; Zhuang, Y.; Chen, Y.; Chen, L. Study on the Relationship between Structure of Synthetic Organic Chemicals and Their Biodegradability. *Environ. Chem.* **1995**, *14*, 354–367.

17. Bremner, D. H.; Carlo, S. D. Chakinala, A. G.; Cravotto, G. Minuteeralisation of 2, 4-Dichlorophenoxyacetic Acid by Acoustic or Hydrodynamic Cavitation in Conjunction with the Advanced Fenton Process. *Ultrason. Sonochem.* **2008**, *15*, 416–419.

18. Song, Y. L.; Li, J. T.; Chen, H. Degradation of C.I. Acid Red 88 Aqueous Solution by Combination of Fenton's Reagent and Ultrasound Irradiation. *J. Chem. Technol. Biotechnol.* **2009**, *84*, 578–583.

19. Gogate, P. R.; Pandit, A. B. A Review of Imperative Technologies for Wastewater Treatment II: Hybrid Methods. *Adv. Environ. Res.* 2004, *8*, 553–597.

20. Namkung, K. C.; Burgess, A. E.; Bremner, D. H.; Staines, H. Advanced Fenton Processing of Aqueous Phenol Solutions: A Continuous System Study Including Sonication Effects. *Ultrason. Sonochem.* 2008, *15*, 171–176.

21. Ioan, I.; Wilson, S.; Lundanes, E.; Neculai, A. Comparison of Fenton and Sono-Fenton Bisphenol A Degradation. *J. Hazard. Mater.* 2007, *142*, 559–563.

22. Papadaki, M.; Emery, R. J.; Hassan, M. A. A.; Bustos, A. D.; Metcalfe, I. S.; Mantzavinos, D. Sonocatalytic Oxidation Processes for the Removal of Contaminants Containing Aromatic Rings From Aqueous Effluents. *Sep. Purif. Technol.* 2004, *34*, 35–42.

23. Sun, J. H.; Sun, S. P.; Sun, J. Y.; Sun, R. X.; Qiao, L. P.; Guo, H. Q.; Fan, M. H. Degradation of Azo Dye Acid Black 1 Using Low Concentration Iron of Fenton Process Facilitated by Ultrasonic Irradiation. *Ultrason. Sonochem.* 2007, *14*, 761–766.

24. Chakinala, A. G.; Bremner, D. H.; Gogate, P. R.; Namkung, K. C.; Burgess, A. E. *Multivariate Analysis of Phenol Minuteeralisation by Combined Hydrodynamic Cavitation and Heterogeneous Advanced Fenton Processing. Appl. Catal. B: Environ.* **2008**, *78*, 11–18.

25. Sivakumar, M.; Pandit, A. B. Ultrasound Enhanced Degradation of Rhodamine B: Optimization with Power Density. *Ultrason. Sonochem.* **2001**, *8*, 233–240.

26. Gogate, P. R.; Mujumdar, S.; Thampi, J.; Wilhelm, A. M.; Pandit, A. B. Destruction of Phenol Using Sonochemical Reactors: Scale Up Aspects and Comparison of Novel Configuration With Conventional Reactors. *Sep. Purif. Technol.* **2004**, *34*, 25–34.

27. Sadia Ameen; ShaheerAkhtar, M.; Hyung-KeeSeo, Hyung Shik Shin. Advanced ZnO–Graphene oxide Nanohybrid and Its Photocatalytic Applications. *Mater. Lett.* **2013**, *100*, 261–265.

28. Ahmad, M.; Ahmed, E.; Hong, Z. L.; Xu, J. F.; Khalid, N. R.; Elhissi, A.; Ahmed, W. A Facile One-Step Approach to Synthesizing ZnO/Graphene Composites for Enhanced Degradation of Methylene Blue Under Visible Light. *Appl. Surf. Sci.* **2013**, *274*, 273–281.

29. Chen, Ya-Li.; Zhang, Chun-E.; Deng, Chao.; Fei, Peng.; Minuteg Zhong, Su, Bi-Tao.; Preparation of ZnO/GO Composite Material with Highly Photocatalytic Performance via an Improved Two-Step Method. *Chin. Chem. Lett.* 2013, *24*, 518–520.

30. Benxia Li.; Tongxuan Liu.; Yanfen Wang.; Zhoufeng Wang. ZnO/Graphene-Oxide Nanocomposite with Remarkably Enhanced Visible-Light-Driven Photocatalytic Performance. *J. Colloid Interface Sci.* 2012, *377*, 114–121.

31. Xiaoning Wang.; Jinping Jia.; Yalin Wang. Degradation of C.I. Reactive Red 2 Through Photocatalysis Coupled with Water Jet Cavitation. *J. Hazard. Mater.* 2011, *185*, 315–321.

32. Bagal, M. V.; Gogate, P. R. Degradation of Diclofenac Sodium Using Combined Processes Based on Hydrodynamic Cavitation and Heterogeneous Photocatalysis. *Ultrason. Sono.* **2014**, *21*, 1035–1043.

33. Behnajady, M. A.; Modirshahla, N.; Shokri, M.; Vahid, B. Investigation of the Effect of Ultrasonic Waves on the Enhancement of Efficiency of Direct Photolysis and Photooxidation Processes on the Removal of a Model Contaminant from Textile Industry. *Global NEST J.* **2008**, *10*, 8–15.

34. Hummers, W. S.; Offeman, R. E. Preparation of Graphitic Oxide. *J. Am. Chem. Soc.* 1958, *80*, 1339–1339.

35. Stankovich, S.; Dikin, D. A.; Piner, R. D.; Kohlhaas, K. A.; Kleinhammes, A.; Jia, Y.; Wu, Y.; Nguyen, S. T.; Ruoff, R. S. Synthesis of Graphene-Based Nanosheets via Chemical Reduction of Exfoliated Graphite Oxide. *Carbon* 2007, *45*, 1558–1565.

36. Xu, C.; Wang, X.; Zhu, J. Graphene - Metal Particle Nanocomposites. *J. Phys. Chem. C.* 2008, *112*, 19841–19845.

37. Li Pan-pan.; Men Chuan-ling.; Li Zhen-peng.; Cao Minute; An Zheng-hua. Study of Graphene Doped Zinc Oxide Nanocomposite as Transparent Conducting Oxide Electrodes for Solar Cell Applications. *J. Shanghai Jiaotong Univ. Sci.* **2014**, *19*, 378–384.

38. Mahamuni, N. N.; Adewuyi, Y. G. Advanced Oxidation Processes (AOPs) Involving Ultrasound for Waste Water Treatment: A Review with Emphasis on Cost Estimation. *Ultrason. Sonochem.* **2010**, *17*, 990–1003.

39. Gul, S.; Ozcan-Yildirium, O. Degradation of Reactive Red 194 and Reactive Yellow 145 Azo Dyes by O_3 and H_2O_2/UV-C Processes. *Chem. Eng. J.* **2009,** *155,* 684–690.
40. Wu, C.H. Effects of Operational Parameters on the Decolourization of C.I. Reactive Red 198 in UV/TiO$_2$-Based Systems. *Dyes Pigments* **2008,** *77,* 31–38.
41. Adewuyi, Y.G. Sonochemistry: Environmental Science and Engineering Applicatons. *Ind. Eng. Chem. Res.* **2001,** *40,* 4681–4715.

CHAPTER 9

SYNTHESIS OF GRAPHENE–TIO$_2$ NANOCOMPOSITES FOR DECOLORIZATION OF MALACHITE GREEN USING HYDRODYNAMIC CAVITATION BASED HYBRID TECHNIQUES

M. SURESH KUMAR[1], S. H. SONAWANE[1*], MEIFANG ZHOU[2], MUTHUPANDIAN ASHOKKUMAR[2], and VIKAS MITTAL[3]

[1]Department of Chemical Engineering, National Institute of Technology, Warangal 506004, Telangana, India

[2]Department of Chemistry, University of Melbourne, VIC 3010, Australia

[3]Department of Chemical Engineering, The Petroleum Institute, Abu Dhabi, UAE

*Corresponding author. E-mail: shirishsonawane09@gmail.com

CONTENTS

ABSTRACT

In this work, graphene–TiO$_2$ nanocomposites (G–TiO$_2$) were synthesized by a sol–gel method using titanium isopropoxide as Ti-precursors and reduced graphene oxide (RGO) and used as photocatalyst for the decolorization of malachite green (MG). Graphene oxide (GO) was prepared by the modified Hummers–Offeman method in presence of ultrasonic irradiation. The prepared GO and graphene–TiO$_2$ was characterized using TEM, XRD. Decolorization of MG has been achieved using hydrodynamic cavitation and in combination with other advanced oxidation processes. The hydrodynamic cavitation was first optimized in terms of different operating parameters such as operating inlet pressure, and pH of the operating medium to get the maximum degradation of MG. In the hybrid techniques, combination of hydrodynamic cavitation with H$_2$O$_2$, G–TiO$_2$, and G–TiO$_2$/H$_2$O$_2$ have been used to get the enhanced degradation efficiency through hydrodynamic cavitation device. The maximum extent of MG decolorization as 99.4% has been observed using hydrodynamic cavitation in conjunction with G–TiO$_2$/H$_2$O$_2$ under the optimized conditions at inlet pressure 2 bar and pH value 8.5 in 90 min. The MG decolorization has been analyzed using UV–visible spectrophotometer.

9.1 INTRODUCTION

Photocatalytic decomposition of various organic compounds in aqueous solutions has been widely studied and many nanomaterials have been developed as photocatalyst for this technology.[1-4] TiO$_2$ has been intensively investigated as a photocatalyst for environmental cleanup and solar energy conversion. However, TiO$_2$ can only decompose aromatic organics into CO$_2$ and H$_2$O under UV-illumination and suffers from a barrier in responding to visible light at wavelengths higher than 387 nm due to a large band gap of 3.2 eV. As a result, only 3–5% of the solar energy that reaches onto the earth surface can be utilized. The common strategies for extending the absorption threshold of TiO$_2$ to visible light region include doping, coupling, or anchoring with other organic or inorganic elements such as nitrogen, carbon, halogen, and metals into the titania lattice.[5-11]

In the past a few years, graphene as a novel carbonaceous nanomaterial has attracted more and more interests due to its unique and excellent performance in chemical, structural, electrical, thermal, mechanical properties, and potential application.[12-16] At present, there have been several methods presented in the literature for the preparation of graphene, for example, micromechanical exfoliation,[17] chemical vapor deposition,[18] and epitaxial growth.[19]

Several attempts in using graphene oxide (GO) or reduced graphene oxide (RGO) for modification of TiO$_2$ for photocatalytic degradation of organics has been reported.[20,21] Zhang et al.[22] used a commercial TiO$_2$(P25) and GO to obtain a TiO$_2$–graphene nanocomposite. Nguyen-Phan et al.[23] prepared a TiO$_2$(P25)–GO composite using a simple colloidal blending method. Liang et al.[24] reported a graphene/TiO$_2$ nanocrystal hybrid fabricated by directly growing TiO$_2$ nanocrystals onto GO sheets.

Wastewater from the textile industry containing dyes causes serious environmental problem due to their intense color and potential toxicity. About 10–20% of the total dyestuff used in the dyeing process is released into the environment.[25,26] The wastewater containing colored solution is the source of aesthetic pollution, eutrophication, and perturbations in aquatic life essentially due to their organic nature. Among all types of dye used in the textile and paper industry around 50–70% dyes are of Azo class.[27–29] Different traditional treatment techniques applied in industrial wastewater such as coagulation/flocculation, membrane separation, adsorption process, but these methods generates secondary pollutants, while biological treatment has limitation to implement as some of dyes show biological resistance.[20] Advanced oxidation processes (AOPs) are defined broadly as those aqueous phase oxidation processes, which are based primarily on the generation and attack of the hydroxyl radicals resulting in the destruction of the target pollutant. Some of the AOPs, which have shown considerable promising effect for wastewater treatment applications, which includes cavitation process, Fenton chemistry, and photocatalytic oxidation.[30,31] Usually a combination of two or more AOPs has been found to be more efficient for wastewater treatment[32–37] as compared to individual oxidation process. Cavitation based hybrid AOPs have been widely investigated for wastewater treatment applications, but focus was on the use of ultrasonically induced cavitational process which have limitation to scale up. Research has been done and development of efficiency of hybrid systems for wastewater treatment using cavitation coupled with various AOPs such as UV photocatalysis, addition of H$_2$O$_2$, cavitation coupled with photo Fenton, etc. The recent literature on wastewater treatment using photocatalysis and cavitation coupled with photocatalysis has been given here.

In this work, hydrodynamic cavitation and photocatalyst for UV–visible radiation is a novel combination for treatment of malachite green (MG) have been studied which were not attempted earlier. Further the photocatalyst graphene–TiO$_2$ was prepared by sonochemical approach is an additional novelty of the work. The exfoliation of the graphene and addition of the TiO_2 using ultrasound technique shows the better performance than any other available technique.

9.2   MATERIALS AND METHODS

9.2.1   MATERIALS AND REAGENTS

Natural graphite powder was used for GO synthesis. All other reagents H_2SO_4 (SDFCL, Mumbai 98 wt%), $KMnO_4$ (SRL Pvt Ltd., 99.5 wt%), $NaNO_3$ (Qualigens fine chemicals, 98 wt%), H_2O_2 (Finar Chemicals, Ahmadabad, 30% w/v), CH_3COOH (SDFCL, Mumbai 99.5%), Hydrazine hydrate (SD Fine Chemical, Mumbai), HCl (Himedia Lab, Thane), C_2H_5OH (Chengshu, Yangyuan Chemical, China, 99.9 wt%), Titanium (IV) isopropoxide (Spectrochem Pvt Ltd., Mumbai). MG dye (molecular weight: 927 g/mol; molecular formula: $C_{52}H_{52}N_4O_{12}$), Sodium hydroxide (NaOH). The experiments were carried out within the temperature range of 30–35°C. All the solutions were prepared with tap water as a dissolution medium. The concentration of dye was kept constant in all the cases at 539 μM (500 ppm). The concentration of H_2O_2 was kept 5390 μM and added externally.

9.2.2   SONOCHEMICAL SYNTHESIS OF GRAPHENE OXIDE (GO)

The GO was prepared from graphite powder by using modified Hummer method[38,39] in the presence of ultrasonic irradiations using an Ultrasonic Horn (Dakshin ultrasonic probe sonicator, 50 Hz frequency, 230 V). In this method graphite powder (1 g) and $NaNO_3$ (1 g) was added to 46 mL H_2SO_4 solution (98%) in an ice bath. Prepared mixture was ultrasonicated for 10 min. Further, gradual addition of 5 g $KMnO_4$ to the as prepared mixture was accomplished and then it was further ultrasonicated for 30 min at 35°C. To the prepared solution, 100 mL DI water was gradually added and again ultrasonicated for 5 min after which 8 mL H_2O_2 was added to the resulting solution. The obtained solution was filtered, washed with 150 mL HCl solution then heat treated in dry oven at 90°C for 12 h to obtain graphite oxide powder. Out of graphite oxide powder, 200 mg was mixed in 200 mL DI water (mg/mL) stirring for 30 min and ultrasonicated for 2 h. The resulting solutions were filtered and washed several times with hot water and kept in a dry oven for 6 h to achieve graphene oxide powder.

9.2.3   PREPARATION OF GRAPHENE–TIO$_2$ NANOCOMPOSITE

GO was prepared by a modified Hummers method and the reduction of exfoliated GO was obtained by hydrothermal reaction[40] using hydrazine hydrate. Typically, GO (100 mg) was loaded in a 250 mL round bottom flask with 100 mL deionized water and subjected to ultrasonic treatment for 2 h, yielding a homogeneous yellow-brown dispersion. Hydrazine hydrate (1.00 mL) was then added in and the solution was heated at 100°C for 24 h. The RGO was gradually precipitated as a black solid. This product was separated by filtration and washed with ethanol and water several times, and then dried at 80°C.

For a typical synthesis of G–TiO$_2$, RGO powder, cetyl trimethyllammonium bromide (CTAB, 0.5 g) and 30 mL ethanol were placed in a 100 mL beaker with stirring. After 30 min, titanium isopropoxide (11 mL) was added dropwise into the reactor. Then, 20 mL deionized water was added into the mixed solution. The suspension was sonicated for 2 h and dried at 80°C. The solid was annealed in a muffle furnace at 500°C from ambient temperature, maintained for 5 min and then naturally cooled down to room temperature. In graphene–TiO$_2$ samples, graphene loading was kept 5 wt%.

9.2.4   TIO$_2$ NANOPARTICLE SYNTHESIS

TiO$_2$ was prepared by a modified sol–gel route.[41] Titanium isopropoxide (25 mL) was added to acetic acid (48 mL) with stirring. Water (150 mL) was added to the mixture dropwise with vigorous stirring. (The titanium isopropoxide, acetic acid, and water are in 1:10:100 molar ratios.). The solution was stirred for 8 h to get a clear transparent sol and allowed to dry at 100°C, after which calcined at 600°c in air for 2 h at ramp rate of 5°C/min.

9.2.5   CHARACTERIZATION OF MATERIALS

Graphene–TiO$_2$ nanocomposite was characterized by using powder X-ray diffractometer (Phillips PW 1800). Graphene–TiO$_2$ nanocomposite microstructure and morphology was studied by using field emission scanning electron microscopy (FESEM). Graphene–TiO$_2$ nanocomposite samples have been analyzed for FTIR spectrometer in KBr medium at room temperature in the region of 4000–40 cm^{-1}.

9.2.6 HYDRODYNAMIC CAVITATION REACTOR SET UP

The experimental setup is shown in Figure 9.1. The setup includes a holding tank of 15 L volume, a centrifugal pump of power rating 0.5 HP, control valve placed in the by-pass line, and flanges to accommodate the cavitating device in the main line and a by-pass line to control the flow through the main line. The suction pipe of the pump is dipped into the tank and discharge from the pump branches into two lines; the main line and a bypass line. The main line consists of a flange which houses the cavitating device which either orifice. The main line flow rate was adjusted by control valve which is provided in the bypass line to control the liquid flow through the main line. Both the main line and bypass line terminate well inside the tank below the liquid level to avoid any induction of air into the liquid due to the falling liquid jet. Diameter of the mail pipeline is 12.7 mm (0.5 inch), and orifice diameter is 2 mm.

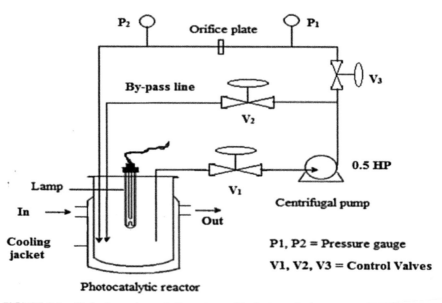

FIGURE 9.1   Hydrodynamic cavitation set up with photocatalysis.

The materials of construction of the entire system except cavitating device are stainless steel (SS316), whereas cavitating device, orifice is made up of brass. UV–visible lamp (9 W) is placed centrally in the feed tank for the UV–visible irradiation. It was used while studying the hybrid techniques involving photocatalytic process.

9.2.7 HYDRODYNAMIC CAVITATION (HC)

All the experiments were performed with 5 L aqueous solution of MG with an initial concentration of 500 ppm (539 µM). Initially operating parameters such as inlet pressure kept constant of 2 bar and pH was optimized at pH of 8.5 which is natural pH of the aqueous solution of MG by varying the pH from 6 to 10 (i.e., 6, 7, 8, 8.5, 9, and 10). As the MG was the basic dye with the natural pH 8.5 and its decolarization favors in basic media and no significant effect on decolorization observed further increasing the pH from 8.5 to 10. The dye solution pH was adjusted by adding the NaOH and H$_2$SO$_4$ as per the requirement.

9.2.8 HYDRODYNAMIC CAVITATION BASED HYBRID TECHNIQUES

All the experiments were performed treating 5 L aqueous solution of MG with initial concentration of 500 ppm. The pump discharge pressure to the cavitating device of 2 bar and the optimal pH 8.5, was used.

Hydrodynamic cavitation combined with addition of H$_2$O$_2$ was employed at the fixed molar ratio of MG: H$_2$O$_2$ as 1:10 with concentration of H$_2$O$_2$ as 5390 µM. Furthermore, combination hydrodynamic cavitation, H$_2$O$_2$ and photocatalytic process was employed at the fixed molar ratio of MG: H$_2$O$_2$ as 1:10. UV assembly was placed centrally in a feed tank for UV irradiation during. Photocatalytic process was employed using graphene–TiO$_2$ photocatalyst which was synthesized. The amount of catalyst used in all experiments was 200 mg/L.

9.2.9 ANALYTICAL METHODS

Hydrodynamic cavitation based decolorization of MG was carried out at different conditions using fixed solution volume of 5 L and for the duration of 90 min. The samples were collected at regular interval of 15 min for UV-spectrophotometer analysis. The initial concentration of MG was kept constant in all the cases at 539 µM (500 ppm). The temperature of the solution during experiments was kept constant in all the cases at about 35°C using the water circulation jacket provision. The absorbance of MG was monitored using UV-Spectrophotometer (Shimadzu-1800) and then the concentration of dye was calculated by analyzing the absorbance of dye solution at the wavelength of 618 nm. The concentration of MG was then calculated using the calibration curve prepared for MG.

9.3 RESULTS AND DISCUSSION

9.3.1 CHARACTERIZATION OF PHOTOCATALYST

Figure 9.2 shows the XRD pattern of the GO. As shown in Figure 9.2, there are some obvious diffraction peaks of 2θ values of 13.7°, 28.83°, 41.4°, and 48.4° for graphite were observed from the GO XRD pattern which can be assigned to the (002), (004), (102), and (106) planes of carbon crystal, respectively (JCPDS No. 79-1470). Furthermore, the diffraction peak of 2θ value of 9.94° was observed for the GO which can be assigned to the (001) crystal plane.[42]

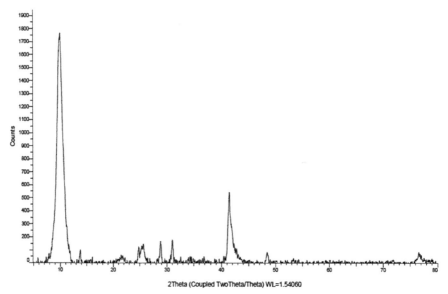

2Theta (Coupled TwoTheta/Theta) WL=1.54060

FIGURE 9.2   X-ray diffraction of the graphene oxide synthesis using sonochemical approach.

Figures 9.3 and 9.4 display the XRD pattern of prepared TiO_2 and graphene–TiO_2 composite, respectively. TiO_2 and graphene–TiO_2 showed the mixed crystalline phases of anatase and rutile. The peaks at 2θ value of 24.3°, 37.6°, 47.2°, 53.3°, 54.2°, 61.8°, 67.9°, 74.2°, and 75.1° were indexed to (101), (004), (200), (105), (211), (204), (116), (220), and

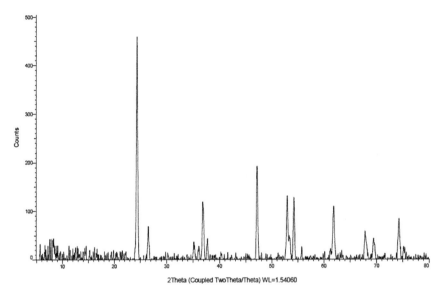

FIGURE 9.3 XRD pattern of TiO$_2$ nanoparticles.

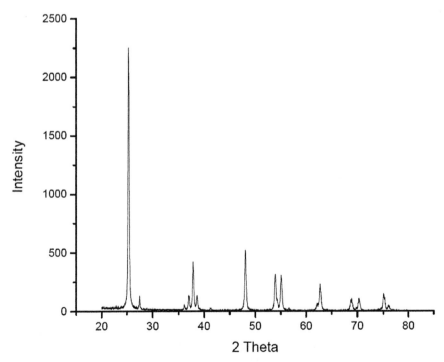

FIGURE 9.4 XRD pattern of graphene–TiO$_2$ nanocomposite.

(215) crystal planes of anatase TiO_2, respectively. In addition, characteristic diffraction peaks at 26.5° and 35.1° were also observed which were attributed to the (110) and (101) crystal planes of rutile TiO_2, respectively. A comparison of TiO_2 and graphene–TiO_2 suggested that addition of graphene would not change the crystalline structure of TiO_2. For graphene–TiO_2, no graphene peak is observed from XRD. The main characteristic of graphene was at 24.5°[43] so the peaks for graphene might be shielded by the strong peak of anatase TiO_2 at 24.3°. Besides, the mass ratio of graphene in the graphene–TiO_2 composite was very low, thus the diffraction intensity of graphene was weak. Similar results have been reported by the other investigation.[40]

From the XRD patterns of TiO_2, the crystallite size of the synthesized TiO_2 nanoparticles were estimated using Scherrer equation[44]

$$d = \frac{k\lambda}{\beta \cos \theta}$$

where
  "d" is crystallite size in nanometer,
  "k" is shape factor constant, which is 0.89,
  "β" is the full width at half maximum (FWHM) in radian, 0.026
  "λ" is the wave length of the X-ray which is 0.154 nm for Cu target Kα
    radiation, and
  "θ" is the Bragg diffraction angle which is 12.2°
By using the above data the estimated crystalline size for synthesized TiO_2 is 302 nm.

FESEM image of GO for evaluation of their morphologies is shown in Figure 9.5. It was found that many exfoliated layers of GO sheets in the image. Figure 9.6 shows TEM images of graphene–TiO_2 nanocomposite and demonstrates that TiO_2 nanoparticles were dispersed uniformly on the graphene sheets.

9.3.2 HYDRODYNAMIC CAVITATION

Initially, the study of decolorization of MG has been carried out for the duration of 90 min using hydrodynamic cavitation at constant inlet pressure of 2 bar for obtaining the optimal pH. The Initial pH of the solution of MG was

S-4800 15.0kV 9.7mm x5.00k SE(U) 11:18   10.0um

FIGURE 9.5   FESEM image of the graphene oxide synthesis using sonochemical approach.

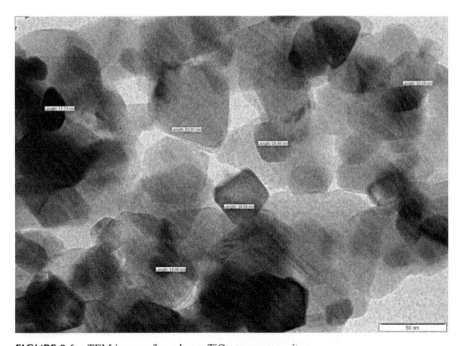

FIGURE 9.6   TEM image of graphene–TiO$_2$ nanocomposite.

optimized by varying the pH from 6 to 10 (i.e., 6, 7, 8, 8.5, 9, and 10) by adding the H_2SO_4 solution and NaOH solution as per the requirement. It has been observed that, the extent of decolorization of MG increases under basic conditions, when the pH of the solution has been reduced 10–6. However, pH 9 has been taken as the optimum because at pH 10 MG has pH transition state where the green color transferred to color less. Figure 9.7 shows UV–visible absorbance values for the MG for time intervals of 15 min of total operation time 90 min. From Figure 9.8, percentage of the MG decolorization increased by increasing pH of dye solution. From Figure 9.9 MG decolorization by hydrodynamic cavitation follows the pseudo first-order reaction. Decolorization percentages of pH values 6, 8, 8.5, 9, and 10 are 90, 91.4, 93.4, 95.8, and 97.8, respectively.

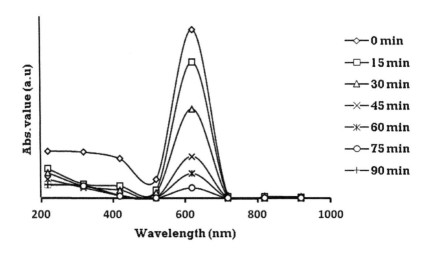

FIGURE 9.7   UV–visible absorbance of MG at different time intervals.

Enhancement in the extent of decolorization of MG at high pH is due to the fact that, under basic conditions the oxidation potential of hydroxyl radicals increase, whereas the recombination reaction of hydroxyl radicals decreases. On the other hand, under basic conditions of the state of MG molecule changes from ionic to molecular state, thereby causing it to locate at the cavity water surface, where the concentration of the hydroxyl radicals are high and hence the decolorization rate increases.

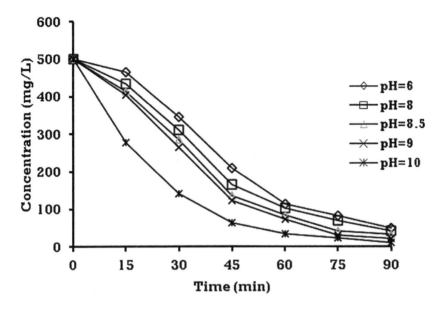

FIGURE 9.9   First-order decolorization of MG at diferent pH solution.

9.3.3   COMBINATION OF HYDRODYNAMIC CAVITATION AND H$_2$O$_2$

In case of HC, main mechanism for the destruction of organic pollutant is the reaction of ˙OH radicals with the pollutant molecules causing oxidation of the same. Hence, the additional supplement of ·OH radicalshould enhance the rate of decolorization of MG. Hydrogen peroxide is a commonly available oxidizing agent which can be used for teatment of waste water due to its high oxidation potential. The following are the reactions expected to take place during the degradation of MG using hybrid process of HC and H$_2$O$_2$.

$$H_2O_2 + (HC) \rightarrow 2 \cdot OH$$
$$H_2O + (HC) \rightarrow H^{\cdot} + {}^{\cdot}OH$$
$${}^{\cdot}OH + {}^{\cdot}OH \rightarrow H_2O_2$$
$${}^{\cdot}OH + H_2O_2 \rightarrow HO^{\cdot}_2 + H_2O$$
$${}^{\cdot}OH + HO^{\cdot}_2 \rightarrow H_2O + O_2$$
$$HO^{\cdot}_2 + H_2O_2 \rightarrow {}^{\cdot}OH + H_2O + O_2$$
$$MG + {}^{\cdot}OH \rightarrow Intermediates$$
$$MG + H_2O_2 \rightarrow Intermediates + H_2O$$

Intermediates + OH$^{\cdot}$ → CO_2 + H_2O
Intermediates + H_2O_2 → CO_2 + H_2O.

The experiments using the hybrid process of HC and H_2O_2 were carried out at 2 bar inlet pressure and solution pH of 9, 1:10 molar ratio of MG and H_2O_2. It has been observed that addition of H_2O_2 increasing the decolrization of MG. From Figure 9.10 decolorization of MG follows the pseudo first-order reaction. Percentage of decolorization of HC and HC+H_2O_2 are 93.4 and 98.8, respectively.

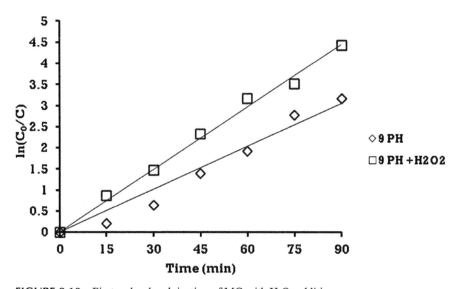

FIGURE 9.10   First-order decoloization of MG with H_2O_2 addition.

9.3.4   COMBINATION OF HC AND H_2O_2 AND PHOTOCATALYTIC PROCESS

Heterogeneous photocatalysis is one of the promising techniques for the degradation of various dyes. Graphene–TiO_2 has been used as a photocatalyst. The experiments using the hybrid process of HC and photocatalytic process combined with H_2O_2 were carried out at 2 bar inlet pressure and solution pH of 9, 1:10 molar ratio of MG and H_2O_2. It has been that addition of H_2O_2 increasing the decolrization of MG. From Figure 9.11 decolorization of MG follows the pseudo first-order reaction. Percentages of decolorization of HC, HC+H$_2$O$_2$, HC+H$_2$O$_2$+photocatalytic process are 93.4, 98.8, and 99.4%, respectively.

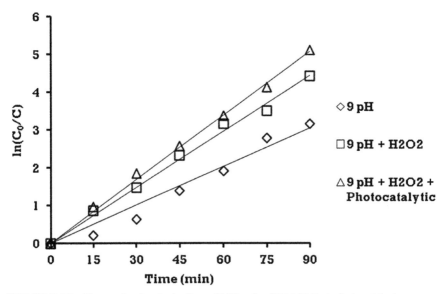

FIGURE 9.11 First-order decolorization of MG using HC + H$_2$O$_2$ + photocatalysis.

9.4 CONCLUSION

In this work graphene–TiO$_2$ nanocomposite has been successfully prepared through a novel and effective ultrasound assisted hydrolysis of titanium isopropoxide with RGO via a reduction of hydrazine hydrate and the same used as the catalyst for the photocatalytic process. Decolorization of MG dye was carried out using hydrodynamic cavitation hybrid techniques. Effect of pH of dye solution at 2 bar inlet pressure, addition of H$_2$O$_2$ and photocatalyst were studied. The decolorization of MG was found to be the pH dependant and acidic medium was found to be favorable for the higher decolorization. The dye solution at pH 9 was optimized at where 95.8% decolorization takes place.

The addition of H$_2$O$_2$ enhances the decolorization rate due to additional radicals available for the oxidation of dye and 98.8% of decolorization was achieved. Furthermore, combination of HC+H$_2$O$_2$+photocatalytic process enhances the decolorization upto 99.4%.

KEYWORDS

- graphene
- TiO_2
- nanocomposites
- photocatalyst
- hydrodynamic cavitation
- AOPs

REFERENCES

1. Alfano, O. M.; Bahnemann, D.; Cassano, A. E.; Dillert, R.; Goslich, R. Photocatalysis in Water Environments Using Artificial and Solar Light. *Catal. Today* **2000,** *58,* 199–230.

2. Chong, M. N.; Jin, B.; Chow, C. W. K.; Saint, C.; Recent Developments in Photocatalytic Water Treatment Technology: A Review. *Water Res.* **2010,** *44,* 2997–3027.

3. Malato, S.; Fernandez-Ibanez, P.; Maldonado, M. I., Blanco, J.; Gernjak, W. Decontamination and Disinfection of Water by Solar Photocatalysis: Recent Overview and Trends. *Catal.Today* **2009,** *147,* 1–59.

4. Ullah, R.; Sun, H.; Wang, S.; Ang, H. M.; Tade, M. O. Wet-Chemical Synthesis of $InTaO_4$ for Photocatalytic Decomposition of Organic Contaminants in Air and Water with UV–vis Light. *Ind. Eng. Chem. Res.* **2012,** *51,* 1563–1569.

5. Sun, H.; Wang, S.; Ang, H. M.; Tade, M. O.; Li, Q. Halogen Element Modified Titanium Dioxide for Visible Light Photocatalysis. *Chem. Eng. J.* **2010,** *162,* 437–447.

6. Kumar, S. G.; Devi, L.G. Review on Modified TiO_2 Photocatalysis under Uv/visible Light: Selected Results and Related Mechanisms on Interfacial Charge Carrier Transfer Dynamics. *J. Phys. Chem. A.* **2011,** *115,* 13211–13241.

7. Han, F.; Kambala, V. S. R.; Srinivasan, M.; Rajarathnam, D.; Naidu, R. Tailored Titanium Dioxide Photocatalysts for the Degradation of Organic Dyes in Wastewater Treatment: A Review. *Appl. Catal. A.* **2009,** *359,* 25–40.

8. Zhang, J.; Wu, Y.; Xing, M.; Leghari, S. A. K.; Sajjad, S.; Development of Modified N Doped TiO_2 Photocatalyst with Metals, Nonmetals and Metal Oxides. *Energy Environ. Sci.* **2010,** *3,* 715–726.

9. Sun, H.; Ullah, R.; Chong, S.; Ang, H. M.; Tade, M. O.; Wang, S. Room-Light-Induced Indoor Air Purification Using an Efficient Pt/N–TiO_2 Photocatalyst. *Appl. Catal. B.* **2011,** 108, 127–133.

10. Pal, M.; U. Pal, U.; Silva Gonzalez, R., Sanchez Mora, E.; Santiago, P. Synthesis and Photocatalytic Activity of Yb Doped TiO_2 Nanoparticles under Visible Light. *J. Nano Res.* **2009,** *5,* 193–200.

11. Pelaez, M.; Nolan, N. T.; Pillai, S.C.; Seery, M. K.; Falaras, P.; Kontos, A. G.; Dunlop, P. S. M.; Hamilton, J. W. J.; Byrne, J. A.; O'Shea, K., Entezari, M. H.; Dionysiou, D.

D. A Review on the Visible Light Active Titanium Dioxide Photocatalysts for Environmental Applications. *Appl. Catal. B.* **2012**, *125*, 331–349.

12. Zhang, Y.; Tan, Y. W.; Stormer, H. L.; Kim, P. Experimental Observation of the Quantum Hall Effect and Berry's Phase in Graphene. *Nature* **2005**, *438*, 201–204.

13. Peigney, A.; Laurent, Ch.; Flahaut, E.; Bacsa, R. R.; Rousset, A. Specific Surface Area of Carbon Nanotubes and Bundles of Carbon Nanotubes. *Carbon* **2001**, *39*, 507–514.

14. Lee, C.; Wei, X.; Kysar, J. W.; Hone, J. Measurement of the Elastic Properties and Intrinsic Strength of Monolayer Graphene. *Science* **2008**, *321*, 385–388.

15. Ghosh, S.; Calizo, I.; Teweldebrhan, D.; Pokatilov, E. P.; Nika, D. L; Balandin, A. A.; Bao, W.; Miao, F.; Lau, C. N. Extremely High Thermal Conductivity of Graphene: Prospects for Thermal Management Applications in Nanoelectronic Circuits. *Appl. Phys. Lett.* **2008**, *92*, 1–3.

16. Lu, Y. H.; Chen, W.; Feng, Y. P.; He, P. M. Tuning the Electronic Structure of Graphene by an Organic Molecule. *J. Phys. Chem. B.* **2009**, *113*, 2–5.

17. Novoselov, K. S.; Geim, A. K.; Morozov, S. V.; Jiang, D.; Zhang, Y.; Dubonos, S. V.; Grigorieva, I. V.; Firsov, A. A. Electric Field Effect in Atomically Thin Carbon Films. *Science* **2004**, *306*, 666–669.

18. Reina, A.; Jia, X. T.; Ho, J.; Nezich, D.; Son, H. B.; Bulovic, V.; Dresselhaus, M. S.; Kong, J.; Large Area, Few-Layer Graphene Films on Arbitrary Substrates by Chemical Vapordeposition. *Nano Lett.* **2009**, *9*, 30–35.

19. Sutter, P. W.; Flege, J. I.; Sutter, E. A.; Epitaxial Graphene on Ruthenium. *Nat. Mater.* **2008**, *7*, 406–411.

20. Xiang, Q.; Yu, J.; Jaroniec, M. Graphene-Based Semiconductor Photocatalysts. *Chem. Soc. Rev.* **2012**, *41*, 782–796.

21. An, X.; Yu, J. C. Graphene-Based Photocatalytic Composites. *RSC Adv.* **2011**, *1*, 1426–1434.

22. Zhang, H.; Lv, X.; Li, Y.; Wang, Y.; Li, J. P25-Graphene Composite as a High Performance Photocatalyst. *ACS Nano.* **2010**, *4*, 380–386.

23. Nguyen-Phan, T. D.; Pham, V. H.; Shin, E. W.; Pham, H. D.; Kim, S.; Chung, J. S.; Kim, E. J.; Hur, S. H. The Role of Graphene Oxide Content on the Adsorption-Enhanced Photocatalysis of Titanium Dioxide/Graphene Oxide Composites. *Chem. Eng. J.* **2011**, *170*, 226–232.

24. Liang, Y. Y.; Wang, H. L.; Casalongue, H. S.; Chen, Z.; Dai, H. J. TiO₂ Nanocrystals Grown on Graphene as Advanced Photocatalytic Hybrid Materials. *Nano Res.* **2010**, *3*, 701–705.

25. Akpan, U. G.; Hameed, B. H Parameters Affecting the Photocatalytic Degradation of Dyes Using TiO₂-Based Photocatalysts: A Review. *J. Hazard. Mater.* **2009**, *170*, 520–529.

26. Cho, I. H.; Zoh, K. D. Photocatalytic Degradation of Azo Dye (Reactive Red 120) in TiO₂/UV System: Optimization and Modeling Using a Response Surface Methodology (RSM) Based on the Central Composite Design. *Dyes Pigm.* **2007**, *75*, 533–543.

27. Saharan, V. K.; Pandit, A. B.; SatishKumar, P. S.; Anandan, S. Hydrodynamic Cavitation as an Advanced Oxidation Technique for the Degradation of Acid Red 88 dye. *Ind. Eng. Chem. Res.* **2011**, *51*, 1981–1989. doi:10.1021/ie200249k.

28. Madhavan, J.; Grieser, F.; Ashokkumar, M. Degradation of Orange-G by Advanced Oxidation Processes. *Ultrason. Sonochem.* **2010**, *17*, 338–343.

29. Song, Y.L.; Li, J.T.; Chen, H. Degradation of C.I. Acid Red 88 Aqueous Solution by Combination of Fenton's Reagent and Ultrasound Irradiation. *J. Chem. Technol. Biotechnol.* **2009**, *84*, 578–583.

30. Bremner, D. H.; Carlo, S. D.; Chakinala, A. G.; Cravotto, G. Mineralisation of 2, 4-Dichlorophenoxyacetic Acid by Acoustic or Hydrodynamic Cavitation in Conjunction with the Advanced Fenton Process. *Ultrason. Sonochem.* **2008**, *15*, 416–419.

31. Song, Y. L.; Li, J. T.; Chen, H. Degradation of C.I. Acid Red 88 Aqueous Solution by Combination of Fenton's Reagent and Ultrasound Irradiation, *J. Chem. Technol. Biotechnol.* **2009**, *84*, 578–583.

32. Gogate, P. R.; Pandit, A. B.; A Review of Imperative Technologies for Wastewater Treatment II: Hybrid Methods. *Adv. Environ. Res.* 2004, *8*, 553–597.

33. Namkung, K. C.; Burgess, A. E.; Bremner, D. H.; Staines, H. Advanced Fenton Processing of Aqueous Phenol Solutions: A Continuous System Study Including Sonication Effects. *Ultrason. Sonochem.* 2008, *15*, 171–176.

34. Ioan, I.; Wilson, S.; Lundanes, E.; Neculai, A. Comparison of Fenton and Sono-Fenton Bisphenol A Degradation. *J. Hazard. Mater.* 2007, *142*, 559–563.

35. Papadaki, M.; Emery, R. J.; Hassan, M. A. A.; Bustos, A. D.; Metcalfe, I. S.; Mantzavinos, D. Sonocatalytic Oxidation Processes for the Removal of Contaminants Containing Aromatic Rings from Aqueous Effluents. *Sep. Purif. Technol.* **2004**, *34*, 35–42.

36. Sun, J. H.; Sun, S. P.; Sun, J. Y.; Sun, R. X.; Qiao, L. P.; Guo, H. Q.; Fan, M. H. Degradation of Azo Dye Acid Black 1 Using Low Concentration Iron of Fenton Process Facilitated by Ultrasonic Irradiation. *Ultrason. Sonochem.* 2007, *14*, 761–766.

37. Chakinala, A. G.; Bremner, D. H.; Gogate, P. R.; Namkung, K. C.; Burgess, A. E. *Multivariate Analysis of Phenol Mineralisation by Combined Hydrodynamic Cavitation and Heterogeneous Advanced Fenton Processing. Appl. Catal. B: Environ.* **2008**, *78*, 11–18.

38. Hummer, Jr. W. S.; Offeman, R. E. Preparation of Graphitic Oxide. *J. Am. Chem. Soc.* **1958**, *80*, 1339.

39. Deosarkar, M. P.; Pawar, S. M.; Sonawane, S. H.; Bhanvase, B. A. Process Intensification of Uniform Loading of SnO_2 Nanoparticles on Graphene Oxide Nanosheets Using a Novel Ultrasound Assisted in situ Chemical Precipitation Method. *Chem. Eng. Process.* **2013**, *70*, 48–54.

40. Shizhen, L.; Hongqi, S.; Shaomin, L.; Shaobin, W. Graphene Facilitated Visible Light Photodegradation of Methylene Blue over Titanium Dioxide Photocatalysts. *Chem. Eng. J.* **2013**, *214*, 298–303.

41. Michael, K. S.; Reenamole, G.; Patrick, F.; Suresh, C. P.; Silver Doped Titanium Fioxide Nanomaterials for Enhanced Visible Light Photocatalysis. *J. Photochem. Photobiol A.* **2007**, *189*, 258–263.

42. Stengl, V.; Bakardjieva, S.; Grygar, T. M.; Bludska, J.; Kormunda, M. TiO_2–Graphene Oxide Nanocomposite as Advanced Photocatalytic Materials. *Chem. Cent. J.* **2013**, *7*, 41.

43. Cheng, P.; Yang, Z.; Wang, H.; Cheng, W.; Chen, M.; Shangguan, W.; Ding, G. TiO_2–Graphene Nanocomposite for Photocatalytic Hydrogen Production from Splitting Water. *Int. J. Hydrogen Energy.* **2012**, *37*, 2224–2230.

44. Bethi, B.; Sonawane, S. H.; Rohit, G. S.; Holkar, C. R.; Pinjari, D.V.; Bhanvase, B. A.; Pandit, A. B. Investigation of TiO_2 Photocatalyst Performance for Decolorization in the Presence of Hydrodynamic Cavitation as Hybrid AOP. *Ultrason. Sonochem.* **2016**, *28*, 150–160.

CHAPTER 10

CERIA-BASED NANOCRYSTALLINE OXIDE CATALYSTS: SYNTHESIS, CHARACTERIZATION, AND APPLICATIONS

ANUSHKA GUPTA[#], V. SAI PHANI KUMAR[#], MANJUSHA PADOLE, MALLIKA SAHARIA[$], K. B. SRAVAN KUMAR[$], and PARAG A. DESHPANDE[*]

Quantum and Molecular Engineering Laboratory, Indian Institute of Technology Kharagpur, Kharagpur 721302, West Bengal, India

[*]*Corresponding author. E-mail: parag@che.iitkgp.ernet.in*

CONTENTS

[#]These authors have equal contribution.

[$]These authors have equal contribution.

ABSTRACT

This work summarizes the recent developments in the field of synthesis, characterization, and activity of ceria-based catalysts. Development of solution combustion technique has proved to be a boon for the synthesis of ceria-based materials. The synthesis technique has provided us with a fast and single step technique for the synthesis of nanocrystalline ceria. Further, substitution of a foreign metal ion has been possible following this technique which has brought about significant improvements in the catalytic activities of ceria-based materials when applied to CO oxidation, NO_x reduction, the water–gas shift reaction, catalytic combustion reactions, and C—C coupling reactions. A detailed report on the activity of the catalysts and mechanism of the reactions over these catalysts is presented here.

10.1   INTRODUCTION

Heterogeneous catalysis has proved its importance as the oft-used catalysis technique for industrial applications. The keen interest of the community is apparent by the extent of its application in process industries and the fundamental research on the development of such systems being carried out by the investigators all over the world. Heterogeneous catalysis offers an ease on industrial scale of operation by elimination of separation and purification processes required during the processing. However, this necessitates an effective contact of the solid catalyst with the fluid reactant phase. Noble metals like platinum and palladium, although very reactive on molecular scale,[1,2] can quickly become ineffective due to ineffective fluid solid contact. To overcome this, supported catalysts have been developed which utilize a fine dispersion of the metals over a high surface area support. Reactive systems throughout the domain of heterogeneous catalysis have been benefited by supported catalysis. However, due to environmental interests, the most notable instance comes from the exhaust catalysis where the removal of CO and NO_x has been achieved in automobile systems using supported catalytic systems.[3,4] Although the classical understanding of heterogeneous catalysis describes the support to be inert toward the reaction, recent observations have proposed the involvement of the support for the reaction via the so-called metal-support interactions.[3,4] Such catalysts, which are the systems of interest of this chapter, have proved to be vital for the modern understanding and development of the catalytic systems.

One of the most popular supports investigated for numerous redox reactions is the ceria-based materials. Ceria due to its oxygen storage capacity (OSC) offers active lattice oxygen for the oxidation steps and due to the presence of oxide ion vacancies, new adsorption sites are formed in the solid thereby making the catalyst more active. Substitution of a metal ion, rather than its impregnation in metallic form, increases both adsorption of gases and creation of oxide ion vacancies in the material. Several methods have been reported for the synthesis of ceria-based solid solutions viz., sol–gel method, co-precipitation, wet impregnation, electrochemical deposition, hydrothermal treatment, dry mixing method, etc.[4–11] The activity of the catalyst depends on the synthesis method, composition of precursors and reaction conditions used. The solid solutions synthesized from co-precipitation method form aggregates easily and they are difficult to be re-dispersed in any solvent.[12] The drawbacks of conventional synthesis techniques are limited control over the composition, grain size, and crystallinity of the nanoparticles formed.[12] Solution combustion technique has made the synthesis of such compounds feasible with high purity and crystallinity. It is a simple process and it produces highly sinterable ceramic powders.[5] As a result, numerous studies have appeared recently reporting the synthesis, characterization, and applications of ceria-based materials for heterogeneous catalysis synthesized by solution combustion technique.[4–6,10,14–19] The catalysts have been tested and found active for reactions like CO oxidation, NO_x reduction, the water–gas shift reaction, hydrogen combustion, hydrocarbon combustion, and C—C coupling reaction. Ceria can also be used as an electrolytic material in solid oxide fuel cells (SOFC's) to generate electric energy from the chemical energy of a fuel.[5] Transition or rare earth metals doped ceria serves as an effective material for the SOFC electrolyte due to its oxygen ion conductivity character. This is due to the oxygen ion vacancy formation ability and the reducibility of Ce^{4+} to Ce^{3+}.[5,6] This chapter focuses on detailing the recent development of such systems including their synthesis, characterization, applications, computational modeling, and mechanistic investigations with particular focus on the systems synthesized by solution combustion technique.

10.2 SYNTHESIS OF CERIA-BASED NANOCRYSTALLINE OXIDES

10.2.1 CO-PRECIPITATION

In this method, the mixed oxides of ceria and transition metals are synthesized by using the nitrates of cerium and dopant metals as the precursors. The

precursors are dissolved in distilled water as per the stoichiometric requirements to obtain the desired concentrations of ceria solid solutions.[7-13] Then,
the required volume of ammonia solution is added dropwise to the solid
solution of ceria under stirring. The resultant mixture is filtered, washed, and
dried at 100°C for overnight. The final solid mass is calcined at 700°C for
5–6 h to obtain the ceria-based mixed oxide.

10.2.2   SOL–GEL METHOD

In sol–gel method, the nitrate precursor of cerium, buffer $CO(NH_2)_2$ and
polyvinylpyrrolidone (PVP) as a dispersant are dissolved in a deionized water under magnetic stirring.[8] The chemical reaction starts at room
temperature and reaction temperature increases as the reaction progresses.
After one hour of the reaction, ammonia solution is added dropwise and the
appearance of reaction mixture changes from purple to pale yellow. This
indicates the formation of ceria hydrosol, which is further dried and calcined
to obtain the compound.

10.2.3   DRY MIXING METHOD

The oxides of cerium and dopant metal are physically mixed in a high energy
ball mill using various diameter balls.[13] The critical speed of the ball mill
and diameter of the balls used in the process plays an important role on the
size of the nanocatalysts synthesized. The final powder after mixing is the
desired oxide catalyst.

10.2.4   HYDROTHERMAL SYNTHESIS

The nitrate solutions of cerium and metal dopants are dissolved in distilled
water. Ammonia solution is added slowly into it by maintaining the desired
pH level.[9-13] Then the reaction mixture is transferred into an autoclave which
is maintained at a high temperature of 200°C for certain time (~24 h) and
then the reaction mixture is cooled to room temperature. The resulting products are washed several times with distilled water and dried in a hot air oven
for overnight at 100°C. The samples are then calcined at 1000°C for 24 h
in atmospheric air.[11] The final powdered samples after calcination are the
required catalysts.

Although there are several techniques available for the synthesis of ceria-based oxide catalysts, combustion synthesis has proved to be an effective technique to achieve nanocrystallinity. Further, the technique is also a vital single-step technique for the synthesis of multicomponent solid solutions. Therefore, we focus on solution combustion technique as the method of synthesis.

10.2.5 SOLUTION COMBUSTION SYNTHESIS

The precursors for the synthesis of a typical $Ce_{1-x}M_xO_{2-\delta}$ nanocatalysts are ceric ammonium nitrate, oxalyl dihydrazide (ODH) fuel, which is synthesized from diethyl oxalate and hydrazine hydrate, and a respective salt of the metal. The salts are typically nitrate, chloride or oxalate compounds of the metal. There is a variety of organic compounds which can used as a fuel for the conventional and modified solution combustion synthesis of nanocatalysts including ODH, glycine, urea, citric acid, tartaric acid, etc.[14,15] Modified solution combustion method uses several fuel combinations for the reaction which leads to selective phases and sizes of the catalysts. The flame temperatures required for the solution combustion are lower for modified solution combustion method than the conventional one leading to a reduction in the crystallite size. Typical metallic derivatives used for the preparation of metal ion substituted ceria nanocatalysts are palladium chloride $(PdCl_2)$, tin (II) chloride $(SnCl_2)$, tin oxalate (SnC_2O_4), tetraammine platinum (II) nitrate $(Pt(NH_3)_4(NO_3)_2)$, zirconium nitrate $(Zr(NO_3)_4.5H_2O)$, iron (III) nitrate nonahydrate $(Fe(NO_3)_3.9H_2O)$, aluminum nitrate $(Al(NO_3)_3)$, etc.[14–19] Various precursors used for the synthesis of metal ion substituted ceria via solution combustion method with the size of crystallites are given in Table 10.1. Stoichiometric amounts of nitrates of ceria and transition metal nitrates are dissolved in a minimum amount of H_2O along with the fuel. The resulting clear solution, obtained after dissolving metal precursors with the fuel, is heated slowly to raise the temperature by means of a hot plate. This hot solution is transferred to the muffle furnace where it is heated at high temperatures (typically 350–500°C). During heating, froth formation and vaporization of H_2O takes place thereby yielding a solid material. The residual solid material after combustion is the required nanocrystalline catalyst which is further dried and calcined at 500°C. The whole process of synthesis gets completed within fifteen minutes (Figs. 10.1 and 10.2 show the schematic and representative synthesis procedure). Thus, the solution combustion technique is an extremely efficient and fast technique for the synthesis of nanocrystalline oxide materials. When compared to other solid-state techniques, it requires much lesser time as well as much lesser temperatures thereby providing an extremely efficient control over the size of the final product. It is noteworthy to mention that the principles of propellant chemistry are involved to determine the stoichiometry between the oxidant and fuel in the solution combustion synthesis method, in which oxidant (O) to fuel (F) O/F ratio plays a key role due to the exothermic nature of the reaction.[5,6] The energy liberated is maximum if the O/F ratio is unity and no carbon residue is left after the completion of the reaction.

TABLE 10.1   List of Precursors and Fuels Used in the Synthesis of Metal Ion Substituted Ceria by Solution Combustion Technique.

Compound	Precursors	Fuel	Crystallite Size (nm)	Ref.
$Ce_{0.9}Pr_{0.1}O_{1.95}$	$(Ce(NO_3)_3.6H_2O)$, $(Pr(NO_3)_3.6H_2O)$	Citric acid $(C_6H_8O_7)$	20	[5]
$Ce_{0.9}La_{0.1}O_{1.95}$	$(Ce(NO_3)_3.6H_2O)$,	Citric acid	20–50	[6]
$Ce_{0.9}Sc_{0.1}O_{1.95}$	$(La(NO_3)_3.6H_2O)$,			
$Ce_{0.9}Yb_{0.1}O_{1.95}$	$(Sc(NO_3)_3.6H_2O)$, $(Yb(NO_3)_3.6H_2O)$			
$MnO_x–CeO_2$	$(Mn(NO_3)_3.4H_2O)$ $(Ce(NO_3)_3.6H_2O)$	Glycine $(C_2H_5NO_2)$, Citric acid	–	[20]
$MnCeO_x$	$Mn(NO_3)_2$, $(Ce(NO_3)_3.6H_2O)$, Cordierite honeycomb	Citric acid	–	[21]
$Ir–CeO_2$	Ceric ammonium nitrate (CAN) $((NH_4)_2Ce(NO_3)_6)$, $((NH_4)_2IrCl_6)$	Glycine	38	[22]
CeO_2	$(Ce(NO_3)_3.6H_2O)$	Urea	26–55	[23]
CeO_2	$(Ce(NO_3)_3.6H_2O)$	Urea	35	[24]
$Gd_2Si_2O_7$:Ce	Gd_2O_3, fumed silica, $(Ce(NO_3)_3.6H_2O)$	Glycine	35–60 nm	[25]
$Ce_{0.9}Er_{0.1}O_{1.95}$	$(Ce(NO_3)_3.6H_2O)$	Citric acid	15–30	[10]
$Ce_{0.9}Nd_{0.1}O_{1.95}$	$(Er(NO_3)_3.5H_2O)$			
$Ce_{0.9}Pr_{0.1}O_{1.95}$	$(Nd(NO_3)_3.6H_2O)$			
$Ce_{0.9}Y_{0.1}O_{1.95}$	$(Pr(NO_3)_3.6H_2O)$ $(Y(NO_3)_3.6H_2O)$			

TABLE 10.1 *(Continued)*

Compound	Precursors	Fuel	Crystallite Size (nm)	Ref.
$Ce_{0.6}Mn_{0.3}Fe_{0.1}O_2$	$(Ce(NO_3)_3.6H_2O)$ $Mn(NO_3)_2$ $Fe(NO_3)_2.9H_2O$	Glycine	1–10 μm	[26]
$Ce_{0.98}Pd_{0.02}O_{2-\delta}$	CAN, palladium chloride $(PdCl_2)$	Oxalyl dihydrazide (ODH)	30–40	[14]
$Ce_{0.83}Zr_{0.015}Pd_{0.02}O_{2-\delta}$	CAN, zirconium nitrate $(Zr(NO_3)_4.5H_2O)$, $PdCl_2$	ODH	20–40	[14]
$Ce_{0.83}Zr_{0.015}Pt_{0.02}O_{2-\delta}$	CAN, $Zr(NO_3)_4.5H_2O$, tetraamine platinum (II) nitrate $((NH_3)_4Pt(NO_3)_2)$	ODH	20–40	[14]
$Ce_{0.65}Fe_{0.33}Pt_{0.02}O_{2-\delta}$	CAN, iron (III) nitrate nonahydrate $(Fe(NO_3)_3 .9H_2O)$, H_2PtCl_6	Diethylenetriamine (DETA) $(C_4H_{13}N_3)$, hydrazine hydrate	3–5	[1]
$Ce_{0.78}Sn_{0.2}Pt_{0.02}O_{2-\delta}$	CAN, tetraamine platinum (II) nitrate, SnC_2O_4.	L-tartaric acid $(C_4H_6O_6)$	4–8	[16]
$Ce_{1-x}Fe_xVO_4$ $(x = 0.01–0.1)$	CAN, ammonium vanadate (NH_4VO_3), Iron (III) nitrate nonahydrate $(Fe(NO_3)_3.9H_2O)$	ODH	10– 0	[27]
$Ce_{0.73}Ti_{0.25}Pd_{0.02}O_{2-\delta}$	CAN, $PdCl_2$, $TiO(NO_3)_2$	Glycine	20	[28]
$Cu_xCe_{1-x}O_{2-\delta}$ $(x = 0.05–0.1)$	CAN, $Cu(NO_3)_2.3H_2O$	ODH	15–25	[29]
$Ce_{1-x}Pt_xO_{2-\delta}$ $(x = 0.01–0.02)$	CAN, H_2PtCl_6	ODH	39–45	[30]
$Ce_{1-x}Rh_xO_{2-\delta}$ $(x\ 0.01–0.02)$	CAN, $RhCl_3._xH_2O$	ODH	45–50	[30]
$Ce_{1-x}Pt_{x/2}Rh_{x/2}O_{2-\delta}$ $(x = 0.01–0.02)$	CAN, H_2PtCl_6, $RhCl_3._xH_2O$	ODH	40–45	[30]

10.3 CHARACTERIZATION OF CERIA-BASED NANOCRYSTALLINE OXIDES

The synthesized nanocrystalline oxide material requires characterization using a number of techniques. X-ray diffraction (XRD), X-ray photoelectron spectroscopy (XPS), temperature-programmed reduction (TPR), transmission electron microscopy (TEM), and Brunauer–Emmett–Teller (BET)

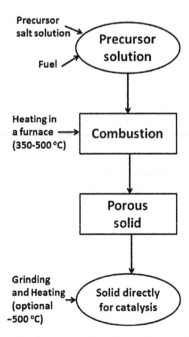

FIGURE 10.1   A schematic for the protocol for the solution combustion technique.

FIGURE 10.2   A representative synthesis procedure for the solution combustion technique.

analysis are typically done for complete structural characterization. Crystal structure and phases present in the catalyst are determined using the XRD analysis. Oxidation state of metal ions in catalysts is determined using XPS. TEM is used for the morphology and size analysis of nanocatalysts. The structural details can further be confirmed by Raman spectroscopy. The specific surface areas of the catalysts are measured using BET technique. We briefly describe the important findings from each of the above techniques below.

XRD can be used as the signature of the ceria-based materials thereby confirming the synthesis of a single-phase compound. The diffraction peaks of CeO_2 and metal ion-substituted CeO_2 were identical corresponding to the cubic fluorite structure. Since, the substituted metal ions occupy the lattice positions of cerium ions, no additional peaks were expected on substitution. Due to the fine dispersion of these metal ions, characteristic ceria peaks were observed in the XRD of metal-ion substituted compounds. However, the lattice parameters changed with the metal ion substitution because of the difference in ionic radii of Ce^{4+} and M^{2+}.

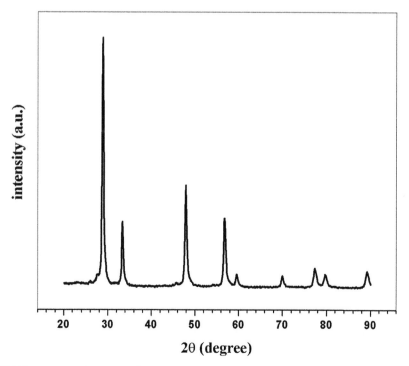

FIGURE 10.3   XRD pattern of $Ce_{0.85}Zr_{0.15}O_{2-\delta}$

The diffraction peaks of Sn-substituted CeO_2 (up to 50%) and CeO_2 were observed to be identical indicating stable cubic fluorite structure of CeO_2 but the lattice parameter was decreased on Sn substitution.[16] This was in agreement with the investigations conducted by Ayastuy et al.[32] for the Sn substituted CeO_2. There were no observable shifts in the XRD patterns for rare earths doped (Er, Nd, Pr, and Y) ceria nanoparticles indicating the uniform cubic fluorite structure of bulk ceria.[10] Typical Rietveld refined XRD patterns for a ceria-zirconia supported nanocatalysts is shown in Figure 10.3. The diffraction peaks in Figure 10.3 for $Ce_{0.85}Zr_{0.15}O_{2-\delta}$ were indexed to the perfect cubic fluorite structure of ceria with a space group of *Fm3m* and no peaks corresponding to zirconia were present. This denotes the complete dispersion of zirconium ion (Zr^{4+}) in place of cerium ion (Ce^{4+}). Table 10.2 shows the Rietveld refined lattice parameters for metal ion substituted ceria for different metal ions. Similarly, no significant diffraction peaks for copper

TABLE 10.2   Rietveld Refined Lattice Parameters for Various Metal Ion Doped Ceria Nanocrystals.

Compound	R_{Bragg}	R_F	χ^2	Ref.
$Ce_{0.85}Zr_{0.15}O_2$	4.08	3.42	5.03	[14]
$Ce_{0.83}Zr_{0.15}Pd_{0.02}O_{2-\delta}$ (before reaction)	6.12	4.11	7.47	[14]
$Ce_{0.83}Zr_{0.15}Pd_{0.02}O_{2-\delta}$ (after reaction)	5.63	3.82	6.29	[14]
$Ce_{0.83}Zr_{0.15}Pt_{0.02}O_{2-\delta}$ (before reaction)	6.56	4.77	7.89	[14]
$Ce_{0.83}Zr_{0.15}Pt_{0.02}O_{2-\delta}$ (after reaction)	6.27	4.54	7.5	[14]
$Ce_{1-x}Sn_xO_2$ ($x = 0.1–0.5$)	1.39–3.46	1.24–2.82	1.06–4.82	[16]
$Ce_{0.78}Pd_{0.02}Sn_{0.2}O_{2-\delta}$	1.06	0.43	1.8	[16]
$Ce_{0.78}Sn_{0.2}Pt_{0.02}O_{2-\delta}$ (before reaction)	0.98	0.72	1.02	[19]
$Ce_{0.78}Sn_{0.2}Pt_{0.02}O_{2-\delta}$ (after WGS reaction)	1.05	1.01	0.95	[19]
$Ce_{0.78}Sn_{0.2}Pt_{0.02}O_{2-\delta}$ (reoxidized at 400°C after WGS reaction)	1.31	0.69	1.14	[19]
$CeAlO_3$			3.72	[17]
$Ce_{0.75}Ti_{0.25}O_2$	2.93			[28]
$Ce_{0.83}Pd_{0.02}Ti0_{.15}O_{2-\delta}$	2.90			[28]
$Ce_{0.67}Fe_{0.33}O_{2-\delta}$ (before reaction)	0.75	0.49	1.41	[1]
$Ce_{0.65}Fe_{0.33}Pt_{0.02}O_{2-\delta}$ (before reaction)	0.71	0.57	1.32	[1]
$Ce_{0.65}Fe_{0.33}Pt_{0.02}O_{2-\delta}$ (after reaction)	0.85	0.58	1.12	[1]
$Ce_{0.9}Fe_{0.1}O_{1.88}$	1.71	1.78	2.34	[31]
$Ce_{0.88}Fe_{0.1}Pd_{0.02}O_{1.86}$	1.89	0.78	1.63	[31]

TABLE 10.2 *(Continued)*

Compound	R_{Bragg}	R_F	χ^2	Ref.
$Ce_{1-x}Pd_xVO_4$ $(x = 0.02–0.1)$	6.97–8.91	3.76–5.60		[34]
$Ce_{0.95}Cu_{0.05}O_{2-\delta}$	1.39	1.12		[29]
$Ce_{0.9}Cu_{0.1}O_{2-\delta}$	1.71	1.41		[29]
$Ce_{0.88}Cu_{0.1}Pd_{0.02}O_{1.88}$	1.93	0.92	1.31	[31]
$Ce_{0.99}Pt_{0.01}O_{2-\delta}$	1.31	0.819		[30]
$Ce_{0.99}Rh_{0.01}O_{2-\delta}$	1.25	0.853		[30]
$Ce_{0.99}Pt_{0.005}Rh_{0.005}O_{2-\delta}$	0.538	0.560		[30]
$Ce_{0.99}Pt_{0.005}Rh_{0.005}O_{2-\delta}$ (heated at 800°C)	0.649	0.570		[30]
$Ce_{0.98}Pt_{0.01}Rh_{0.01}O_{2-\delta}$	0.5	0.630		[30]
$Ce_{0.98}Pt_{0.01}Rh_{0.01}O_{2-\delta}$ (heated at 800°C)	0.568	0.650		[30]
$Ce_{0.9}Mn_{0.1}O_{1.95}$	1.36	1.15	1.56	[31]
$Ce_{0.88}Mn_{0.1}Pd_{0.02}O_{1.93}$	1.65	1.35	1.12	[31]
$Ce_{0.9}Co_{0.1}O_{1.95}$	1.24	1.42	1.80	[31]
$Ce_{0.88}Co_{0.1}Pd_{0.02}O_{1.93}$	1.14	1.18	1.87	[31]
$Ce_{0.9}Ni_{0.1}O_{1.90}$	1.33	0.73	0.96	[31]
$Ce_{0.88}Ni_{0.1}Pd_{0.02}O_{1.88}$	1.02	0.96	0.98	[31]
$Ce_{0.9}La_{0.1}O_{1.95}$	1.28	1.43	0.94	[31]
$Ce_{0.9}Y_{0.1}O_{1.95}$	2.11	1.85	1.77	[31]
$Ce_{0.98}Pd_{0.02}O_{1.98}$	1.42	1.17	0.98	[31]
$Ce_{0.98}Pt_{0.02}O_{1.98}$	1.74	1.45	2.02	[31]

or oxides of copper were observed[29] in case of Cu substituted CeO_2 and sintered $Ce_{1-x}Cu_xO_{2-\delta}$ at 600°C. However, the incorporation of Cu ions into CeO_2 lattice led to the reduction of lattice parameter. Metal ion substitution creates oxygen vacancies in the lattice. The oxide ion vacancies further change the lattice parameter. A reduction of lattice parameter for Mn-substituted ceria was also reported,[31] but the reverse effect was observed in case of Fe and La-substituted ceria. The change in lattice parameter was due to the difference in ionic radii of dopant and the parent ion. Similarity of the diffraction patterns of pure CeO_2 and transition (Fe, Cu, Co, Mn, Ni, Pd, Pt, and Ru) and rare earth metal (La and Y) ion substituted CeO_2 showed a fine ionic dispersion of metal ions in the ceria lattice. The foreign metal ions occupy the same lattice positions as the cerium ions thereby making stable multicomponent nanocrystalline solid solutions.

Transition metals exhibit multiple oxidation states. Further, Ce in ceria also tends to reduce from +4 state to +3 state by reversible oxygen exchange. Therefore, it is important to determine the electronic structure of the synthesized compounds which can be done using XPS. The binding energies of the electrons occupying different orbitals are determined from characteristic XPS peaks and the amount of metal ion in different oxidation states is calculated from the area under the XPS curves corresponding to the different oxidation states. The XPS analysis for different metal ion substituted CeO_2 compounds were given in Table 10.3 along with relative percentage of metal content before and after the reactions in which they were used. The existence of both Ce^{4+} and Ce^{3+} states was observed for the metal-ion substituted ceria due to the presence of redox couples. The reduction of Ce^{4+} to Ce^{3+} was observed with a corresponding change in the oxidation state of the substituted metal ion. Changes in the oxidation states were often observed during reaction owing to these redox couples. Mn and Co had +3 oxidation state in the as prepared $Ce_{1-x}M_xO_{2-\delta}$ while they existed in +2 state after reduction in H_2 at 600°C in $Ce_{0.88}Mn_{0.1}Pd_{0.02}O_{2-\delta}$ and $Ce_{0.88}Co_{0.1}Pd_{0.02}O_{2-\delta}$ where as Pd and Ni appeared in metallic state in the reduced samples from +2 state in as prepared ones.[31] The oxidation states of rare earth metal ions La and Y were same (+3) in the prepared samples and H_2 reduced ones. However, cerium ion underwent a reduction from +4 states to +3 oxidation state after H_2 reduction. These dynamics were accurately captured by XPS analysis.

An important advantage of the solution combustion technique was the crystallization of the products in nanoscopic dimensions. This was confirmed by XRD and TEM analysis. For example, the XRD and TEM analyses revealed that the sizes of the Pd-substituted $Ce_{0.78}Sn_{0.2}Pd_{0.02}O_{2-\delta}$ and unsubstituted $Ce_{0.8}Sn_{0.2}O_2$ were in the range of 5–12 nm.[16] Figures 10.4a and 10.4b show the TEM images of CeO_2 and alumina doped CeO_2 with varying amounts of alumina content. Electron diffraction pattern are also given in Figure 10.4a for pure CeO_2. The presence of small particles less than 10 nm and bigger particles of size 40 nm were observed from Figure 10.4a and this clearly revealed the polydisperse nature of ceria nanoparticles. However, higher crystallinity was achieved with the formation of mixed hybrid oxide of CeO_2, Al_2O_3, and $CeAlO_3$ and the crystallite size was around 3.1 nm. Scherrer analysis and TEM images (Fig. 4a,b) for hybrid metal oxide CeO_2-Al_2O_3-$CeAlO_3$ clearly showed the average size of the crystallites in the range of 2.3 nm. The mixture of fuels in the synthesis may led to smaller particles of sub-10 nm.[15] Similar results have been obtained for other of $Ce_{1-x}M_xO_{2-\delta}$ compounds.

TABLE 10.3 Binding Energies and Relative Intensities of Various Metal Ion Substituted Ceria Compounds in Different Oxidation States in as Prepared Samples and After Reaction at Higher Temperatures.

Compound	Identity	Oxidation State	Relative Intensity (%)	Binding Energy (eV)	Ref.
$Ce_{0.83}Zr_{0.15}Pd_{0.02}O_{2-\delta}$	Before reaction	Pd^0	–	335	[14]
		Pd^{2+}		337.5	
		Pd^{4+}		338.4	
$Ce_{0.83}Zr_{0.15}Pd_{0.02}O_{2-\delta}$	After reaction	Pd^0	–	340.6	[14]
		Pd^{2+}		342.9	
		Pd^{4+}		343.5	
$Ce_{0.83}Zr_{0.15}Pt_{0.02}O_{2-\delta}$	Before reaction	Pt^0	24	71	[14]
		Pt^{2+}	47	72	
		Pt^{4+}	29	74.4	
$Ce_{0.83}Zr_{0.15}Pt_{0.02}O_{2-\delta}$	After reaction	Pt^0	34	74.2	[14]
		Pt^{2+}	78	75.2	
		Pt^{4+}	10	77.6	
$Ce_{0.98}Pt_{0.02}O_{2-\delta}$	Before reaction	Pt^{2+}	9.7	72	[15]
		Pt^{4+}	90.3	74	
$Ce_{0.98}Pt_{0.02}O_{2-\delta}$	After reaction	Pt^{2+}	19.4	72	[15]
		Pt^{4+}	80.6	74	
$Ce_{0.83}Zr_{0.15}Pt_{0.02}O_{2-\delta}$	Before reaction	Pt^{2+}	0	72	[15]
		Pt^{4+}	100	74	
$Ce_{0.83}Zr_{0.15}Pt_{0.02}O_{2-\delta}$	After reaction	Pt^{2+}	0	72	[15]
		Pt^{4+}	100	74	
$Ce_{0.83}Ti_{0.15}Pt_{0.02}O_{2-\delta}$	Before reaction	Pt^{2+}	33.8	72	[15]
		Pt^{4+}	66.2	74	
$Ce_{0.83}Ti_{0.15}Pt_{0.02}O_{2-\delta}$	After reaction	Pt^{2+}	57.8	72	[15]
		Pt^{4+}	42.2	74	
$Ce_{0.83}Zr_{0.15}Pd_{0.02}O_{2-\delta}$	Before reaction	Pd^{2+}	–	237.5	[15]
$Ce_{0.83}Zr_{0.15}Pd_{0.02}O_{2-\delta}$	After reaction	Pd^{2+}	–	238	[15]
$Ce_{0.6}Sn_{0.4}O_2$	Before reaction	Sn^{4+}	–	487	[16]
$Ce_{0.6}Sn_{0.4}O_2$	After reaction	Sn^{4+}	–	487	[16]
		Sn^{2+}		486.6	
$Ce_{0.78}Sn_{0.2}Pd_{0.02}O_{2-\delta}$	Before reaction	Sn^{4+}	–	487	[16]
$Ce_{0.78}Sn_{0.2}Pd_{0.02}O_{2-\delta}$	After reaction	Sn^{2+}	–	486.6	[16]
$Ce_{0.9}Sn_{0.1}O_2$	Before reaction	Ce^{4+}	–	882.7	[16]
$Ce_{0.8}Sn_{0.2}O_2$	Before reaction	Ce^{4+}	–	901	[16]
CeO_2 by solution combustion synthesis (SCS)	Soot conversion reaction	Ce^{4+}	4.4–19.8	882.2–916.9	[24]
		Ce^{3+}	8.6–27.5	882.9–902.1	

TABLE 10.3   *(Continued)*

Compound	Identity	Oxidation State	Relative Intensity (%)	Binding Energy (eV)	Ref.
CeO_2 powder by SCS	Soot conversion reaction	Ce^{4+}	8.64–28.22	881. 67–888. 27	[23]
		Ce^{3+}	5.43–11.63	900.17–916.06	
$Ce_{0.9}Mn_{0.1}O_{2-\delta}$	As prepared	Mn^{3+}	–	653.6	[31]
$Ce_{0.9}Mn_{0.1}O_{2-\delta}$	Reduced by H_2 at 600°C	Mn^{2+}	–	642	[31]
MnO_x–CeO_2		Mn^{2+}	6.9–34.1	640.57–640.84	[20]
		Mn^{3+}	34.5–66.1	641.61–642.16	
		Mn^{4+}	26.4–31.4	643.04–643.30	
$Ce_{0.88}Mn_{0.1}Pd_{0.02}O_{2-\delta}$	As prepared	Pd^{2+}	–	338.2	[31]
$Ce_{0.88}Mn_{0.1}Pd_{0.02}O_{2-\delta}$	Reduced by H_2 at 600°C	Pd^0	–	335.1	[31]
$Ce_{0.9}Co_{0.1}O_{2-\delta}$	As prepared	Co^{3+}	–	795.4 and 780.3	[31]
$Ce_{0.9}Co_{0.1}O_{2-\delta}$	Reduced by H_2 at 600°C	Co^{2+}	–	796.9 and 781.2	[31]
$Ce_{0.9}Ni_{0.1}O_{2-\delta}$	As prepared	Ni^{2+}	–	854.5	[31]
$Ce_{0.88}Ni_{0.1}Pd_{0.02}O_{2-\delta}$	As prepared	Ni^{2+}	–	872.5	[31]
$Ce_{0.9}Ni_{0.1}O_{2-\delta}$	Reduced by H_2 at 600°C	Ni^{2+}	–	852.2	[31]
$Ce_{0.88}Ni_{0.1}Pd_{0.02}O_{2-\delta}$	Reduced by H_2 at 600°C	Ni^{2+}	–	852 and 857.8	[31]
$Ce_{0.9}Y_{0.1}O_{2-\delta}$	As prepared and reduced sample	Y^+	–	158	[31]
Ir–CeO2		Ir	–	61.2	[22]
Au10PrCe by micro emulsion synthesis	Fresh catalyst	Pr^{4+}	–	930.39–968.13	[33]
		Pr^{3+}		928.63–953.66	
$Ce_{0.67}Fe_{0.33}O_{2-\delta}$	As prepared and reduced	Fe^{3+}	–	710.4	[1]
$Ce_{0.65}Fe_{0.33}Pt_{0.02}O_{2-\delta}$	As prepared and reduced	Fe^{3+}	–	710.4	[1]
$Ce_{0.65}Fe_{0.33}Pt_{0.02}O_{2-\delta}$	As prepared	Ce^{4+}	100	882.7	[1]
		Ce^{3+}	0		
$Ce_{0.65}Fe_{0.33}Pt_{0.02}O_{2-\delta}$	Reduced	Ce^{4+}	90	882.7	[1]
		Ce^{3+}	10	889.1	
$Ce_{0.65}Fe_{0.33}Pt_{0.02}O_{2-\delta}$	As prepared and reduced	Pt^{4+}	–	74.3	[1]
$Ce_{0.98}Pd_{0.02}VO_4$		Pd^{2+}	–	337.7	[34]
$Ce_{0.98}Pd_{0.02}VO_4$		V^{5+}	–	517.2	[34]

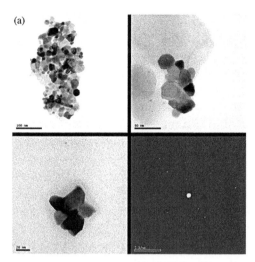

FIGURE 10.4a   A TEM micrograph of pure ceria nanoparticles[17] (Reproduced from Deshpande, P. A.; Aruna, S. T.; Madras, G. CO oxidation by CeO2–Al2O3–CeAlO3 hybrid oxides Catal. Sci. Technol. 2011, 1, 1683–691 with permission from the Royal Society of Chemistry.)

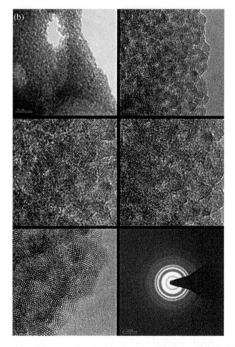

FIGURE 10.4b   A TEM micrograph of aluminum substituted hybrid CeO_2.[17] Reproduced from Deshpande, P. A.; Aruna, S. T.; Madras, G. CO oxidation by CeO_2–Al_2O_3–$CeAlO_3$ hybrid oxides Catal. Sci. Technol. 2011, 1, 1683–691 with permission from the Royal Society of Chemistry.

10.4 CATALYTIC APPLICATIONS OF CERIA-BASED NANOCRYSTALLINE OXIDES

10.4.1 OXIDATION OF CARBON MONOXIDE

CO is considered to be toxic to humans and animals because of its high affinity for hemoglobin. An increase in concentration of CO in the environment is primarily because of transportation, power plants, and industrial activities. High concentration of CO can trigger serious respiratory problems. CO is sparingly soluble in water which limits its elimination by aqueous treatment. Combustion of CO is also not feasible as its concentration in atmosphere is low (50–5000 ppm) which results in incomplete combustion. Ceria-based catalysts are used to concentrate the pollutants at its surface which allows the oxidation reaction to proceed at a sufficient rate. Different noble metal (Pt, Pd, Rh, etc.) catalysts and oxides have been reported for CO oxidation.[35–38] Because of good adsorption characteristics and high OSC, ceria was found to be a good CO oxidation catalysts. When the oxygen pressure in gas phase decreases (below the stoichiometry), ceria allows the metal to work in transient condition.[39–41] An increase in CO conversion with increasing temperature and complete CO conversion was achieved at 450°C temperature with CeO_2 and SnO_2 solid solution as reported by Ayastuy et al.[32] Yao and Yu Yao [40] in their study investigated the impact of ceria and alumina supported catalysts on the kinetics of the reaction. It was observed that the transition metal (metal ion) supported ceria catalysts exhibited less activation barrier and more stable catalytic activity due to dual site mechanism in which CO and O_2 did not compete for the same sites. Kang et al.[42] showed that ceria had an exceptional activity for CO oxidation than either alumina or silica. Effect of ceria on CO oxidation was confirmed in several other studies.[43–46] Manuel et al.[47] and Shekhtman et al.[48] observed two types of active sites on ceria supports having differences in activation barriers. Sites located at the metal-support interface were highly reactive providing 23 kJmol^{-1} activation barrier whereas sites located on the support were comparatively less reactive with the activation barrier of 44 kJmol^{-1}. Sensitivity of metal state to the carbon monoxide oxidation reaction was also observed.

Along with different metal atoms, phase, size, microstructure, and morphology of material also plays an important role. Recently, Deshpande et al.[17] investigated the effect of all these material properties on the CO oxidation reaction. Hybrid oxides like CeO_2-Al_2O_3-$CeAlO_3$ have been synthesized using modified solution combustion technique. These hybrid oxide catalysts were observed to be catalytically active for CO oxidation reaction showing 90% of conversion in most of their synthesized compounds. Out of all hybrid oxides, that with crystallites of size 3 nm showed the highest catalytic activity and lowest activation energy. CO oxidation reaction was also carried out by Baidya et al.[16–28] using Sn, Zr, and Pd doped ceria based catalyst. In their studies, rates of CO oxidation were observed higher over Pd^{2+} ion substituted $Ce_{1-x}Pd_xO_{2-\delta}$ compared to Pd^{2+} ion in CeO_2. Rate of CO oxidation over Pd doped $Ce_{0.78}Sn_{0.2}Pd_{0.02}O_{2-\delta}$ was observed 10 times higher compared to $Ce_{0.73}Ti_{0.25}Pd_{0.02}O_{2-\delta}$ because of higher OSC.

The importance of noble metal ionic catalysts for CO oxidation was propounded by Hegdeet al.[3] Amongst various metal ion-substituted compounds, Pd ion substituted $Ce_{1-x}Ti_xO_2$ showed the highest rate of CO conversion and lowest activation energy. Electronic interaction between the noble metal ions and the lattice and creation of redox sites led to the specific adsorption which enhanced the activity. Gupta et al.[19] carried out the CO oxidation reaction over Pt doped $Ce_{0.78}Sn_{0.2}Pt_{0.02}O_{2-\delta}$ in presence and absence of oxygen. In the presence of stoichiometric oxygen, complete conversion was observed at room temperature whereas 80% conversion was observed at 180°C in absence of feed oxygen utilizing the activated lattice oxygen. Mahadevaiah et al.[1] also carried out the reaction in presence and absence of feed oxygen over $Ce_{0.67}Fe_{0.33}O_{2-\delta}$ and $Ce_{0.65}Fe_{0.33}Pt_{0.02}O_{2-\delta}$. Higher activity of CO oxidation was observed by substitution of Pt ion in $Ce_{0.67}Fe_{0.33}O_{1.835}$. Complete conversion was observed in presence of oxygen at 125°C for $Ce_{0.65}Fe_{0.33}Pt_{0.02}O_{1.785}$ with activation energy of 56.3 kJmol^{-1} and at 450°C for $Ce_{0.67}Fe_{0.33}O_{1.835}$ with activation energy of 15.2 kJmol^{-1}. Roy et al.[18–49] also observed complete conversion over Ti and Pd doped CeO_2 in his study. Pd doped on $CeVO_4$ showed higher catalytic activity over pure $CeVO_4$ for CO oxidation in presence of oxygen.[50] $Ce_{0.98}Pd_{0.02}VO_4$ showed considerably lower activation energy than reported for impregnated Pd/$CeVO_4$, Pd substituted catalyst like Pd/CeO_2-ZrO_2, Pd/SiO_2, and Pd/CeO_2/Al_2O_3.[34,51,52] Figure 10.5 shows a typical CO oxidation profile over metal ion substituted ceria catalyst.

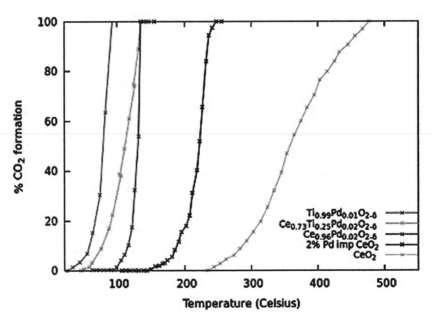

FIGURE 10.5 CO oxidation activity different ceria-based catalysts.[3] (Reprinted with permission from Hegde, M. S.; Madras, G.; Patil, K. C. Noble Metallic Catalysts. *Acc. Chem. Res.* **2009,** *42,* 704–712.© 2009 American Chemical Society.)

10.4.2 REDUCTION OF NITROGEN OXIDES

Like CO oxidation, efforts are being made to develop catalysts for reducing NO_x emissions for a wide range of air to fuel ratios. The reactions can be represented as:

$$2NO\ (g) + CO\ (g) \rightarrow N_2O\ (g) + CO_2\ (g)$$
$$2NO\ (g) + CO\ (g) \rightarrow N_2O\ (g) + CO_2\ (g)$$
$$N_2O\ (g) + CO\ (g) \rightarrow N_2\ (g) + CO_2\ (g)$$

Although desired, it is difficult to get 100% selectivity for N_2. Various studies have shown the presence of N_2O in the reaction system. NO reduction by CO has been reported using different oxide catalysts. It has been observed that NO reduction by CO is unaffected by the presence of oxygen. The incorporation of titanium into ceria ($Ti_{0.75}Ce_{0.25}O_2$) improves the activity of selective catalytic reduction (SCR) performance with more than 90% NO conversion in the temperature range of 240–440°C.[13] The availability of oxygen vacancies on the surface of ceria-titania led to the formation of NO_2

and nitrates thereby strengthening the SCR performance. Higher catalytic activity of MnO_x–CeO_2 catalysts than that of clean MnO_2 using glycine as a fuel has been reported by Andreoli et al.[20] for the NH_3-selective catalytic conversion of NO_x. Complete conversion of NO_x over MnO_x–CeO_2 catalysts at a temperature of 150°C was achieved with N_2 yield of more than 60%. The catalyst $Ce_{0.98}Pd_{0.02}O_{2-\delta}$ developed by Baidya et al.[28] showed high activity as shown in Figure 10.6 They also observed 100% N_2 selectivity on $Ce_{0.73}Ti_{0.25}Pd_{0.02}O_{2-\delta}$. Their study showed that the dissociation of NO can be easily achieved by increasing oxide ion vacancy sites in ceria nanocatalyst. These vacancies are created due to CO oxidation by lattice oxygen and an increase in the NO adsorption took place on oxide ion vacancy. Roy and Hegde[50] have analyzed the effect of Pd, Pt, and Rh ion doped CeO_2 catalyst for reduction of NO by CO. Pd doped CeO_2 was found to be a good catalyst in their studies. Hecker and Bell[55] had previously studied and reported Rh to be a good ionic catalyst when supported on Al_2O_3 for NO reduction. NO reduction reaction by CO on $CeVO_4$ showed 50% of NO conversion at 650°C whereas it shows 50% conversion at 280°C and 100% conversion at 350°C in case of $Ce_{0.98}Pd_{0.02}VO_4$. All these results therefore suggest that Pd substitution in ionic form enhances the catalytic activity of the catalysts.

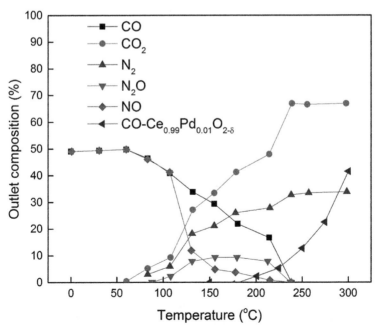

FIGURE 10.6   TPR profile of NO + CO + O_2 reaction over the catalyst $Ce_{0.98}Pd_{0.02}O_{2-\delta}^2$.

10.4.3   WATER–GAS SHIFT (WGS) REACTION

Water–gas shift (WGS) reaction is used for hydrogen production which can be used for running fuel cells.

$$CO\ (g) + H_2O \rightarrow CO_2\ (g) + H_2\ (g)$$

During hydrogen production, formation of methane and coke are also observed over the catalyst at high temperatures and low steam content. This process decreases the efficiency of hydrogen production in WGS. Jain and Maric[56] have shown an increase in catalytic activity of Pt(1 wt%)/ceria nano-catalyst synthesized by reactive spray deposition technology (RSDT) method for WGS reaction. Complete conversion of CO, that is, 100% was achieved at a temperature of 250°C without any methane formation. The increase in hydrogen production by steam reforming and autothermal reforming processes using Ce–ZrO$_2$ catalysts were reported by Nahar and Dupont.[57] They reviewed the catalytic activity of ceria-zirconia supported catalysts for the hydrogen production by steam reforming, autothermal reforming, catalytic partial oxidation and dry reforming of alkanes and hydrocarbons. Gold deposited on praseodymium (5 and 10%) doped ceria via impregnation synthesis was demonstrated to have a high catalytic activity for hydrogen production and CO conversion in WGS.[33] This was due to the reversible exchange of lattice oxygen during reduction and oxidation on the gold depos-ited nanocatalyst. The catalytic activity of Ir–CeO$_2$ catalyst synthesized by single step solution combustion technique for the production of hydrogen from the steam reforming of methane was demonstrated by Postole et al.[22] The iridium-doped CeO$_2$ showed a high catalytic activity for the steam reforming of methane and it served as a suitable catalyst for the effective production of H$_2$. Hilaire et al.[58] developed a ceria based catalysts alterna-tive to conventional Cu–ZnO WGS catalyst. Studies have been carried out to enhance the activity and efficiency of WGS conversion process by adding Pt, Au, or Cu in small amounts in ceria based catalysts.[59–65] Bunlueistein et al.[59] have shown that the bi-functional redox mechanism between Pt and ceria was responsible for high WGS activity. However, many studies[66–69] have also reported deactivation of ceria based catalysts during the reaction because of loss of oxygen storage ability, sintering of the noble metal parti-cles at high temperatures, and formation of surface carbonates. Gupta et al.[19] synthesized Ce$_{0.78}$Sn$_{0.2}$Pt$_{0.02}$O$_{2-\delta}$ catalyst in which the OSC of the catalyst was reversible even after subjecting the catalyst to high temperature. Deactiva-tion of the catalyst was not observed even after long exposure to the excess of CO_2 in the feed gas. Overall, 5000 ppm methane formation was reported with $Ce_{0.98}Pt_{0.02}O_{2-\delta}$ at 350°C.[70] But methane formation was not observed in their studies using $Ce_{0.78}Sn_{0.2}Pt_{0.02}O_{2-\delta}$. Sn substitution suppressed the formation of methane as catalyst's oxygen storing capacity (OSC) was enhanced because of amphoteric nature of Sn. Similarly, Pt substitution in $Ce_{0.67}Fe_{0.33}O_{2-\delta}$ enhanced the catalytic activity for WGS reaction over Pt impregnated on $Ce_{0.67}Fe_{0.33}O_{2-\delta}$ due to synergistic interaction of the Pt ion with Ce and Fe ions as can be seen from Figure 10.7. Importance of Pt doping on CeO_2 was also investigated by Deshpande et al.[15] It was inferred that the activity of Pt-substituted compounds is higher than that of Pd substituted ceria nanocatalysts.

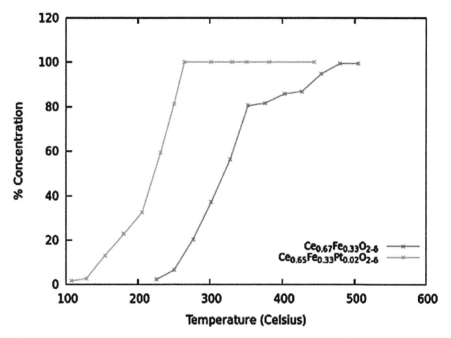

FIGURE 10.7   WGS activity of $Ce_{0.67}Fe_{0.33}O_{2-\delta}$ and $Ce_{0.67}Fe_{0.33}Pt_{0.02}O_{2-\delta}$.[1]

10.4.4   HYDROGEN COMBUSTION

Hydrogen used in fuel cell has many advantages like high energy density, low operating temperatures and fast response to load changes. However, due to incomplete consumption of H_2 in the fuel cell, it becomes harmful to environment. Excess supply of H_2 further results in higher amount of unreacted

H_2 in the system. Recycling of the exhaust stream from fuel cell process increases the complexity of the system[71] and after recycling, some residual H_2 is present in the outlet stream. Therefore, it is necessary to remove residual H_2 from the fuel cell process. Many studies have reported the synthesis of catalysts which give 100% H_2 combustion.[3,71] H_2 is also a good reducing agent for SCR[71-74] of NO and hydrogenation reactions. Catalytic hydrogen combustion (CHC) is exothermic and the rates of catalytic hydrocarbon combustion were observed to be higher over Pd substituted CeO_2 than over Pt impregnated or substituted in CeO_2 and Al_2O_3.[15] CHC reaction has also been studied over Pd and Pt noble metal substituted on ZrO_2, TiO_2, Fe_2O_3, $CeVO_4$, $FeVO_4$, and CeO_2.[71,75,76] Changes in electronic structure of the compounds before and after the reactions were studied using XPS. The activity of the same metal ion was found to be different over different supports because of different mechanism observed over different supports. $Ce_{0.98}Pd_{0.02}O_{2-\delta}$, $Ce_{0.98}Pt_{0.02}O_{2-\delta}$, $Ce_{0.83}Zr_{0.15}Pd_{0.02}O_{2-\delta}$, and $Ce_{0.83}Zr_{0.15}Pt_{0.02}O_{2-\delta}$ showed high activity for H_2 combustion and complete conversion was observed below 200°C for all the catalysts when O_2 was used in stoichiometric amount. With an increase in O_2 concentration, an increase in reaction rates was observed and complete conversion was observed below 100°C.

10.4.5   HYDROCARBON COMBUSTION

Energy produced by hydrocarbon combustion is utilized by human beings in their day-to-day lives. Majority of the world's energy needs are met by the combustion of fossil fuels, including methane, coal, and petroleum. Energy is produced by combustion of hydrocarbons. The combustion reaction proceeds as follows:

$$C_xH_y + ZO_2 \rightarrow_x CO_2 + \frac{y}{2}H_2O$$

where $z = x + 1/4\ y$. The fuel processing systems due to their inefficiency reject unburnt hydrocarbons which are harmful and must be eliminated from the atmosphere. This can be done by catalytic combustion of hydrocarbons. Ceria-based catalyst has been tested for this activity. Complete hydrocarbon (C_2H_2 and C_2H_4) oxidation takes place at lower temperatures on $Ce_{0.98}Pd_{0.02}O_{2-\delta}$[18] than on $Ti_{0.99}Pd_{0.01}O_{1.99}$.[49] Similar to CO oxidation, chemisorption of oxygen in the oxide ion vacancy of the catalyst takes place in hydrocarbon oxidation. Activation energy obtained from Arrhenius plot for C_2H_2 and C_2H_4 are 71 kJmol^{-1} and 90.3 kJmol^{-1} over $Ce_{0.98}Pd_{0.02}O_{2-\delta}$ and

70 kJmol^{-1} and 53.4 kJmol^{-1} over $Ti_{0.98}Pd_{0.01}O_{1.99}$ which was much smaller than the activation energies of 92–109 kJmol^{-1},[77] over other reported conventional catalysts.

10.4.6 CARBON–CARBON COUPLING REACTIONS (C–C COUPLING)

C—C bond formation is observed in many compounds important in biological, pharmaceutical, and material processing. Due to this, there is a need to develop mild and general methods for C–C coupling. The synthesis of these bonds involve nucleophillic aromatic substitution reactions which necessitate the use of electron deficient aryl halides. Several coupling reactions have been developed with different substrates. Most prevalent among these methods are the palladium (0) catalyzed cross coupling reactions namely Heck reaction,[78] Sonogashira–Miyaura reaction,[79] Suzuki–Miyaura reaction,[80] and Hartwig–Buchwald coupling.[81] These C–C coupling reactions allow a one-step synthesis of aromatic olefins which are used extensively as biologically active compounds, natural products, pharmaceuticals, and precursors of conjugated polymers. Products formed from C—C coupling reactions are the intermediates for various pharmaceutical products showing antimicrobial activity and antitumor effects. This makes the above reactions important. C—C coupling reactions proceed in the presence of homogeneous as well as heterogeneous catalysts. These reactions are catalyzed by palladium (Pd) supported on supports such as charcoal, mesoporous carbon, magnesium oxide, silica, alumina or titania, silica, and basic supports such as basic zeolites, mixed oxide, and flourapatite. Apart from palladium, Suzuki–Miyaura reaction has been catalyzed by several other transition metal complexes. Iron–MCPA complex has been recently reported to catalyze the reaction.[82] Fe halides have been reported as catalysts for the reaction.[83] Cu has also been used for catalyzing the reaction.[83] Further bimetallic catalysts consisting of Cu along with Pd, Pt, and Ru have also been tested and found active for the reaction. Combined Cu/Pd catalyst was found to show the highest activity among all the combinations. Ni complex has also been used for the synthesis of biaryls using Suzuki–Miyaura reaction.[84] But no study has been conducted on these C—C coupling reactions before investigation by Hegdeet al.[3,85] where they investigated the substitution of noble metal in reducible oxides such as CeO_2 which showed higher catalytic activity than the homogeneous catalytic reactions. Overall, 100% conversions of the reactants was observed as shown in Figure 10.8.

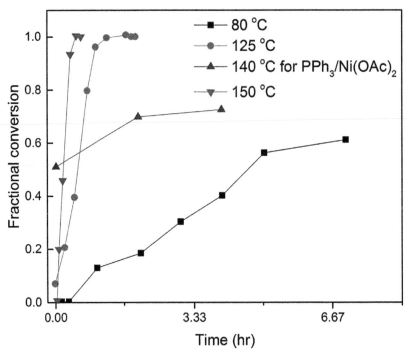

FIGURE 10.8 Activity of Pd-substituted ceria for reaction of iodobenzene and methyl acrylate.

10.4.7 PHOTOCATALYSIS

Photocatalysis creates strong oxidation agents to breakdown organic matter to carbon dioxide and water in the presence of a photocatalyst, light, and water. Photocatalytic activity of the catalyst has been used for various applications.[86–91] TiO_2,[92] rare earth metal based pervoskites[93] and $LaNiO_3$[94] have been reported as photocatalytically active material. The reason for the photocatalytic activity was the generation of hydroxyl radicals which were highly reactive. Recently, Fe substituted $CeAlO_3$, synthesized by Deshpande et al.[95] showed a band gap of 3.3 eV which was close to that of commercial Degussa P-25 TiO_2 catalyst with a value of 3.2 eV. Therefore, they investigated the photocatalytic activity of $CeAlO_3$ for the degradation of dyes present in water. They considered four anionic (orange G (OG), remazol brilliant blue R (RBBR), ACG, indigo carmine (IC)), and four cationic (acriflavin (AF), methylene blue (MB), methyl violet (MV), rhodamine 6G (R6G)) dyes and studied the kinetics of the reaction. In their study, a reduced band gap of

$CeAlO_3$ was observed due to formation of nanocrystallites which generated an electron–hole pair through excitation of electron from valence band to the conduction band. Because of this electron hole pair, degradation of dyes occurred through subsequent formation of hydroxyl and superoxide radicals. For low concentrations (25 and 50 ppm) of orange G, almost complete conversion was observed whereas for higher concentrations, 90% of conversion was observed. For RBBR and ACG at lower concentration (up to 50 ppm), complete degradation of dye was observed and the following trend of cationic dyes for rate of degradation was observed in their study: acriflavin > rhodamine 6G > methylene blue > methyl violet.

10.4.8 HUMIDITY SENSOR

Humidity sensors are used in industrial processes and human comfort. Organic polymers or metal oxides are used as humidity sensors.[96] Humidity sensors based on organic polymers have numerous challenges due to their weak mechanical strength and poor physical and chemical stabilities. Recently, Parvatikar et al.[97] investigated the humidity sensor based on pure CeO_2 which showed poor sensitivity as compared to CeO_2 doped with Ba studied by Zhang et al.[98] Hu et al.[99] studied the response of humidity sensor by synthesizing Mn doped CeO_2. The samples were first dispersed in ethanol and then dropped on top of pre-deposited Au electrodes on the surface of Si wafer. The dimension of the film was 5 mm in length, 3 mm in width, and ~20 μ in thickness. Ohmic contact for the planar film with copper wires was made by silver paste. The measurements were carried out by putting the sensors in an airproof glass vessel with a volume of 2 L. A hygroscope was placed into the vessel to monitor the relative humidity during the experiments. The sensors resistance was measured by an external testing circuit. The resistance was observed to decrease more rapidly with increasing humidity. The studies conclusively demonstrated the potential of metal ion doped ceria for humidity sensor applications.

10.5 DFT STUDIES ON CERIA-BASED SOLID SOLUTIONS

The enhancement of properties of CeO_2 like OSC and the dielectric properties upon the doping of Ceria with other metals have been extensively studied using first principles.[100–107] Density functional theory with local density approximation (LDA) to the exchange correlation energy has been widely employed in these studies. Ultrasoft pseudo potentials were used for representing the interactions between the valence electrons and the nucleus and core electrons of the atoms. Periodic calculations using plane wave basis sets were done to represent the actual compound. DFT+U calculations which involve the addition of Hubbard-like, localized term (U) to the LDA density function were deemed more suitable for calculations of ceria in reduced form.[100–104] Hu and Metiu reported the use of U value of 5.5 eV in their study of effect of dopant on the energy of oxygen-vacancy formation.[105] The outcome of the calculations primarily was focused on the electronic structure of doped CeO_2, analysis of bond lengths and its distributions and energies and charges of atoms. These outcomes have provided insights into changes in the properties upon the doping of Pt, Ru, Zr, Nb, Ta, Mo, or W into ceria.

Vacancy studies on surface of doped ceria by creating oxygen vacancies have been performed by Hu and Metiu.[105] The change in energy of vacancy formation with dopant, though not significant shows an approximate linear relationship between the energy of oxygen-vacancy formation and the Bader valence of the dopant.[105] The Bader valence is the number of electrons on the atom of the dopant in gas phase, minus the number of electrons on the dopant, calculated by the Bader method. The DFT calculations showed that dopants Mo, Nb, Ta, Ru, Zr, and W cause the oxygen atoms to come closer to them away from the cerium atoms.[105]

DFT studies on Pt doped Ceria explained its high hydrogen storage capacity, the Pt ion resulting in the splitting of H_2 to protons.[106] Spillover of hydrogen on to oxygen in the lattice has been reported in Pt/CeO_2 system because of the presence of Pt, where hydrogen being closer than cutoff distance of oxygen is considered spillover.[106] In this study of hydrogen spillover on CeO_2/Pt, Waghmare et al. have considered the (100) surface of CeO_2, substituting the cerium in the second layer with platinum ($Ce_{15}PtO_{31}$) and creating an oxygen vacancy to maintain Pt in its + 2 oxidation state.[106] The substitution of Pt for Ce is proven to be energetically favorable in the presence of oxygen vacancy by about 1.22 eV, with a loss of 11.83 eV by creating an oxygen vacancy in bulk ceria, and a gain of 10.61 eV by substituting Pt in the presence of an oxygen vacancy in ceria.[106] The presence of oxygen vacancy has been proved experimentally in Pt–ion-doped ceria from EXAFS analysis.[108] The above two arguments make the model used by Dutta et al., more realistic.[106] A vacuum of 15–17 Å was taken above the top layer to neglect the slab–slab interactions in this model. Different initial configurations from one hydrogen atom to seven hydrogen atoms on platinum ion have been considered. Hydrogen atoms were placed in a layer at a distance of 1.8 Å from the Pt ion and with a H—H distance of 0.74 Å.[106]

The configurations having odd number of hydrogen (5 H and 7 H) have been found to be the most stable in the arrangement with hydrogen atoms arranged in chain with platinum as the center, whereas the configurations having even number of hydrogen (4 H and 6 H) have been found to be most stable in the arrangement where hydrogen atoms arranged in two mutually perpendicular lines with Pt as the center. The energy release per hydrogen calculated as [E $(Ce_{15}PtO_{31} + nH)$—E $(Ce_{15}PtO_{31})$—nE(H)]/n was the highest for 2 H case with an energy release of 3.465 eV per hydrogen and lowest of 2.501 eV for 3 H case making 2 H case the most stable and 3 H case the least stable.[106] The net positive charge for surface Ce atoms was correlated with spillover of hydrogen atoms on to Ce ions. As one goes from 1 H case to the 4 H case, the dissociated hydrogen atoms increase and in the 4 H case, there was a spillover of hydrogen onto Ce thus resulting in negative charge transfer into the Ce. This has been explained with a Hirshfeld atomic charge analysis showing a decrease of charge of surface Ce from 1 H case to the 4 H case.[106] The net positive charge for surface Ce atoms again increased for 5 and 6 H's, and was the least for the 7 H case, because of maximum spillover to Ce in the 7 H case. In the 6 H case, spillover of hydrogen atoms was only on to oxygen atoms resulting in a maximum charge flow out of oxygen because of lesser electro-negativity of hydrogen with respect to oxygen. This has been reflected in the charge analysis of oxygen atoms showing a net minimum charge for 6 H case and a net maximum charge for the 4 H case.[106]

Waghmare et al. have also studied the effect of doping metals in ceria on OSC, doping of metals like Fe, Co, Pd, and Pt led to local distortion in structure resulting in strain in the lattice of ceria. In this study, they used the Plane-wave self-consistent field (PWscf) method of density functional theory to find the internal structure of doped compounds. Plane wave basis with energy cutoff of 30 Ry and ultrasoft pseudo potentials for representing the ionic cores were considered in the study. A model of $2 \times 2 \times 2$ super cell (96 sites), which contained 32 formula units of CeO_2 was used. When doped with other metals, 4 out of 32 Ce atoms were replaced by metal dopant (M) atoms which correspond to 12.5 % substitution, and the oxygen vacancies were created near the dopant atom to maintain charge neutrality. From the density functional theory calculations on palladium and platinum doped ceria, a bond valence sum (BVS) of 1.83 for Pt and 3.99 for Ce in $Ce_{32}Pt_1O_{63}$, and 1.92 for Pd and 3.99 for Ce in $Ce_{32}Pd_1O_{63}$ was observed indicating that Pt and Pd ions were under bonded, which was responsible for the easier reducibility of Pt and Pd ions compared to Ce ion in the doped ceria. This reducibility for Pd and Pt ions has been observed in the WGS reaction mechanism where the hydrogen initially adsorbs on Pt and Pd ions and then spills over to the oxygen atoms adjacent to the Pd and Pt ions forming hydroxyl groups on the surface doped ceria.[107] The Mn—O bond lengths observed were 2.003 Å for Mn in six coordination number and 1.897 Å and 1.987 Å for a co-ordination number of four. In the bulk calculations of doped Ceria, Mn was reported to be present in the 4+2 co-ordination state when doped in ceria with short bond distances between 1.8 and 2.1 Å and the longer ones higher than 2.8 Å. These bond lengths show a large deviation from the 2.34 Å bond length of Ce—O in pure ceria.[31] The BVS of oxide ions in $Ce_{28}Mn_4O_{62}$ is reported to be in the range of 1.55–2.25, the oxide ions with lower BVS were more susceptible toward reduction. Mn doping in ceria leading to oxide ions of BVS less than two was the reason for the enhanced OSC upon doping. Similar effects of increase in the oxide ions with low BVS and increase in the range of Ce—O bond length distribution have been observed upon the doping of ceria with Fe, Co, Pd, and Pt. Dopant metal ions in the simulated structure exhibit both short and long M—O bonds. The longer M—O bonds indicate higher reducibility of the dopant ion compared to the Ce ion. The presence of dopant ion also results in longer Ce—O bonds leading to a synergistic reduction of both longer Ce—O and M—O bonds leading to higher OSC compared to CeO_2 without any dopant ions. The presence of longer oxygen bonds decreases the oxygen bond valence well below two, leading to the activation of lattice oxygen, which in turn was responsible for the high OSC property observed in the doped ceria. With Pd substitution, the optimized structure results in still longer M—O and Ce—O bonds, which explains the higher OSC for Pd substituted solid solution. These effects were further magnified when the Fe, Mn, and Co were doped in ceria which was already substituted with palladium ($Ce_{27}Fe_4Pd_1O_{61}$, $Ce_{27}Mn_4Pd_1O_{61}$, and $Ce_{27}Co_4Pd_1O_{61}$), which was the reason for tremendous increment in OSC at even low temperatures of these compounds when compared to undoped CeO_2.

10.6 MECHANISTIC INSIGHTS INTO THE WORKING OF CERIA-BASED CATALYSIS

10.6.1 NO + CO REACTION

For the CO+NO reaction carried out over the catalysts $Ce_{1-x}M_xO_{2-\delta}$ where M represents Pt^{+2}, Pd^{+2}, and Rh^{+3}, a dual site adsorption was suggested because noble metals are potential adsorbents of NO and CO, which would get adsorbed via molecular adsorption and ceria has oxide ion vacancies, "v" where NO can dissociatively get adsorbed. Hence, the following bi-functional mechanism was found to fit the experimental data well. The mechanism proposed was as follows:[50]

$$CO + M^{n+} \Leftrightarrow CO_M^{n+}$$

$$NO + M^{n+} \Leftrightarrow NO_M^{n+}$$

$$NO + \text{``V''} \Leftrightarrow N \text{``O''}$$

$$CO_M^{n+} + \text{``O''} \rightarrow CO_2 + M^{n+} + \text{``V''}$$

$$NO_M^{n+} + N \text{``O''} \rightarrow N_2O + \text{``O''} + M^{n+}$$

$$N_2O + \text{``V''} \rightarrow N_2 + \text{``O''}$$

Steps 1 and 2 represent the molecular adsorption of NO and CO on the metal ion followed by dissociative chemisorption of NO in step 3 on the oxide ion vacancy. Further, in step 6, N_2O was reduced to N_2 in the oxide ion vacancy.

10.6.2 THE WATER–GAS SHIFT REACTION

WGS has been an extensively studied reaction as it was one of the reactions involved in Fischer–Tropsch's process[109] and steam reforming of hydrocarbons.[110] Three mechanisms were proposed, two of them being based on the famous Langmuir–Hinshelwood mechanism and the Eley–Rideal mechanism, respectively. The following elemental steps describe the mechanisms.

1) Eley–Rideal mechanism:
$$CO + S_M \Leftrightarrow CO_M$$
$$CO_M + H_2O \rightarrow CO_2 + H_2 + S_M$$

2) Langmuir–Hinshelwood
$$CO + S_M \Leftrightarrow CO_M$$
$$S_M + H_2O \Leftrightarrow H_2O_M$$
$$CO_M + H_2O \rightarrow CO_2 + 2S_M + H_2$$

In the Eley–Rideal mechanism, CO adsorbs on the metal ion (M) and steam (H_2O) reacts with the adsorbed CO. Upon completion of the reaction, the reactants get desorbed thus regenerating the active site of the catalyst. On the other hand, in the Langmuir–Hinshelwood mechanism, all the reacting species (CO and H_2O) get adsorbed on the metal ion site. As no significant conversion was observed over unsubstituted ceria, adsorption over the oxide ion vacancies was neglected. However, due to evidence of the dependence of the support in WGS, a mechanism based on the utilization of dual sites, was also proposed. Reduced ceria has oxide ion vacancies and the presence of more anionic vacancies due to substitution of ionic metals could provide for an oxidizing environment. Therefore, the following mechanism was proposed.[1,14]

3) Novel dual site mechanism:

$$CO + M^{n+} \Leftrightarrow CO_M^{n+}$$
$$NO + M^{n+} \Leftrightarrow NO_M^{n+}$$
$$NO + \text{"V"} \Leftrightarrow N \text{ "O"}$$
$$CO_M^{n+} + \text{"O"} \rightarrow CO_2 + M^{n+} + \text{"V"}$$
$$NO_M^{n+} + N \text{ "O"} \rightarrow N_2O + \text{"O"} + M^{n+}$$
$$N_2O + \text{"V"} \rightarrow N_2 + \text{"O"}$$

The subscript "M" represents the adsorption site of the metal over the catalyst. Thus CO_M represents CO molecules adsorbed over metal ion. "v" represents the oxide ion vacancy and "o" the intermediate species formed over the oxide ion vacancy. CO is adsorbed over the metal ion and water gets dissociatively adsorbed on the oxide ion vacancy. These two adsorbed species react on the surface to form the products which upon getting desorbed, restores the catalyst. The effect of H_2 and CO_2 on conversions was included in the kinetic modelling and along with the argument of CO providing a reducing environment due to the oxidized states of the metal ions in the catalyst, it was proposed that this is why CO would get oxidized to CO_2 and H_2 could be adsorbed on the metal site. The optimization routines showed that the Eley–Rideal mechanism showed higher activation energies and an analysis of the rate constants obtained via the Langmuir–Hinshelwood mechanism gave unrealistic values because of which, the mechanisms A and B were discarded. It was also observed, that C best described the kinetics.

10.6.3 NO REDUCTION BY H_2

A bi-functional mechanism[18,28,49] where in adsorption and reaction take place on both the metal site and ionic vacancy was proposed for NO–H_2 reaction over nano-$Ce_{0.98}Pd_{0.02}O_{2-\delta}$.

$$H_{-2} + 2S_{Pd} \Leftrightarrow 2H_{Pd}$$
$$NO + S_{Pd} \Leftrightarrow NO_{Pd}$$
$$NO + \text{``V''} \Leftrightarrow N\text{``O''}$$
$$NO_{Pd} + N\text{``O''} \rightarrow N_2O + S_{Pd} + \text{``O''}$$
$$H_{Pd} + O \rightarrow \text{``O''} H + S_{Pd}$$
$$H_{Pd} + \text{``O''}H \rightarrow H_2O + S_{Pd}$$
$$N_2O + \text{``V''} \rightarrow N_2 + \text{``O''}$$

Steps 1–3 represent the adsorption of H_2, NO on the Pd site and NO on the oxide ion vacancy, respectively. This was also a dual site mechanism involving the utilization of both oxide ion vacancies and the metal ions.

10.7 CONCLUSIONS

It can be seen from the foregoing discussion that combustion synthesis is an efficient technique for the synthesis of nanocrystalline oxides. Ceria-based materials have proven their importance for a large number of heterogeneous reactions including CO oxidation, NO_x reduction, the water–gas shift reaction, hydrogen combustion, hydrocarbon combustion, and C—C coupling reactions. Nanocrystallinity achieved due to the synthesis procedure imparts important characteristics to the catalyst thereby increasing their catalytic activity. Substitution of an aliovalent metal ion improves the activity of the catalysts by providing new adsorption and defect sites in the catalysts. Enhanced activities have been observed over such catalyst when compared to conventional impregnated catalysts and metal-free catalysts. Both experimental as well as theoretical investigations have proven the existence dual site mechanism for the reactions taking place over such catalysts. Current reports emphasize the use of noble metal ions for effective catalysis. Optimization of the use of base metals for improved conversion and economics provides a new direction for research into the development of metal ion substituted ceria systems.

KEYWORDS

- **ammonium nitrate**
- **oxalyl dihydrazide**
- **crystal structure**
- **flourapatite**
- **homogeneous**

REFERENCES

1. Mahadevaiah, N.; Singh, P.; Mukri, B. D.; Parida, S. K.; Hegde, M.S. *Appl. Catal. B Environ.* **2011,** *108–109, 117–126.*
2. Roy, S.; Marimuthu, A.; Hegde, M. S.; Madras, G. *Catal. Commun.* **2008,** *9,* 101–105.
3. Hegde, M. S.; Madras, G.; Patil, K. C. *Acc. Chem. Res.* **2009,** *42,* 704–712.
4. Deshpande, P. A.; Hegde, M. S.; Madras, G. *AIChE.* **2010,** *56* (5), 1315–1324.
5. Esther, J. C.; Siddheswaran, R.; Pushpendra, K.; Karl, C. M.; Rajarajan, K.; Jayavel, R. *Mater. Chem. Phys.* **2015,** *151,* 22–28.
6. Esther, J. C.; Siddheswaran, R.; Pushpendra, K.; Siva, S., V.; Rajarajan, K. *Ceram. Int.* **2014,** *40,* 8599–8605.
7. Nousir, S.; Maache, R.; Azalim, S.; Agnaou, M.; Brahmi, R.; Bensitel, M. *Arab. J. Chem.* **2015,** *8,* 222–227.
8. He, H. W.; Wu, X. Q Ren, W.; Shi, P.; Yao, X.; Song, X. T. *Ceram. Int.* **2012,** *38S,* S501–S504.
9. Kim, J. R.; Lee, K. Y.; Suh, M. J.; Ihm, S. K. *Catal. Today.* **2012,** *185,* 25–34.
10. Esther Jeyanthi, C.; Siddheswaran, R.; Rostislav Medlin, Karl Chinnu, M.; Jayavel, R.; Rajarajan, K. *J. Alloys Compd.* **2014,** *614,* 118–125.
11. Qiang, D.; Shu, Y.; Chongshen, G.; Takeshi, K. Tsugio, S. *RSC Adv.* **2012,** *2,* 12770–12774.
12. Zhou, H. Pi.; Si, R.; Song, W. G.; Yan, C. H. *J. Solid State Chem.* **2009,** *182,* 2475–2485.
13. Liu, Y.; Yao, W.; Cao, X.; Weng, X.; Wang, Y.; Wang, H.; Wu, Z. *Appl. Catal. B Environ.* **2014,** *160–161,* 684–691.
14. Deshpande, P. A.; Hegde, M. S.; Madras, G. *Appl. Catal. B Environ.* **2010,** *96,* 83–93.
15. Deshpande, P. A.; Madras, G. *Appl. Catal. BEnviron.* **2010,** *100,* 481–490.
16. Baidya. T.; Deshpande, P. A.; Madras, G.; Hegde, M. S. *J. Phys. Chem. C.* **2009,** *113,* 4059–4068.
17. Deshpande, P. A.; Aruna, S. T.; Madras, G. *Catal. Sci. Technol.* **2011,** *1,* 1683–691.
18. Roy, S.; Marimuthu, A.; Hegde, M. S.; Madras, G. *Appl. Catal. B Environ.* **2007,** *71,* 23–31.
19. Gupta, A.; Hedge, M. S. *Appl. Catal. B Environ.* **2010,** *99,* 279–288.
20. Andreoli, S.; Deorsola, F. A.; Pirone, R. *Catal. Today.* **2015,** 253, 199–206.
21. Huang, Q.; Yan, X.; Li, B.; Chen, Y.; Zhu, S.; Shen, S. *J. Rare Earth.* **2013,** *31* (2), 124.

22. Postole, G.; Nguyen, T. S.; Aouine, M.; Gelin, P.; Cardenas, L.; Piccolo, L. *Appl. Catal. B Environ.* **2015,** *166–167,* 580–591.

23. Miceli, P.; Bensaid, S.; Russo, N.; Fino, D. *Chem. Eng. J.* **2015,** *278,* 190–198.

24. Piumetti, M.; Bensaid, S.; Russo, N.; Fino, D. *Appl. Catal. B Environ.* **2015,** *165,* 742–751.

25. Shinde, S.; Pitale, S.; Singh, S. G.; Ghosh, M.; Tiwari, M.; Sen, S.; Gadkari, S. C.; Gupta, S. K. J. *Alloys Compd.* **2015,** *630,* 68–73.

26. Zhu, C.; Nobuta, A.; Ju, Y. W.; Ishihara, T.; Akiyama, T. *Int. J. Hydrogen Energy.* **2013,** *38,* 13419–13426.

27. Deshpande, P. A.; Madras, G. *Chem. Eng. J.* **2010,** *158,* 571–577.

28. Baidya,; T.; Marimuthu, A.; Hedge, M. S.; Ravishankar, N.; Madras, G. *J. Phys. Chem. C.* **2007,** *111,* 830–839.

29. Gayen, A.; Priolkar, K. R.; Shukla, A. K.; Ravishankar, N.; Hedge, M. S. *Mater. Res. Bull.* **2005,** *40,* 421–431.

30. Gayen, Baidya. A.; Biswas, T. K.; Roy, S.; Hedge, M. S. *Appl. Catal. A Gen.* **2006,** *315,* 135–146.

31. Gupta, A.; Waghmare, U. V.; Hedge, M. S. *Chem. Mater.* **2010,** *22,* 5184–5198.

32. Ayastuy, J. L.; Iglesias-Gonzalez, A.; Gutierrez-Ortiz, M. A. *Chem. Eng. J.* **2014,** *244,* 372–381.

33. Ilieva, L.; Petrova, P.; Ivanov, I.; Munteanu, G.; Boutonnet, M.; Sobczac, J. W.; Lisowski, W.; Kaszkur, Z.; Markov, P.; 87Venezia, A. M.; Tabakova, T. *Mater. Chem. Phys.* **2015,** *157,* 138–146.

34. Bellakki, M. B.; Baidya, T.; Shivakumara, C.; Vasanthacharya, N. Y.; Hedge, M. S.; Madras, G. *Appl. Catal. B Environ.* **2008,** *84,* 474–481.

35. Yu Yao, Y. F. *J. Catal.* **1984,** *87,* 152–162.

36. Ertl G. *Science.* **1991,** *254,* 1750–1755.

37. Barbier Jr, J.; Duprez, D. *Appl. Catal. B.* **1993,** *3,* 61–83.

38. Heck, R. M.; Farrauto, R. J. *Catalytic Air Pollution Control;* 2nd ed.; Wiley: New York; 2002, Chap. 6, pp 69–129.

39. Barbier, J. Jr,; Duprez, D. *Appl. Catal. B.* **1994,** *4,* 105–140.

40. Yao, H. C.; Yu Yao, Y. F. *J. Catal.* **1984,** *86,* 254–265.

41. Harrison, B.; Diwell, A. F.; Hallett, C. *Platinum Met. Rev.* **1988,** *32,* 73–83.

42. Kang, M.; Song, M. W.; Lee, C. H. *Appl. Catal. A.* **2003,** *251,* 143–156.

43. Barbier, Jr, J.; Duprez, D. *Appl. Catal. B.* **1993,** *3,* 61–83.

44. Nibbelke, R. H.; Campman, M. A. J.; Hoebink, J. H. B. J.; Marin, G. B. *J Catal.* **1997,** *171,* 358–378.

45. Oh, S. H.; Eickel, C. *J. Catal.* **1988,** *112,* 543–555.

46. (a) Serre, C.; Garin, F.; Belot, G.; Maire, G. *J. Catal.* **1993,** *141,* 1–8. (b) Serre, C.; Garin, F.; Belot, G.; Maire, G. *J. Catal.* **1993,** *141,* 9–20.

47. Manuel, I.; Chaubet, J.; Thomas, C.; Colas, H.; Matthess, N.; Djga-Ma- riadassou, G. *J. Catal.* **2004,** *224,* 269–277.

48. Shekhtman, S. O.; Goguet, A.; Burch, R.; Hardacre, C.; Maguire, N. *J. Catal.* **2008,** *253,* 303–311.

49. Roy, S.; Marimuthu, A.; Hegde, M. S.; Madras, G. *Appl. Catal. B Environ.* **2007,** *73,* 300–310.

50. Roy, S.; Hegde, M. S. *Catal. Commun.* **2008,** *9,* 811–815.

51. Bekyarova, E.; Fornasiero, P.; Kaspar, J.; Graziani, M. *Catal. Today.* **1998,** *45,* 179–182.

52. Cant, N. W.; Hicks, P. C.;Lenon, B. S. *J. Catal.* **1978,** *54,* 372–383.

53. Cho, B. K.; Snaks. *J. Catal.* **1986,** *115,* 486.
54. Cho, B. K. *J. Catal.* **1992,** *138,* 255.
55. Hecker, W. C.; Bell, A. T. *J. Catal.* **1983,** *84,* 200.
56. Jain, R.; Maric, R. *Appl. Catal. A Gen.* **2014,** *475,* 461–468.
57. Nahar, G.; Dupont, V. *Renew. Sust. Energ. Rev.* **2014,** *32,* 777–796.
58. Hilaire, S.; Wang, X.; Luo, T.; Gorte, R. J.; Wagner, J. *Appl. Catal. A.* **2004,** *258,* 271.
59. Bunlueistein, T.; Gorte, R. J.; Graham, G. W. *Appl. Catal. B.* **1998,** *15,* 107–114.
60. Fu, Q,; Saltsburg, H.; Flytzani-Stephanopoulous, M. *Science* 2003, *301,* 935.
61. Fox, E.; Lee, A.; Wilson, K.; Song, C. *Top. Catal.* **2008,** *49,* 89–96.
62. Yeung, C.; Tsang, C. *Catal. Lett.* **2009,** *128,* 349–355.
63. Schumacher, N.; Boisen, A.; Dahl, S.; Gokhale, A. A.; Kandoi, S.; Grabow, L. C.; Dumesic, J. A.; Mavrikakis, M.; Chorkendorff, J. *J. Catal.* **2005,** *229,* 265–275.
64. Serrano-Ruiz, J. C.; Huber, G. W.; Sánchez-Castillo, M. A.; Dumesic, J. A.; Rodríguez-Reinoso, F.; Sepúlveda-Escribano, A. *J. Catal.* **2006,** *241,* 378–388.
65. Si, R.; Flytzani-Stephanopoulos, M. *Angew. Chem. Int. Ed.* **2008,** *47,* 2884–2887.
66. Liu, X.; Ruettinger, W.; Xu, X.; Farrauto, R. *Appl. Catal. B.* **2005,** *56,* 69.
67. Zalc, J. M.; Sokolovskii, V.; Löffler, D. G. *J. Catal.* **2002,** *206,* 169–171.
68. Fu, Q.; Deng, W.; Saltsburg, H.; Flytzani-Stephanopoulos, M. *Appl. Catal. B.* **2005,** *56,* 57–68.
69. Goguet, A.; Burch, R.; Chen, Y.; Hardacre, C.; Hu, P.; Joyner, R. W.; Meunier, F. C.; Mun, B. S.; Thompsett, D.; Tibiletti, D. *J. Phys. Chem. C.* **2007,** *111,* 1692–1693.
70. Sharma, S.; Deshpande, P. A.; Hegde, M. S.; Madras, G. *Ind. Eng. Chem. Res.* **2009,** *48,* 6535–6543.
71. Deshpande, P. A.; Madras, G. *Phys. Chem. Chem. Phys.* **2011,** *13,* 708–718.
72. Pomonis P. J.; Efstathiou, A. M. *J. Catal.* **2002,** *209,* 456.
73. Costa, C. N.; Savva, P. G.; Fierro J. L. G.; Efstathiou, A. M. *Appl. Catal. B.* **2007,** *75,* 147.
74. Yu, Q.; Richter, M.; Li, L.; Kong, F.; Wu G.; Guan, N. *Catal. Commun.* **2010,** *11,* 955.
75. Deshpande, P. A.; Polisetti, S.; Madras, G. *AIChE J.* **2011,** *57,* 2928–3245.
76. Parthasarathi, B.; Hegde, M. S.; Patil, K. C. *Curr. sci.* **2001,** *80,* 1576–1578.
77. Papadakis, V. G.; Pliangos, C. A.; Yentekakis, I. V.; Verykios, X. E.; Vayenas, C. G. *Catal. Today.* **1996,** *29,* 71–75.
78. Heck, R. F.; Nolley Jr, J. P. *J. Org. Chem.* 1972, *37,* 2320–2322.
79. Nicolaou, K. C.; Bulger, P. G.; Sarlah, D. *Angrew. Chem. Int. Ed.* **2005,** *44,* 4442–4489.
80. Miyaura, N.; suzuki, A. *Chem Rev.* **1995,** *95,* 2457–2483.
81. Yang, B. H.; Buchwald, S. L. *J .Organo,met.Chem.* 1999, *576,* 125–146.
82. Dong, L.; Wen, J.; Qin, S.; Yang ,N.; Yang, H.; Su, Z.; Yu ,X.; Hu, C. *ACS Catal.* **2012,** *2,* 1829–1837.
83. Thathagar, M. B.; Beckers, J.; Rothenberg, G. *J. Am. Chem. Soc.* **2002,** *124,* 11858–11859.
84. Zim, D.; Lando, V. R, Dupont, J. *Org. Lett.* **2001,** *3,* 3049–3051.
85. Sanjaykumar, S. R.; Mukri, B. D.; Patil, S.; Madras, G.; Hegde, M. S. *J. Chem Sci.* **2011,** *123,* 47–54.
86. Konstantinou, I. K.; Albanis, T. A. *Appl. Catal. B.* **2003,** *42,* 319–335.
87. Malato, S.; Blanco, J.; Vidal, A.; Richter, C. *Appl. Catal. B.* **2002,** *37,* 1–15.
88. Chahbane, N.; Lenoir, D.; Souabi, S.; Collins, T. J.; Schramm, K. W. *Clean Air Soil Water.* **2007,** *35,* 459–464.

89. Zhao, S.; Li, J.; Wang, L.; Wang, X. *Clean Soil Air Water*. **2010,** *38,* 268–274.

90. Mahata, P.; Aarthi, T.; Madras, G.; Natarajan, S. *J. Phys. Chem. C.* **2007,** *111,* 1665–1674.

91. Saha, D.; Mahapatra, S.; Guru Row, T. N.; Madras, G. *Ind. Eng. Chem. Res.* **2009,** *48,* 7489–7497.

92. Chen, P.; Zhang, X. *Clean Soil Air Water.* **2008,** *36,* 507–511.

93. Feteira, A.; Sinclair, D. C.; Lanagan, M. T. *J. Appl. Phys.* **2007,** *101,* 1–7.

94. Li, Y.; Yao, S.; Wen, W.; Xue, L.; Yan, Y. *J. Alloys Comp.* **2010,** *491,* 560–564.

95. Deshpande, P. A.; Aruna, S. T.; Madras, G. *Clean Soil Air Water* **2011,** *39* (3), 259–264.

96. Fei, T.; Jiang, K.; Liu, S.; Zhang, T. *Sens. Actuators B.* **2014,** *190,* 523–528.

97. Parvatikar, N.; Jain, S.; Bhoraskar, S. V.; Ambika Prasad, M. V. N. *J. Appl. Plym. Sci.* **2006,** *102,* 5533.

98. Zhang, Z. W.; Hu, C. G.; Xiong, Y. F.; Yang, R. S.; Wang, Z. L. *Nanotechnology* **2007,** *18,* 465504.

99. Hu, C. H.; Xia, C. H.; Wang, F.; Zhou, M.; Yin, P. F.; Han, X. Y. *Bull. Mater. Sci.* **2011,** *34,* 1033–1037.

100. Ganduglia-Pirovano, M. V.; Da Silva, J. L. P.; Sauer, J. *Phys. Rev. Lett.* **2009,** *102,* 026101.

101. Li, H. Y.; Wang, H. F.; Gong, X. Q.; Guo, Y. L.; Guo, Y.; Lu, G.; Hu, P. *Phys. Rev. B.* **2009,** *79,* 193401.

102. Loschen, C.; Carrasco, J.; Neyman, K. M.; Illas, F. *Phys. Rev. B.* **2007,** *75,* 035115.

103. Andersson, D. A.; Simak, S. I.; Johansson, B.; Abrikosov, I. A.; Skorodumova, N. V. *Phys. Rev. B.* **2007,** *75,* 035109.

104. Nolan, M. Grigoleit, S.; Sayle, D. C.; Parker, S. C.; Watson, G. W. *Surf. Sci.* **2005,** *576,* 217–229.

105. Hu, Z.; Metiu. H. *J. Phys. Chem. C.* **2011,** *115,* 17898–17909.

106. Dutta, G.; Waghmare, U. V.; Baidya, T.; Hedge. M. S. *Chem. Matter.* **2007,** *19,* 6430–6436.

107. Binet, C.; Daturi, M.; Lavalley, J. C. *Catal. Today.* **1999,** 50, 207–225.

108. Bera, P.; Priolkar, K. R.; Gayen, A.; Sarode, P. R.; Hedge, M. S.; Emura, S.; Kumashiro, R.; Jayaram, V.; Subbanna, G. N. *Chem. Matter.* **2003,** *15,* 2049–2060.

109. Dry, R. E.; Shingles, T.; Boshoff, L. J. *J. Catal.* **1972,** *25,* 99–104.

110. Gre noble, D. C.; Estadt, M. M. *J. Catal.* **1981,** *67,* 90–102.

CHAPTER 11

PHASE MANIPULATION IN TiO$_2$ FOR BETTER PHOTOCATALYTIC AND ELECTROCATALYTIC EFFECT

K. PARAMESWARI[*] and C. JOSEPH KENNADY

Department of Chemistry, School of Science and Humanities, Karunya University, Coimbatore 641114, India

[*]*Corresponding author. E-mail: parameswari@karunya.edu*

CONTENTS

ABSTRACT

Phase manipulation of TiO_2 was carried out by coating TiO_2 on titanium substrate by the thermal decomposition and anodization methods using different precursors. The electrodes were studied for their surface morphology, phase formation and corrosion resistance. The electrocatalytic efficiency of the electrodes was studied for the electroreduction of fumaric to succinic acid. The product formed was isolated and confirmed by melting point measurement. The product was also confirmed by nuclear magnetic resonance (NMR) spectroscopic technique. Phase of TiO_2 was found to be anatase and efficiency of electrocatalytic activity was high in the electrode obtained by the thermal decomposition of $TiCl_3/HNO_3$ compared to the other electrodes prepared by other methods and precursors.

11.1　INTRODUCTION

TiO_2 films were prepared by a variety of deposition techniques such as sol–gel process, sputtering techniques, thermal plasma, and electron beam evaporation. The preparation of titania is very much interesting, since titania exhibits polymorphism (anatase, rutile, and brookite) and the phase formation is highly dependent on the preparation conditions. In fact, anatase has higher photocatalytic activity than rutile, because of a difference in Fermi energy. The charge carrier in anatase thin film has a higher mobility than that of rutile. The preparation variables affecting the nature of TiO_2 on titanium substrate and its electrocatalytic properties are also not well studied. The aim of this work is to investigate the structural and electrochemical properties of Ti/TiO_2 electrodes prepared from the four following methods. The anatase and rutile phases are the main focus of this investigation.

11.2　EXPERIMENTAL METHODS

11.2.1　INTRODUCTION

Titanium dioxide-coated titanium (Ti/TiO_2) electrodes are used in a wide range of applications such as electrocatalysts, photocatalysis, and dye-sensitized solar cells.[1-5] The material properties of TiO_2 nanoparticles are a function of the crystal structure, nanoparticle size, and morphology and, hence, are strongly dependent on the method of synthesis.[5-15] TiO_2 exists in three main phases: anatase, brookite, and rutile. TiO_2 in its anatase form has found to have better electrocatalytic activity compared to the other forms of TiO_2. Hence, the challenge here is to prepare TiO_2 in its anatase form. As a bulk material, rutile is the stable phase; however, solution-phase preparation methods for TiO_2 generally favor the anatase structure.[8,15–18] These observations are attributed to two main effects: surface energy and precursor chemistry. At very small particle dimensions, the surface energy is an important part of the total energy and it has been found that the surface energy of anatase is lower than those of rutile and brookite.[19–22] Surface energy considerations accurately describe the observation of a crossover size of about 30 nm where anatase nanoparticles transform to rutile.[19,21] The crystal structure stability has been explained on the basis of a molecular picture, where the nucleation and growth of the different polymorphs of TiO_2 are determined by the precursor chemistry, which depends on the reactants used.[23–29] A complicating factor in the understanding of nanoparticle formation is the multitude of experimental conditions used for synthesis of the different TiO_2 phases, making it difficult to compare mechanisms. Synthesis of anatase nanoparticles by solution-phase methods in aqueous environments has been reported for a large variety of experimental conditions: using an alkoxide precursor, anatase is formed by reaction in aqueous solutions of a variety of acids or bases. Phase-pure anatase nanoparticles with diameters ranging from 6 to 30 nm are generally prepared from titanium (IV) isopropoxide and acetic acid.[16] When stronger acids are used, a fraction of the product usually consists of brookite nanoparticles.[18,30] Larger anatase particles are difficult to synthesize due to transformation to rutile upon increasing treatment times and/or temperature. Phase-pure brookite particles (0.3–1 μm) have been prepared using amorphous titania as the starting material and hydrothermal treatment with NaOH.[31] The mechanism is related to the formation of a sodium titanate, which subsequently is transformed to pure TiO_2 with the brookite structure. The synthesis of brookite nanoparticles of 5–10 nm has been reported by thermolysis of $TiCl_4$ in aqueous HCl solution.[29] The composition of the reaction product was found to be strongly dependent on the Ti: Cl concentration ratio, and up to 80% pure brookite was reported at Ti:Cl = 17:35. The brookite nanoparticles could be separated by selective precipitation of rutile. At higher Ti:Cl ratio, pure rutile particles were obtained which were generally much larger and rod-shaped. Phase-pure rutile nanoparticles have been prepared from $TiCl_4$ or $TiCl_3$ in HCl solution or from titanium (IV) isopropoxide in nitric acid at pH = 0.5.[26,32–35] Recently, several authors have compared synthesis methods for the three phases, in order to determine the effect of crystal structure on the physical properties.[27,36,37] An advanced understanding of the parameters that deter-mine phase formation is crucial for the successful technological application of nanomaterials. In this work, we determine the influence of the reactant chemistry in the preparation of Ti/TiO$_2$ from four different methods with main goals: (1) to gain insight into the phase formation mechanisms, (2) to provide experimental methods for the controlled synthesis of phase-pure anatase and rutile with control over the nanoparticle dimensions, (3) to study the stability of the TiO$_2$ film on titanium substrate, and (4) to study its elec-trocatalytic activity.

Keeping this in mind, an attempt has been made to prepare stable redox Ti/TiO$_2$ electrocatalysts with good electrocatalytic activity and also to reduce the cost of preparation was carried out in four different methods.

11.2.2 PREPARATION OF TITANIUM DIOXIDE-COATED TITANIUM ELECTRODES

1. Thermal decomposition method using different concentration of TiCl$_3$ at different temperatures and HNO$_3$ as oxidizing agent.
2. Thermal decomposition method using different concentration of tita-nium tetraiospropoxide and citric acid at different temperatures.
3. Thermal decomposition method using different concentration of TiCl$_3$ at different temperatures and H$_2$O$_2$ as oxidizing agent.
4. Anodizing titanium substrate in H$_2$SO$_4$ at different temperatures, current densities, and time.

11.2.2.1 PRELIMINARY TREATMENT OF THE TITANIUM SUBSTRATE

The working electrode is titanium metal strip (Imperial Metal Industries Ltd., IMI-125 grade) and titanium mesh (Pattern of expansion: LWD: 12.5 mm; SWD: 7 mm; strand width: 2 mm; strand thickness: 1 mm) which are sand blasted. Then the samples are rinsed with double distilled water followed by isopropanol. Then the specimen is cleaned (etched) in a 5 wt% oxalic acid solution at 80–98°C for 1 h and then cleaned in deionized water and air dried. This pretreatment procedure gives a rough surface which will enhance the adhesion of TiO$_2$ coating. Ti/TiO$_2$ electrodes were prepared in two stan-dard sizes namely 1 × 5 cm^2 and 9 × 12 cm^2 for cyclic, voltammetric, and preparative scale experiments, respectively.

11.2.2.2   THERMAL DECOMPOSITION METHOD USING TiCl$_3$ AND HNO$_3$

In general, the experimental procedure recommended by Beck et al.[38] is adopted. The Ti/TiO$_2$ electrode is prepared by the following procedure. The precursor is prepared in different concentrations 0.05, 0.10, 0.15, and 0.2 N using TiCl$_3$, HNO$_3$ (5.0% v/v), and isopropyl alcohol (Merck). In this solution the concentration of HNO$_3$ and isopropyl alcohol are kept constant and the coating solutions are prepared by adding HNO$_3$ and isopropyl alcohol slowly cooled to 10°C, maintained at 10°C for 20 min, and then allowed to raise to room temperature. A stable clear yellow solution is obtained. The solution is sprayed over pretreated titanium substrate strips and the adherent film is dried in an oven at 100–110°C. The metal strip is placed in a muffle furnace for 25 min. This procedure is repeated at least six times to get a good coating of TiO$_2$ on titanium metal. These electrodes are essentially prepared in two different standard sizes namely 1 × 5 cm^2 and 9 × 12 cm^2 for cyclic voltammetric and preparative scale experiments, respectively. For cyclic voltammetric experiments a punched tape is usually bound around the electrode strip just exposing 0.078 cm^2 for electrochemical activity. In preparative scale experiments 9 × 8 cm^2 electrode surface is immersed in the electrolyte. The electrode prepared by the above procedure remained electrochemically active. Figures 11.1 and 11.2 show the photographs of the Ti/TiO$_2$ electrode prepared from 0.15 N TiCl$_3$ at 600°Cwith 5.0% v/v HNO$_3$ and the electrode used in cyclic voltammetric studies. The preparation method of Ti/TiO$_2$ from the thermal decomposition of TiCl$_3$ and HNO$_3$ is given in flow diagram.

FIGURE 11.1   Ti/TiO$_2$ electrode prepared from 0.15 N of TiCl$_3$ at 600°C with HNO$_3$ (5% v/v).

FIGURE 11.2 Ti/TiO$_2$ electrode used for cyclic voltammetric studies prepared from 0.15 N of TiCl$_3$ at 600°C with HNO$_3$ (5% v/v).

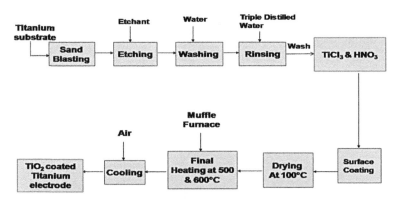

Flow diagram indicating the Ti/TiO$_2$ electrode preparation procedure from TiCl$_3$ and HNO$_3$

11.2.2.3 *THERMAL DECOMPOSITION METHOD USING TITANIUM TETRAISOPROPOXIDE*

Among various methods of preparing oxide electrodes the Pechini method has been also widely used.[39-44] With different molar propositions of titanium tetraisopropoxide 1:2:8 (solution I) (titanium tetraisopropoxide: citric acid: ethylene glycol) and 1:4:16 (solution II) are prepared as it is given in

Table 11.1. The solution is prepared by heating ethylene glycol (Merck) to 60°C with stirring and then adding the titanium tetraisopropoxide (Merck). Finally, citric acid is added. Then the temperature was increased to 90°C and the solution is stirred at this temperature until it turned clear and allowed to cool down to room temperature. The pretreated titanium substrate is dipped in the precursor, after each layer is deposited, the films are thermally treated at 250°C for 10 min in an electric oven, and then the metal strip is placed in a muffle furnace for 15 min. This procedure is repeated with above prepared solution I at 350, 450, 550, and 650°C. The same procedure is repeated with solution II at same four different temperatures. The procedure of preparing Ti/TiO$_2$ from titanium tetraisopropoxide is given in the flow diagram.

TABLE 11.1 The Composition of the Precursor Solutions.

Solution	Titanium Tetraisopropoxide	Citric Acid	Polyethylene Glycol
Solution I	1	2	8
Solution II	1	4	16

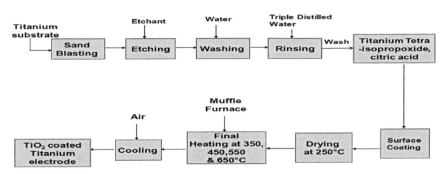

Flow diagram indicating the Ti/TiO$_2$ electrode preparation procedure from TTIP and citric acid

11.2.2.4 THERMAL DECOMPOSITION METHOD USING TiCl$_3$ AND H$_2$O$_2$

TiCl$_3$ solution is prepared as it is given in Section 11.2.2.2 instead of HNO$_3$, here H$_2$O$_2$ is taken. The complete procedure of preparing Ti/TiO$_2$ by the thermal decomposition method using TiCl3 and H$_2$O$_2$ is given in the following flow diagram.

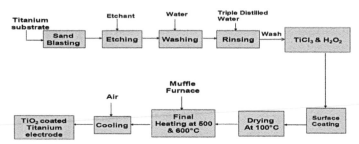

Flow diagram indicating the Ti/TiO$_2$ electrode preparation procedure
from TiCl$_3$ & H$_2$O$_2$

11.2.2.5   PREPARATION OF Ti/TiO$_2$ BY ANODIZATION METHOD

Many studies have been reported in the literature on TiO$_2$ films obtained by
electro oxidation of titanium.[45-47] The anodic oxidation process is conducted in
a dual electrode reaction chamber, in which the pretreated titanium substrate
is used as the anode and a stainless steel sheet of the same size as the cathode.
Two electrodes are dipped in electrolyte H$_2$SO$_4$ and connected externally
through a DC power supply (EPS600). Ti/TiO$_2$ electrodes are prepared with
varying current densities, temperature, and time as it is given in Table 11.2.

TABLE 11.2   Anodization at Different Current Densities, Temperatures, and Time.

Expt. No	Current Density (mA/cm^2)	Temperature (°C)	Time (min)
1	10	30	45
2	10	40	45
3	10	50	45
4	10	60	45
5	10	60	30
6	12	60	45
7	12	70	45

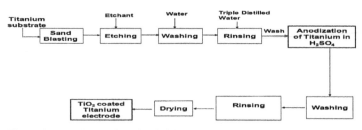

Flow diagram indicating the Ti/TiO$_2$ electrode preparation procedure
by anodization method

The electrodes are characterized for their surface morphology and micro-structure. Surface morphology of Ti/TiO$_2$ electrodes was carried out by scanning electron microscope (SEM) using JEOL 6390 instrument. X-ray patterns were recorded using SHIMADZU-6000 X-ray diffractometer. From these, investigations optimized conditions were obtained and based on these results obtained from SEM and X-ray diffraction (XRD) studies, from each preparation method, one electrode with uniform crystalline size was chosen for corrosion studies, cyclic voltammetric studies, and electrolysis.

11.2.2.6 INSTRUMENTATION

SEM employed for surface studies is JEOL 6390 instrument. The XRD pattern of the oxide layer is recorded using SHIMADZU-6000 X-ray diffrac-tometer. Other instruments employed such as, pH meter and constant current source for preparative electrolysis are of Indian make.

11.2.2.7 CYCLIC VOLTAMMETRIC CELL

Cyclic voltammetric studies are carried out using electrochemical worksta-tion (CHI660C). These experiments are carried out in an undivided glass cell provided with ground joints. A platinum foil, whose electrode area is more than five times of the working Ti/TiO$_2$ electrode, used as the counter electrode. Ti/TiO$_2$ electrode is used as the working electrode and saturated calomel electrode attached with luggin capillary is used as the reference electrode in all reduction experiments. The glass cell also has inlet for passing inert gas and provision for inserting a thermometer. The complete cell is placed in water-bath whose temperature could be controlled to an accuracy of ±1°C. For removal of dissolved oxygen form the electrolyte hydrogen gas is used which is electrochemically generated from a locally built alkaline water electrolyzer. The hydrogen gas is passed through a series of wash bottles containing pyrogallol, 1M sulfuric acid, and conduc-tivity water, respectively before it entered the electrochemical cell. Before each set of experiments, the electrochemical cell is washed with nitric acid, distilled water, and finally with triple distilled water. Purified hydrogen gas is passed into the electrolyte solution for at least 10 min. Whenever gas is passed, the electrolyte is also stirred with a magnetic stirrer. The cyclic voltammetric studies for all the electrodes are carried out under nitrogen inert atmosphere.

11.2.2.8 CORROSION STUDIES

Tafel and impedance plots are recorded using electrochemical workstation (CHI660C). The corrosion resistance behavior of the electrode (coated and uncoated) is evaluated by electrochemical impedance spectroscopy. The specimen of 1.0 cm² area is used as working electrode, the standard calomel electrode (SCE) as reference electrode and the platinum foil as counter electrode. These three electrodes are immersed in 0.1 M H_2SO_4 in a 3-electrode cell assembly. The time interval, often in minutes, is given to attain a steady-state open-circuit potential (OCP) and then the impedance measurements are carried out at the OCP. A sine wave voltage of 10 mv is super imposed on the rest potential. The real part (Z') and the imaginary part (Z") are measured in the frequency range of 100–10 KHz. From the Nyquist plots, the value of (Rct) charge transfer resistance is obtained. The corrosion resistance behavior of Ti/TiO$_2$ coated and the titanium substrate of 1.0 cm² area is evaluated by Tafel polarization technique studies. The cell constant used in this study is same as that used in the impedance measurements studies. The coated specimen of 1.0 cm² area is immersed in 0.1 M H_2SO_4. The time interval, often in minutes, is given to attain a steady-state potential. The electrode is then polarized cathodic to anodic direction and the potentiodynamic polarization has been carried out in the potential range −200 to +200 mv from OCP at a scan rate of 1 mv/s. Parameters such as Ecorr and Icorr values are obtained from the plot of E vs log$i$. From I_{corr}, the corrosion rate millimeter per year (mmpy) is calculated by applying from the Faraday's first law:

$$CR(\text{mmpy}) = \frac{I_{corr}T_yM_{me}\times10^6}{ZFD_{me}}$$

where
I_{corr} is the corrosion current density (A/dm²),
T_y is the year in seconds,
M_{me} is the molar weight of titanium metal (47.87 kg/mol),
F is the Faraday constant (96485 C/mol),
Z is the number of electrons (Z = 4), and
D_{me} is the density of the metal (Ti = 4.506 g/cm³).
$K = T_y/F = 3.2$
The constant factor 3.2 is substituted in the above equation,
CR (mmpy) = 3.2 I_{corr} equivalent weight/density,
Note: For measurement corrosion rate in millimeter/year,
Equivalent weight of titanium = 11.97 g/mol, and
Density of titanium = 4.506 g/cm³ (was taken).

11.2.2.9   PREPARATIVE ELECTROLYSIS

Preparative scale experiments for the reduction of fumaric acid are carried out in a 1 dm^3 (one liter) capacity glass cell. The cell cover used is equipped with openings for introducing the porous diaphragm, thermometer, condenser, and electrodes. About 0.2 dm^3 1M H$_2$SO$_4$ (200 mL) is used as catholyte and anolyte. A 0.72 dm^2 area of Ti/TiO$_2$ electrode is used as cathode and a stainless steel electrode as anode, which is placed inside the porous pot. The electrolyte is stirred by means of magnetic stirrer. The current density and temperature are varied. After completion of electrolysis the product is separated and identified. The mesh type electrode Ti/TiO$_2$ (prepared from TiCl$_3$ and HNO$_3$) is used for preparative electrolysis is given in Figure 11.3.

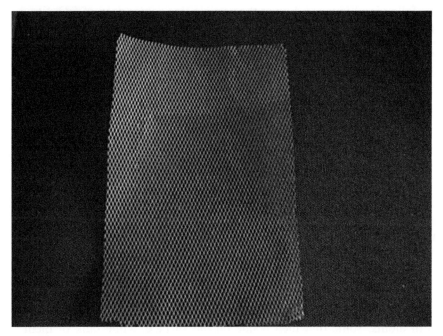

FIGURE 11.3   Ti/TiO$_2$ mesh electrode prepared from 0.15 N of TiCl$_3$ with HNO$_3$ (5% v/v) at 600°C.

11.2.2.10   PRODUCT ANALYSIS

The characterized Ti/TiO$_2$ electrode is used in the electroreduction of fumaric acid to succinic acid and the product is confirmed by NMR spectroscopic technique using a T-60 Varian 60 MHz spectrometer.

11.3   COMPARATIVE STUDY OF Ti/TiO₂ ELECTRODES

The comparative study of Ti/TiO$_2$ prepared by the above given four methods, are done in terms of its particle size, phase formation of TiO$_2$, corrosion resistance in 0.1 M H$_2$SO$_4$, electrocatalytic activity, and the yield of succinic acid obtained by the electroreduction of fumaric acid. This study helps us to get a clear picture about the efficient method of preparation of Ti/TiO$_2$ in terms of above given studies.

11.3.1   COMPARISON OF FILM TiO₂ ON TITANIUM SUBSTRATE IN TERMS OF PARTICLE SIZE

The following Figure 11.4 gives the particle size comparison of TiO$_2$ film on titanium substrate prepared by the four different methods.

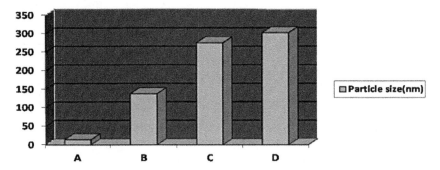

FIGURE 11.4   Comparison of particle size of TiO$_2$ film on titanium prepared from (A) TiCl$_3$ and HNO$_3$, (B) TTIP and Citric acid, (C) TiCl$_3$ & H$_2$O$_2$, and (D) anodization.

The particle size of TiO$_2$ film is 14 nm formed by the thermal decomposition of TiCl$_3$ with HNO$_3$. Particles of the condensed matter are made up of atoms, some of which reside on the surface, while bulk of them is locked in the interior. Surface of a particle not only has planar areas but also the edges and the corners. The atoms on the surface are exposed to the dissimilar surroundings while the interior atoms have homogeneous environment. Since the surface area to volume ratio is small in the bulk state, the proportion of the surface atoms to total atoms is very low and hence, the behavior of the material as a whole is less influenced by the surface atoms. As one makes particles finer and finer, the surface area to volume ratio keeps increasing and the properties related to the surface area continue to vary in accordance with the known relationships. This change in the properties with increasing surface area is fairly intuitive. Further down in this subdivision of matter, a stage is reached when the proportion of surface atoms becomes significant and their thermodynamics is a dominant contributor in the properties of the whole material. At this very low length scale, there is also a great increase in total edge length and the number of corners; and the atoms on the edges and corners enhance the surface energy substantially. If the size of an atom is assumed to be 0.1 nm (which is a fair approximation), Figure 11.5 shows the simplified relationship between surface atoms (as a percentage of total atoms) and the particles size in nanometers. In the TiO_2 film prepared by the thermal decomposition of $TiCl_3$ and HNO_3 at 600°C, 0.15 N $TiCl_3$ the particle size was 14 nm and for such particle size from the graph in Figure 11.5. The surface to total atom was more which contribute to the better electrocatalytic activity of the Ti/TiO_2 electrode. At the same time the particle size of TiO_2 film prepared by other methods as given in Figure 11.4, as the particle size increases the surface to total atoms reduces so the decreased electrocatalytic activity was also observed in the yield of succinic acid obtained by the preparatory electrolysis. (Table 11.3).

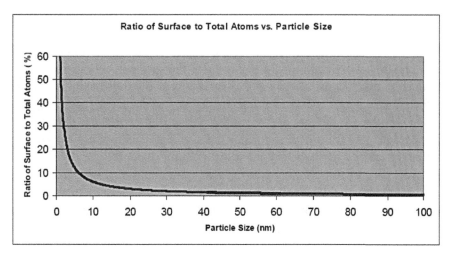

FIGURE 11.5   Surface to total atoms expressed as percentage against particle size in nm.

One can draw the following inferences from the graph:

The relationship between percentage of surface atoms and the particle size is fairly linear in the region 100–20 nm. There is drastic increase in the dominance of surface atoms over interior atoms as we go down from

20 to 1 nm. From the graph, it is inferred that in the Ti/TiO$_2$ electrode prepared from TiCl$_3$ and HNO$_3$ the particle size was 14 nm and the surface to total atoms are more in this electrode which contributes towards better electrocatalytic activity. There is a region approximately between 20 and 2 nm where surface atoms increasingly govern the properties of atoms. This region of few nanometers (varying for different materials) corresponds to atomic clusters consisting of 10–100 atoms and is the domain of real nanotechnology. It is the region flanked by atomic scale on one side and bulk scale on the other. In these atomic clusters, since many of the atoms are on the surface, the surface behavior transforms into the bulk behavior thereby leading to a drastic change in properties. Additionally, when a crystalline material goes down in particle size to nano-region, it assumes closely packed structures with fewer defects such as interstitials, vacancies, and dislocations. This rearrangement in matter at nano-scale brings about alteration in properties like reactivity, adsorption behavior, color, UV absorption, catalytic activity, conductivity, hardness, and microbiological activity. The bulk crystalline state is far from perfect as there are impurities and defects introduced during the crystal phase formation stage. These imperfections cause significant alterations in the properties of atoms. The defects influence chemical, mechanical, optical, electrical, magnetic, and thermal properties of the bulk crystalline matter. However, as we move down the length scale into the nanoscale region by slicing the bigger crystals into progressively finer crystals, more and more atoms come to the surface as compared to the number of atoms embedded inside. This process also brings the inner crystal defects to the surface. High surface energy of the nanoscale matter drives out most of these defects on the surface. Even the embedded defects are not too deep and mild heat can squeeze these out. This leads to far greater perfection and purity in the crystal structure in the nanoparticles. Obviously, the properties of these defect-free crystals are notably different from their bulk counterpart. The second change that occurs in crystals at nanoscale is the density of packing of surface atoms. High surface energy of nano-crystals leads to surface relaxation by contraction and even rearrangement of the atoms in some cases. This leads to formation tightly packed structures. Such nanocrystals are very pure and perfect and resist any ingress of foreign atoms. These modifications in the crystal at nanoscale give rise to their novel properties. The particle size of film TiO$_2$ on titanium substrate by the thermal decomposition of TiCl$_3$ with HNO$_3$ was 14 nm, compared to the film TiO$_2$ formed by other methods. The particle size is one of the very important parameters in deciding the electrocatalytic activity, and it has been confirmed from the cyclic voltammetric studies of the same electrode and

further this finding is proved by the high yield obtained by the electroreduction of fumaric acid from succinic acid. (Table 11.3). Due to the formation of nano particle in the film prepared by the decomposition of $TiCl_3$ with HNO_3, it has better electrocatalytic activity resulting in high yield of succinic acid compared to Ti/TiO₂ prepared by other methods.

TABLE 11.3   Comparative Studies of Ti/TiO₂ Electrodes by Four Different Methods.

Method	Preparation Condition	XRD Structure	Particle Size (nm)	Corrosion Rate (mm/y)	Phase Angle (θ)	Yield of Succinic Acid (%)
Thermal decomposition of $TiCl_3$ and HNO_3	600°C, 0.1 N of $TiCl_3$ with HNO_3	Anatase	14	19.060×10^{-6}	48	93.00
Thermal decomposition of TTIP and citric acid	550°C Solution II	Rutile	138	1.17	24	75.00
Thermal decomposition of $TiCl_3$ and H_2O_2	600°C, 0.1 N of $TiCl_3$ with H_2O_2	Rutile	275	2.31	20	47.00
Anodization of titanium in H_2SO_4	60°C, 30 min, 10A/dm²	Rutile	302	11.85	15	15.00

11.3.2   COMPARISON OF PHASE OF THE FILM TiO₂ ON TITANIUM SUBSTRATE

From the X-ray diffraction studies, the comparison of phase formation of TiO₂ on titanium substrate is given in Table 11.3. The TiO₂ film formed by the decomposition of $TiCl_3$ with HNO_3, 100% anatase phase is formed and in other methods mostly TiO₂ is in its rutile form. Anatase phase is found to have good catalytic activity than rutile and brookite forms of TiO₂. However, the rutile structure is known to be the most stable phase. TiO₂ exists in several polymorphic forms, including amorphous, anatase, and rutile depending on fabricating conditions and further heat treatment.[48,49] Barnard et al. presented the modeling structure and electronic properties of anatase TiO₂ nanoparticles by using a self consistent tight-binding method and density functional theory and they found nonbonding electrons at the edges and corners of the nanoparticles.[50] It has been shown that the anatase phase becomes more stable than the rutile phase when the particle size is reduced below 14 nm.[51] For example, in calcination of amorphous titanium (IV) hydroxide at low temperatures, nanocrystalline anatase particles are originally formed. Further increase in temperature and treatment duration results in that rutile crystallites appear and the anatase–rutile phase transition occurs.[52] There are seven known polymorphs of titania, only three of which (rutile, brookite, and anatase) occur in nature and the structure of anatase and rutile TiO_2 are given in Figure 11.6. The anatase and rutile phases are the focuses of this investigation. Their bases consist of Ti octahedrally coordinated to oxygen. The right side of Figure 11.6 shows the octahedral stacked, so that, each oxygen has three-fold coordination. Among the three different crystal modifications: anatase, rutile, and brookite, the first two being most frequently encountered and used. According to the phase diagram of TiO_2,[53] anatase, and brookite transform into rutile on heating to 700 and 900°C, respectively, while amorphous TiO_2 crystallizes into anatase on heating above 300°C. Accordingly, variation of the parameters of technological processes leads to the formation of TiO_2 coatings with different properties.

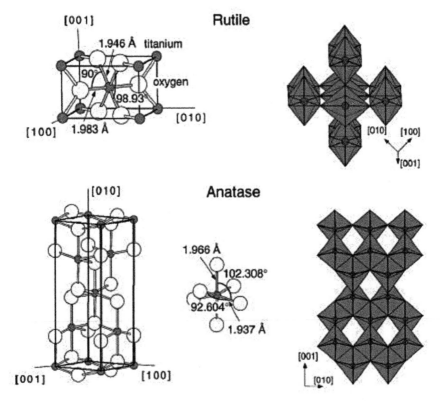

FIGURE 11.6   Structure of anatase and rutile TiO_2.

For the rutile (110) facet, half of the Ti atoms retain their six-fold coordination, while the other half remains in five-fold coordination. The stability decreases for the (100) and (001) facets as the coordination for the Ti atoms falls to five-fold and four-fold, respectively. For anatase, the (101) surface is the most stable with a surface free energy lower than even the rutile (110).[54] As the degree of coordination unsaturation increases, the driving force for chemical and physical adsorption also increases. If there are no adsorptions species available, surface atoms will relax into configurations that maximize their coordination and stabilize the electrostatic interactions.[55,56] In fact, anatase has higher photocatalytic activity than rutile because of a difference in Fermi energy[57] and the charge carrier in anatase thin film has a higher mobility than that of rutile.[58]

For anatase phase the mechanism responsible for better electrocatalytic behavior is a controversial subject. Considerable research is in progress to gain complete understanding into this phenomenon. However, at present there are four hypotheses seeking to explain higher photocatalytic activity of anatase which can be related to the electrocatalytic activity of anatase TiO$_2$. Crystal size, surface area, defect population, and porosity: These factors influence the rate of recombination of the holes and the electrons. Any generic difference in these aspects between rutile and anatase can lead to a significant difference in their electro or photocatalytic activity. Higher Fermi level: The anatase has higher Fermi level over the rutile by 0.1 eV. This leads to a lower oxygen affinity and a higher level of hydroxyl groups on the surface. These hydroxyl groups contribute to the higher electro or photocatalytic activity.

Indirect band gap: The anatase possesses an indirect band gap while the rutile has a direct band gap. In a direct band gap material, the minimum in the conduction band coincides with the maximum in the valence band, thus facilitating an early return of the electron to the valence band. In the indirect band gap materials, the minimum in the conduction band is away from the maximum in the valence band. This enables the excited electron to stabilize at the lower level in the conduction band itself leading to its longer life and greater mobility.[59] As it has been discussed that the Ti/TiO$_2$ electrode prepared by the thermal decomposition of TiCl$_3$ with HNO$_3$ is found to be in anatase phase and it becomes more stable than the rutile phase when the particle size is reduced to14 nm, and the particle size of TiO$_2$ formed by this method was 14 nm with anatase phase in the electrode prepared at 600°C with 0.15 N TiCl$_3$ with HNO$_3$.

11.3.3 COMPARISON OF CORROSION RESISTANCE CAPACITY OF THE FILM TiO$_2$

The corrosion rate (mmpy) and phase angle obtained from Tafel polarization curves and AC-impedance studies for the electrode Ti/TiO$_2$ prepared by the four different methods are compared and are given in Figures 11.7, 11.8, and Table 11.3. In the case of coated sample the anodic reaction is related to TiO$_2$ particle and it is related to titanium in the uncoated one. TiO$_2$ particles increase the contact between coating and H$_2$SO$_4$ solution, so the intensity of anodic reaction will increase. As the TiO$_2$ nano particles of the Ti/TiO$_2$ electrode prepared by the thermal decomposition of TiCl$_3$ and HNO$_3$ was 14 nm, which increases the homogeneity and uniformity of coating and also decrease the defects of coating (Table 11.3). Therefore the nanoparticles control defects on molecular level and will lead to increase corrosion protection properties. When TiO$_2$ coating by anodizing titanium substrate exposed in 0.1 M H$_2$SO$_4$ shows very less corrosion resistance compared to the uncoated but the difference is not much that may be due to the thin TiO$_2$ film on titanium substrate.

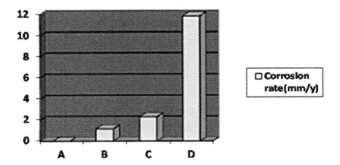

FIGURE 11.7 Comparison of corrosion rate of TiO$_2$ film on titanium in 0.1 M H$_2$SO$_4$ prepared from (A) TiCl$_3$ and HNO$_3$, (B) TTIP and Citric acid, (C) TiCl$_3$ and H$_2$O$_2$, and (D) anodization.

Phase angle value from 40 to 70 indicates that the impedance has a significant contribution in the corrosion resistance properties of the electrode TiO$_2$.[60,61] From Table 11.3 Ti/TiO$_2$ prepared by the thermal decomposition of TiCl$_3$ with HNO$_3$, phase angle value is 48, compared to the other Ti/TiO$_2$ electrode prepared by other methods which again brings the confirmation that the Ti/TiO$_2$ electrode's stability is far superior to the electrodes prepared by other methods. And this finding is further confirmed by the corrosion resistance (mmpy) calculated from Tafel polarization curves (Table 11.3). A comparative study of the yield of succinic acid obtained from the eletrore-duction of fumaric acid at Ti/TiO$_2$ electrodes is given in Figure 11.9.

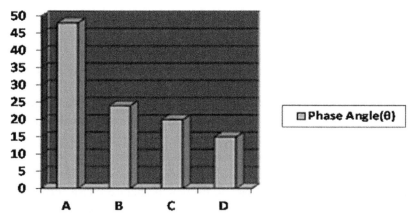

FIGURE 11.8   Comparison of phase angle obtained from AC-Impedance study of TiO$_2$ film on titanium prepared from (A) TiCl$_3$ & HNO$_3$, (B) TTIP and citric acid, (C) TiCl$_3$ & H$_2$O$_2$, and (D) anodization.

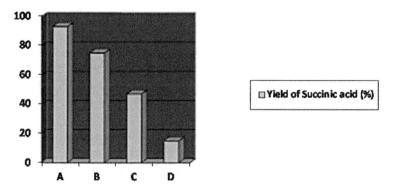

FIGURE 11.9   Comparison of yield of succinic acid from fumaric acid at Ti/TiO$_2$ electrode prepared from (A) TiCl$_3$and HNO$_3$, (B) TTIP and citric acid, (C) TiCl$_3$ & H$_2$O$_2$, and (D) anodization.

A decrease of the maximum absolute value of phase angle of the elec-trode is attributed to the decrease of the corrosion resistance of the metal.[62–66] And these findings were confirmed from the high yield of succinic acid obtained by the preparatory electrolysis carried out for reduction of fumaric acid, using the Ti/TiO$_2$ electrode prepared from the thermal decomposition of TiCl$_3$ and HNO$_3$, compared to the electrodes prepared by other methods. Comparative study of the yield of succinic acid obtained by the preparatory electrolysis is given in Table 11.3.

11.4 CONCLUSION

This research work has optimized the parameters in the preparation procedure which can control the phase formation of TiO$_2$ on titanium substrate in its anatase phase. Anatase phase has good electrocatalytic property compared to other two phases (rutile and brookite) since titania exhibits polymorphism (anatase, rutile, and brookite) and the phase formation is highly dependent on the preparation process. From SEM and X-ray diffraction studies, it is found that the phase formation of TiO$_2$ film on titanium substrate purely depend on the preparation variables, by which anatase phase formation is possible with high acidic pH and strong oxidizing agent like HNO$_3$ and also it depends on temperature factor, by increasing the temperature anatase phase changes to rutile.

This work presents the result of the corrosion resistance of the TiO$_2$ film carried out by Tafel polarization techniques and AC-impedance spectroscopic studies which depend on the particle size, phase, and porous nature of the film. From cyclic voltammetric studies, the Ti/TiO$_2$ electrodes prepared by the four different methods have quasi-reversible electrocatalytic cycle. Out of the above four methods of preparation of Ti/TiO$_2$ electrode, the electrode prepared by the thermal decomposition method from TiCl$_3$ and HNO$_3$ was found to be the best.

1. SEM, XRD studies show that the TiO$_2$ particle size is in nanoscale, the phase formation of TiO$_2$ on the titanium substrate is in anatase form which has better electrocatalytic activity than other forms (rutile and brookite).
2. It exhibits high corrosion resistance in the electrolyte 0.1 M H$_2$SO$_4$.
3. The results obtained by cyclic voltammetric studies clearly reveal that Ti/TiO$_2$ electrode prepared by the thermal decomposition method from TiCl$_3$ and HNO$_3$ is having quasi-reversible electrocatalytic cycle.
4. The findings were further confirmed by the high yield of succinic acid obtained by the electroreduction of fumaric acid using this electrode.

Further the method of preparation of the electrodes proves to be economical. The prepared electrode has got good stability and hence, it is used for the electroreduction of fumaric acid to succinic acid.

11.5   FUTURE RESEARCH SCOPE

There is an immense and extensive scope for future research in the field of heterogeneous catalytic reduction processes.

1. Improvement in the yield of succinic acid by the electroreduction of fumaric acid at Ti/TiO$_2$ electrodes prepared by the thermal decomposition of TTIP and citric acid, from TiCl$_3$ and H$_2$O$_2$ and by anodization method is the target for future research.
2. Ti/TiO$_2$ electrode prepared by the above given methods can be used for reduction of various other organic carbonyl compounds and nitro compounds. A comparative study can be made by using the Ti/TiO$_2$ electrodes prepared by the given four methods for the reduction of organic carbonyl and nitro compounds.
3. Photoelectrochemical measurement of Ti/TiO$_2$ electrodes can be studied by carrying out a set of linear sweep voltammogram measurements with and without UV illumination in 1.0 M H$_2$SO$_4$ and for a titanium substrate and based on the results obtained, these electrodes may be employed for the degradation of pollutants like phenol, bisphenol, etc.

KEYWORDS

- **Ti/TiO$_2$ electrode**
- **phase**
- **organic synthesis**
- **crystal structure**

REFERENCES

1. Hagfeldt, A.; Grätzel, M. *Chem. Rev.* **1995**, *95,* 49.

2. Hoffmann, M. R; Martin, S. T; Choi, W.; Bahnemann, D. W. *Chem. Rev.* **1995**, *95*, 69.

3. Linsebigler, A. L.; Lu, G.; Yates, J. T. *Chem. Rev.* **1995**, *95*, 735.

4. Chen, X.; Mao, S. S. *Chem. Rev.* **2 007**, *107*, 2891.

5. Matijevic, E. *Langmuir* **1986**, *2*, 12.

6. Sugimoto, T. *Adv. Colloid Interface Sci.* **1987**, *28*, 65.

7. Livage, J.; Henry, M.; Sanchez, C. *Prog. Solid State. Chem.* **1988**, *18*, 259.

8. Anderson, M. A.; Gieselmann. M. J.; Xu, Q. *J. Membr. Sci.* **1988**, *39*, 243.

9. Barringer, E. A.; Bowen, H. K. *Langmuir.* **1985**, *1*, 414.

10. Jean, J. H.; Ring, T. A. *Langmuir.* **1986**, *2*, 251.

11. Look, J. L.; Zukoski, C. F. *J. Am. Ceram. Soc.* **1992**, *75*, 1587.

12. Bokhimi, X.; Morales, A.; Novaro, O.; Lopez, T.; Sanchez, E.; Gomez, R. *J. Mater. Res.* **1995**, *10*, 2788.

13. Vorkapic, D.; Matsoukas, T. *J. Colloid Interface Sci.* **1999**, *214*, 283.

14. Penn, R. L.; Banfield, J. F. *Geochim. Cosmochim. Acta.* **1999**, *63*, 1549.

15. Barb´e, C. J.; Arendse, F.; Comte, P.; Jirousek, M.; Lenzmann, F.; Shklover, V.; Gr¨atzel, M. *J. Am. Ceram. Soc.* **1997**, *80*, 3157.

16. Zaban, A.; Aruna, S. T.; Tirosh, B. A.; Gregg, B. A.; Mastai, Y. *J. Phys. Chem. B.* **2000**, *104*, 4130.

17. Zhang, Z.; Wang, C. C.; Zakaria, R.; Ying, J. Y. *J. Phys. Chem. B.* **1998**, *102*, 10871.

18. Oskam, G.; Nellore, A.; Penn, R. L.; Searson, P. *J. Phys. Chem. B.* **2003**, *107*, 1734.

19. Zhang, H.; Banfield, J. *J. Mater. Chem.* **1998**, *8*, 2073.

20. Zhang, H.; Banfield, J. *J. Phys. Chem. B.* **2000**, *104*, 3481.

21. Navrotsky, A. *Geochem. Trans.* **2003**, *4*, 34.

22. Naicker, P. K.; Cummings, P. T.; Zhang, H.; Banfield, J. F. *J. Phys. Chem. B.* **2005**, *109*, 15243.

23. Cheng, H.; Ma, J.; Zhao, Z.; Qi, L. *Chem. Mater.* **1995**, *7*, 663.

24. Yanagisawa, K.; Yamamoto, Y.; Feng, Q.; Yamasaki, N. *J. Mater. Res.* **1998**, *13*, 825.

25. Yanagisawa, K.; Ovenstone, J. *J. Phys. Chem. B.* **1999**, *103*, 7781.

26. Aruna, S. T.; Tirosh, S.; Zaban, A. *J. Mater. Chem.* **2000**, *10*, 2388.

27. Li, J. G.; Ishigaki, T.; Sun, X. *J. Phys. Chem. C.* **2007**, *111*, 4969.

28. Yin, H.; Wada, Y.; Kitamura, T.; Kambe, S.; Murasawa, S.; Mori, H.; Sakata,T.; Yanagida, S. *J. Mater. Chem.* **2001**, *11*, 1694.

29. Pottier, A.; Chaneac, C.; Tronc, E.; Mazerolles, L.; Jolivet, J. P. *J. Mater. Chem.* **2001**, *11*, 1116.

30. Wu, M.; Lin, G.; Chen, D.; Wang, G.; He, D.; Feng, S.; Xu, R. *Chem. Mater.* **2002**, *14*, 1974.

31. Nagase, T.; Ebina, T.; Iwasaki, T.; Hayashi, H.; Onodera, Y.; Chatterjee, M. *Chem. Lett.* **1999**, *9*, 911.

32. Wang, C. C.; Ying, J. Y. *Chem. Mater.* **1999**, *11*, 3113.

33. Yin, H.; Wada,Y.; Kitamura,T.; Kambe, S.; Murasawa, S.; Mori,H .; Sakata, T.; Yanagida, S. *J. Mater. Chem.* **2001**, *11*, 1694.

34. Park, N. G.; Van de Lagemaat, J.; Frank, A. J. *J. Phys. Chem. B.* **2000**, *104*, 8989.

35. Wang, W.; Gu, B.; Liang, L.; Hamilton, W.; Wesolowski, D. *J. Phys. Chem. B.* **2004**, *108*, 14789.

36. Koelsch, M.; Cassaignon, S.; Thanh Minh, C. T.; Guillemoles, J. F.; Jolivet, J. P. *Thin Solid Films.* **2004**, *451*, 86.

37. Tomita, K.; Petrykin, V.; Kobayashi, M.; Shiro, M.; Yoshimura, M.; Kakihana, M. *Angew. Chem. Int. Ed.* **2006,** *45,* 2378.

38. Beck, F.; Gabriel, W.; *Angew. Chem. Int. Ed. Engl.* **1985,** *24,* 771.

39. Beck, F. *Electrochim. Acta.* **1989,** *34,* 81.

40. Ravichandran, C.; Vasudevan, D.;Anantharaman, P. N. *J. Appl. Electrochem.* **1992,** *22,* 1192.

41. Noel, M.; Anantharaman, P. N.; Udupa, H. V. *J. Appl. Electrochem.* **1982,** *12,* 291.

42. Ravichandran, C.; Kennady, C. J.; Chellamal, S.; Anantharaman, P. N. *J. Appl. Electrochem.* **1991,** *25,* 60.

43. Noel, M.; Ravichandran, C.; Anantharaman, P. N. *J. Appl. Electrochem.* **1995,** *25,* 690.

44. Ronconi, C. M.; Pereira, E. C. *J. Appl. Electro.Chem.* **2001,** *31,* 319–323.

45. Sharma, O. R.; De Pillai, C. P.; Giordano, M. C. *Electrochim. Acta.* **1984,** *29,* 111.

46. Torresi, R. M.; Camara, O. R.; Pauli, C. P.; Girordano, M. C. *Electrochim. Acta.* **1987,** *32,* 1291.

47. Delplancke, J. L.; Winand, R. *Electrochim.Acta.* **1988,** *33,* 1539.

48. Ranade, M. R.; Navrotsky, A.; Zhang, H. Z Banfield, J. F.; Elder, S. H.; Zaban, A.; Borse, P. H.; Kulkarni, S. K.; Doran, G. S.; Whitfield, H. J. PNAS 99. *Proc. Natl. Acad. Sci.* U S A **2002,** *99 (Suppl.* 2), 6476–81.

49. Zhang, H.; Banfield, J. F. *J. Chem. Mater.* **2002,** *14,* 4145.

50. Barnard, A. S.; Erdin, S.; Lin, Y.; Zapol, P.; Halley, J. W. Phys. *Rev. B.* **2006,** *73,* 205405.

51. Zhang, H. Z.; Banfield, J. F. *J. Mater. Chem.* **1998,** *8* (9), 2073.

52. Diebold, U.; The Surface Science of Titanium Dioxide. *Surf. Sci. Rep.* **2003,** *48,* 53.

53. Landolt, B. *Numerical Data and Functional Relationships in Science and Technology;* Springer: Berlin, 1984, Group III, 17G, p 413.

54. Ramamoorthy, M .; Vanderbilt, D.; King-Smith, R. D. First-Principles Calculations of the Energetics of Stoichiometric TiO2 Surfaces. *Phys. Rev. B.* **1994,** *49,* 16721.

55. Charlton, G.; Howes, P. B.; Nicklin, C. L.; Steadman, P.; Taylor, J. S. G.; Muryn, C. A.; Harte, S. P.; Mercer, J.; McGrath, R.; Norman, D.; Turner, T. S.; Thornton, G. Relaxation of TiO2 (110)-(1 x 1) Using Surface X-Ray Diffraction. *Phys. Rev. Lett.* **1997,** *78,* 495.

56. LaFemina, J. P. Total Energy Computations of Oxide Surface Reconstructions. *Crit. Rev. Surf. Chem.* **1994,** *3,* 297.

57. Maruska, H. P.; Ghosh, A. K. *Sol. Energy.* **1978,** *20,* 493.

58. Tang, H.; Prasad, K.; Sanjin'es, R.; Schmid, P. E.; L'evy, F. *J. Appl. Phys.* **1994,** *75,* 2042.

59. Bannerji, S.; Muraleedharan,; Tyagi,; Raj. Physics and Chemistry of Photocatalytic Titanium Dioxide, Visualization of Bactericidal Activity Using Atomic Force Microscopy. *Curr. Sci.* **2006,** *90* (10), 25.

60. Ping, Gu,; Yan, Fu,; Pig, Xie,; Deaudoin, J. J. *Characterestics of Surface Corrosion of Reinforcing Steel in Cement Paste by Low Frequency Impedance Spectroscopy. Cem. Concr. Res.* **1994,** *24* (2) 231.

61. Rozenfeld, I. C. *Corrosion Inhibitors*: Chapter 5; Mc Graw-Hill, International Book Company part 2: Pennsylvania Plaza, NY, 1981; p145.

62. Mansfeld, F.; Shih, H. *Detection of Pittig with Electrochemical Impedance Spectroscopy. J. Electrchem. Soc.* **1988,** *135* (4), 1171.

63. Shih, H.; Mansfeld, F. A Pittig Procedure for Impedance Spectroscopy Obtained for Cases of Localized. *Corrosion.* **1989,** *45* (8), 610.

64. Mannnsfeld, F.; Lin, S.; Kim, S.; Shih, H. Corrosion Protection of Al Alloys & Albased Metal Matrix Composites by Chemical Passivation. *Corrosion* **1989,** *45* (8), 615.
65. Macdonald, D. D.; Tantawry, Y. A. EI; Rocha, R. C. Filho; Urquidi Macdonald, M. *Evaluation of Electrochemical Impedance Techniques for Detecting Corrosion on Rebar in Reinforced Concrete, SHRP-10/UFR-91-524*; Strategic Highway Research Programme Report: National Research Council Washington, DC, 1994.

PART III
Biomedical Application of Nanomaterials

BIOSYNTHESIS AND CHARACTERIZATION OF SILVER NANOPARTICLES SYNTHESIZED FROM SEAWEEDS AND ITS ANTIBACTERIAL ACTIVITY

S. THANIGAIVEL, AMITAVA MUKHERJEE, NATARAJAN CHANDRASEKARAN, and JOHN THOMAS*

Centre for Nanobiotechnology, VIT University, Vellore, Tamil Nadu, India

Corresponding author. E-mail: john.thomas@vit.ac.in; th_john28@ yahoo.co.in

CONTENTS

ABSTRACT

This study aims to investigates the use of silver nanoparticles (AgNps) and its antibacterial activity against the disease causing baceterial pathogens, which are most predominantly found in the cultivation of aquaculture fish farming. The green synthesis of AgNps using *Caulerpa scalpelliformis* reported to have antibacterial activity agaisnt the Gram-negative bacterium, such as *Pseudomonas aeruginosa* infection which is identified as a fish pathogen in this study. The reduction of silver ions and its charecterization has been performed to ascertain the efficacy of AgNPs toward the fish pathogen. The formation of AgNPs was observed by the visualization of the color change further confirmed by the UV spectroscopy and scanning electron microscopic studies and the nanoparticles (NPs) size distribution of the AgNps was also studied by the dynamic light scattering (DLS) technique to detrmine the particle behavior and average particle size distribution at the suspension. Then, the in vivo pathogenicity was performed with *Catla catla* fish and treatment of AgNPs was demonstrated in both in vitro and in vivo.

12.1   INTRODUCTION

Bacterial fish disease is a major problem in aquaculture. It poses major challenges to the fish farmers. Fish diseases caused by *Pseudomonas aeruginosa*, a known pathogenic organism, is responsible for considerable economic losses in the commercial cultivation of many edible fishes. *Pseudomonas* infection has been identified as the most predominant bacterial infection among fish *Catla catla* and appears to be a stress-related disease of fresh water fish especially under culture conditions.[1] These microorganisms are responsible for ulcerative diseases including ulcerative syndrome and bacteria hemorrhagic septicemia. Nanoparticles (NPs) play a significant role in the drug delivery systems.[2] Nanocrystalline silver particles have found tremendous applications in the field of high sensitivity biomolecular detection and diagnostics, antimicrobials, and therapeutics.[3] The use of environmentally benign materials like plant leaf extracts[4–7] for the synthesis of silver nanoparticles (AgNps) offers numerous benefits and eco-friendliness and compatibility for biopharmaceutical application and other biomedical applications as they do not cause toxic effects.[8] Green synthesis provides advantages over chemical and physical methods as it is cost effective, environment friendly, and can easily be scaled up for large-scale synthesis. This novel approach of biologically synthesized NPs was formulated and its size was optimized through the dynamic light scattering (DLS) and scanning electron microscope (SEM). This was challenged against the disease causing pathogen. This was given through the intramuscular, immersion, and oral routes. The antibacterial activity of the green synthesized NPs has good antibacterial activity as reported by many studies.[9,10]

The use of plants for biosynthesis is still largely an unexplored area and hence needs to be utilized effectively.[11] It includes various advantages compared to chemical means of synthesis with respect to cost effectiveness, large-scale synthesis, and does not need high temperatures or energy.[12] To date, there are a lot of reports on green synthesis of AgNps which have the above-mentioned advantages. There are reports which state that AgNps are nontoxic within the precisely quantified limit, and are safe inorganic antibacterial agents which have been used for centuries since they are capable of killing about 650 type of diseases caused by the pathogenic microorganisms.[13,14] Silver has been described as being "oligodynamic" in nature because of its ability to exert a bactericidal effect at minute concentrations.[15] AgNps show antifungal and antibacterial activity against the resistant bacteria, thus preventing infections, inaddition to promoting wound healing, and acting as anti-inflammatory.[16] This work aims to improve the aquaculture production by supporting aquatic fish health and its disease resistance toward the excessive use of antibiotics. This kind of NPs-mediated process plays an important role against a commercial antibiotics and these can be used as an alternative source to control bacterial infection and mass mortalities in fish farms.

12.2 MATERIALS AND METHODS

12.2.1 COLLECTION AND MAINTENANCE OF EXPERIMENTAL ANIMALS

Healthy fingerlings of *C. catla* with an average body weight of 12–15 g were collected from a farm located in Walajapet, Vellore district, Tamil Nadu, India, with no record of disease symptoms. They were transported in live condition in 20 L containers containing freshwater with continuous battery powered aeration. They were maintained under lab condition in 1000 L fiberglass tanks with continuous aeration in fresh water.

12.2.2   COLLECTION AND PREPARATION OF PLANT EXTRACTS

Seaweed used in the study, *Caulerpa scalpelliformis*, was collected from coastal regions of Mandapam, Rameswaram, Tamil Nadu, India. The plants were collected from the natural environment. They were washed with treated water, and shade dried. Then the aqueous extract of this seaweed was prepared separately by crushing and soaking them (5 g) in double distilled water (100 mL) for 48 h. They were then filtered and the filtrate was used for the synthesis of NPs.

12.2.3   BACTERIAL CULTURE

P. aeruginosa strain was purchased from Microbial Type Culture Collection (MTCC) and confirmed by the biochemical reaction scheme provided by Bergy's Manual of Systemic Bacteriology. The bacterial culture was processed using standard protocol and then serially diluted. The pathogenicity of *P. aeruginosa* in *C. catla* was tested by two different methods namely intramuscular and immersion. Bacterial culture (15 μL) was injected through intramuscular route. In the case of immersion 1 mL of bacterial culture was added to 1 L of water. Histopathological analysis was carried out and compared with that of the control organs (gills, liver, and skin).

12.2.4   GREEN SYNTHESIS OF SILVER NANOPARTICLES (AgNPs)

This was done by preparing 1 mM silver nitrate which was used for the synthesis of AgNps. *C. scalpelliformis* aqueous extract (10 mL) was added to 90 mL concentration of 1 mM silver nitrate for reduction into Ag^+ ions and kept at room temperature for 5 h.

12.2.5   CHARACTERIZATION OF AgNPs

12.2.5.1   UV–VIS SPECTROSCOPIC STUDY

UV spectrophotometer was employed to detect the presence of silver ions and its absorbance at 420 nm was measured. The formation of Ag^+ ions was monitored by measuring the UV–Vis spectrum of reaction mixture for the absorbance in the ion medium at 5 h after diluting a small aliquot of the sample in distilled water. The analysis was done by UV–Vis spectrophotometer UV-2450 (Shimadzu). The formation of AgNps was analyzed by the reduction reaction between the metal ions and the leaf extract by UV spectrophotometer from 200 to 700 nm.

12.2.5.2 DYNAMIC LIGHT SCATTERING MEASUREMENTS (DLS)

DLS technique was employed for the measurement of particles present in the colloidal suspension. The size of the particles was measured on Brookhaven Instrument model 90 Plus Particle Size Analyzer. This was used to determine the particle size distribution and average size of the NPs in the suspension.

12.2.5.3 SCANNING ELECTRON MICROSCOPE (SEM)

SEM analysis was employed (Hitachi S-4500) to ascertain the exact size of the NP. Thin glass coated NPs were prepared. This was done by placing a very small amount of the sample on the grid. Extra solution was removed using a blotting paper and then the film on the SEM grid was allowed to dry by placing it at room temperature for 5–10 min.

12.2.6 ANTIBACTERIAL ACTIVITY

12.2.6.1 WELL DIFFUSION METHOD

Disc diffusion was done by swabbing the bacterial culture on agar plates. The plates were kept under sterile condition for 10 min. Two wells of 8 mm diameter were made on the agar surface. Extract (30 µL) was added into one of the wells. Into the other well, miliq water was added as a control. The samples were allowed to diffuse. The plates were monitored after 48 h for antibacterial activity and the zone of inhibition was measured at the end of 48 h.[17]

12.2.6.2 ANTIBIOTIC SENSITIVITY TEST

The antibacterial activity was studied by standard disc diffusion method. Fresh overnight cultures of bacterial inoculum (100 µL) were spread on to nutrient agar plates. Sterile paper discs of 6 mm diameter (containing

50 mg/L AgNps) along with standard antibiotic were placed in each plate. The antibiotic test was carried out following the method of Bauer Kir by 1966. The antibiotic disc used in the study was ciprofloxacin. Then antimicrobial suceptiblity discs were placed on the Muller Hinton agar surface containing the bacterial lawn culture, on the plate. The comparative study was performed by placing the AgNPs loaded disc for comparison with the standard antibiotic.

12.3   EXPERIMENTAL INFECTION VIA INTRAMUSCULAR INJECTION

The method of Thomas et al.[1] was followed. Fishes (10 per dosage and tank) were maintained in fiberglass tanks (100 L) containing freshwater at 28°C. The isolated bacterial culture was administered to healthy fishes by intramuscular injection, at the concentration of 10^1 10^2 10^3, 10^4, 10^5, and 10^6 CFU/animal^{-1}. Control fish received only sterile *phosphate-buffered saline* (PBS) buffer. For healthy fish, 25 µL of bacterial suspension from each of these five concentrations was injected intramuscularly using 1 mL insulin syringes. The experiment was conducted in triplicates in each bacterial concentration. Animals were checked twice a day for symptoms and mortality.

12.3.1   TOXICITY PROFILE AND OPTIMUM EXPOSURE OF NANOPARTICLES

Experimental infection was induced to the fishes with the causative bacterium by intramuscular and immersion method. This was followed by the slightly modified method of Thomas et al. and Thanigaivel et al.[1,17,18] The infected fishes with clinical signs were identified and reared separately in sterile freshwater treated with a concentration of 250 mg/L of each AgNPs separately. The fishes were collected at 24, 48, and 72 h of exposure and reared further for 30 days in aquarium tanks containing freshwater. The fishes surviving at the end of 20 and 30 days with the appropriate concentration were counted to ascertain the optimum exposure time to AgNPs and the concentrations of NPs required for fishes to recover from infection caused by the bacteria. In the case of intramuscular injection, 25 µL of NPs was injected intramuscularly, and the fishes challenged with bacteria after third dose.[1]

12.3.2   TREATMENT OF BACTERIAL INFECTION

Treatment of bacterial infection in fishes was performed using standard protocol. AgNP (10–50 µL) was injected by intramuscular route. AgNP (1 mL) added to 1 L of water in the case of immersion method.

12.3.3   HISTOPATHOLOGY

Histopathological analysis of gills, skin, and liver were studied by the standard protocol.[17,18]

12.4   RESULTS AND DISCUSSION

The seaweed extract used in this study was capable of synthesizing NPs, which were used as a reductant in the process of reducing the Ag$^+$ ions to form AgNps. UV–Vis spectroscopy analysis was performed at room temperature with respect to the color changes which apperared during the green synthesis. The UV spectrophotometer at a resolution of 1 nm was used for the study. It was observed that the resonance band occurs at 412 nm and steadily increases without any shift in the peak wavelength. Formation of stable AgNps using the seaweed extract of *C. scalpelliformis* in its aqueous form confirmed the presence of NPs using UV–Vis spectral analysis. The characteristic surface plasmon absorption band observed at 420 nm for AgNps synthesized from 1 mM AgNO3 was studied for the preliminary confirmation of the presence of AgNPs in the colloidal suspesion. AgNps exhibited yellowish brown color in the aqueous solution due to the excitation of surface plasmon vibrations in AgNPs.[29] In this work, the extract was mixed in the aqueous solution of the silver ion complex under dark condition. The mixture changed its color from light brown to yellowish brown due to reduction of silver ion which indicated formation of AgNps and UV–Vis spectroscopy analysis helped in the examination of size and shape of controlled NPs in aqueous suspensions.[19] Figure 12.1 shows the UV–Vis spectra recorded from the reaction medium after 5 h. The absorption spectra of AgNps formed in the reaction medium were found to have an absorbance peak at 450 nm, broadening of peak indicated that the particles were poly dispersed.

DLS was also performed to ascertain the size of NPs in the colloidal suspension, which was found to be 83.5 nm as shown in Figure 12.2. It confirmed that the particles seen in the mixture had a nanometer range of particle size distribution. This result agrees with the work on lime oil nano-emulsion and its size characterization in the treatment of *Pseudomonas* infection in tilapia reported by Thomas et al.[1]

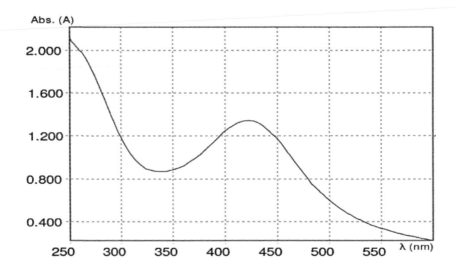

FIGURE 12.1 UV–Vis absorption spectrum of silver nanoparticles synthesized from *Caulerpa scalpelliformis*.

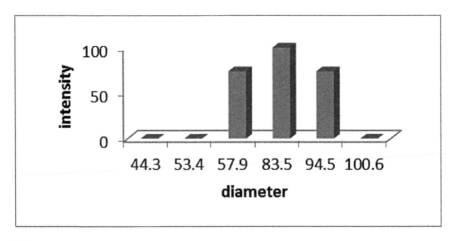

FIGURE 12.2 Particle size distribution by DLS.

The SEM analysis showed the presence of high density AgNps synthesized by the *C. scalpelliformis* extract which further confirmed the development of silver nanostructures (Fig. 12.3).

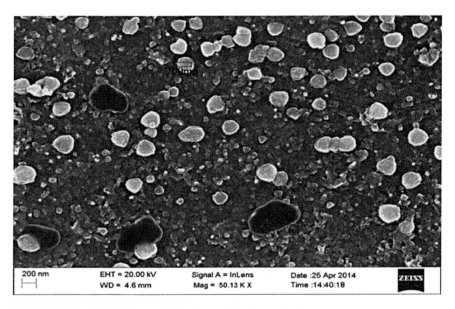

FIGURE 12.3 SEM micrograph for the particle size analysis.

The size of the NPs produced was measured. It confirmed the formation of spherical shaped AgNps with size ranging from 50 to 200 nm which was confirmed by SEM. AgNps synthesized within 10 min had an absorbance at 420 nm and the broadening of the peak indicates the poly dispersion of the particle. The SEM showed that spherical shape NPs with a diameter range 50–100 nm were formed. Our results agree with the report of Panneerselvam et al.[20] The high density AgNps synthesized by the *Andrographis paniculata* shows that SEM image had average size ranging from 35 to 55 nm with inter-particle diameter with sperical shaped NPs. The aggregation of the NPs indicates that they were in the direct contact, but were stabilized by a capping agent.[20] The AgNps synthesized from *Pseudomonas* sp. show the sharpening of peaks. It revealed that the average size of the AgNPs ranged from 20 to 100 nm. The shape of the AgNps was spherical and they aggregated into larger irregular structures with no well-defined morphology.[21] Bharathi et al.[22] analyzed green synthesis of AgNps from *Cleome viscosa* by SEM. They reported that AgNps were uniformly distributed on the surface of the cells. However, it does not indicate that all the NPs are bound to the surface of the cells. This may be because the particles dispersed in the solution may also be deposited onto the surface of the cells.

The susceptibility of *C. catla* to *P. aeruginosa* was tested by intramuscular injection. The highest concentration, 35×10^6 and 35×10^7 viable cells of *P. aeruginosa* per animal caused 80% mortality within 120 and 96 h of post-inoculation, respectively, when animals were injected intramuscularly, whereas the lower concentration of 35×10^3 and 35×10^4 viable cells of *P. aeruginosa* caused 35 and 40% mortality, respectively, after 120 and 96 h of post-injection. The LD_{50} value of *P. aeruginosa* for intramuscular route was determined at different time intervals. It was found to be 1.38×10^5 and 1.41×10^6 CFU per animal after 72 and 96 h of post-infection, respectively. Similar results were reported by Thanigaivel et al.[17,18] The concentration of *Citrobacter freundii* (35×10^6 CFUmL^{-1} of rearing medium) caused 10, 20, 30, 35.2, and 40% mortality through immersion route after 24, 48, 72, 96, and 120 of exposure, respectively. The LC_{50} value of *C. freundii* for immersion route was determined at different time intervals. It was found to be 1.4×10^5, 2.6×10^5, and 3.4×10^6 CFUmL^{-1} of rearing medium, after 24, 48, and 72 h of post-infection, respectively. The symptoms revealed the action of bacterial pathogen on the healthy fishes. Thanigaivel et al.[17,18] reported the pathogenicity of *P. aeruginosa* in *C. catla* which showed ulceration and exophthalmia in experimentally infected fishes.

The antibacterial activity (Fig. 12.4) of AgNps was estimated by the zone of inhibition. In our study, the action of AgNPs against the test organism *P. aeruginosa* showed 17 mm of zone of inhibition along with the standard antibiotic gentamycin. There are many reports of plant extracts which have antibactericidal properties agaisnt various pathogens in humans and animals. This can be effectively used in the synthesis of AgNps as a greener route. The NPs synthesized from plants have many applications for benefit of humans. However, the mechanisim of NP synthesis in plants is quite complex to understand. Only very little information is available about their synthesis, with regard to the reducing agent, proteins, and phenolic precipitation. The most promising area of research is the elucidation of the mechanism of plant-mediated synthesis of AgNps. Although the use of AgNps has some negative impact it can penetrate deep into the Gram-negative bacteria. Other clinical studies revealed the effectiveness of the AgNps. They had comparatively higher antibacterial activity against Gram-negative bacteria than against Gram-positive bacteria, which may be due to the thinner peptidoglycan layer and presence of beta-barrel proteins called porins.[23] Recently, NPs have gained greater significance in the field of biomedicine. The most significant and distinguishing property of NPs was found to be its larger surface area to volume ratio. Surface area corresponds to the various properties, such as the catalytic reactivity and antimicrobial activity. When surface area of the NPs increases, their surface energy will increase as a result of which their biological effectiveness will also increase.[24] Another study revealed that the smaller sized NPs with a larger surface area to volume ratio provide a more effective antibacterial activity even at a very lower concentration. AgNps of many different shapes (spherical, rod-shaped, truncated, and triangular nanoplates) were developed by various synthetic routes. Truncated triangular shaped silver nanoplates were found to show the strongest antibacterial activity. This property could be due to their larger surface area to volume ratios and their crystallographic surface structures. Nanosilver was considered to be a very effective in antibacterial agent against the fish pathogen, *P. aeruginosa*.[25] The well-diffusion method was used to assay the antibacterial activity against certain human and animal pathogens.[26] The in vitro activity of the prepared AgNps was analyzed by disc diffusion and antibiotic sensitivity tests. The efficacy of AgNPs against the test organism was proved by clear zone of inhibition. Similar results were also

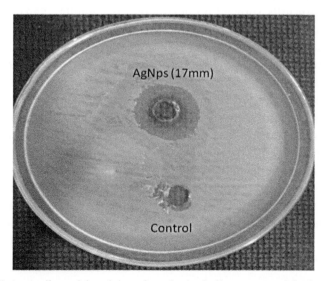

FIGURE 12.4 Antibacterial activity of synthesized silver nanoparticle from *Caulerpa scalpelliformis*.

demonstrated.[17,18,23] This obtained results agree with Shirley et al.[27]. Such clothes are sterile and are used in hospitals to prevent or to minimize the infection with pathogenic bacteria like *Staphylococcus aureus*. The average zones of inhibition expressing a profound inhibitory effect were represented as 35 mm in *P. aerogenosa*, 30 mm in *Klebsiella pneumonia*, 36 mm in *S. aureus*, 40 mm in *Salmonella typhi*, 38 mm in *Salmonella epidermis*, and 34 mm in *E. coli*.[27,28] This type of biologically synthesized NPs has a wide range of applications like biological detections, controlled drug delivery, optical filters, sensor design, etc. With respect to diagnosis, AgNps interact with HIV-1 virus via preferential binding to the gp 120 glycoprotein knobs.[28] Our results revealed that silver NPs can be used for treating bacterial infection in fish and that it was nontoxic at a concentration of 20 ppm. Pathogenicity tests revealed that *P. aeruginosa* showed clinical symptoms namely fin lesion, detachment of scales, exophthalmia, and septicemia both by intramuscular and immersion routes. Histopathological analysis was performed to determine the damages caused to the internal organs which occurred due to the bacterial infection. The results revealed that gills, liver, and skin were damaged. The fish treated with AgNps showed reversal of damage in the internal organs. AgNps exhibited good antibacterial activity against *P. aeruginosa* in treating bacterial infection in fishes.

Histopathological analysis of this study investigated the damage caused to the organs during the pathogenicity and its reversal after treatment. Control gills showed no destruction of lamella. Negative control group which received bacterial suspension showed destruction in the primary and secondary lamella (Fig. 12.5a). In the fish treated with NPs primary and secondary lamellae were seen (Fig. 12.5b). The normal skin of fish showed the presence of basal layers, stratum compactum, and mucus glands. In experimentally infected fish mucus cells and mucus glands were absent showed in Figure 12.5c. In the fish treated with NPs mucus cells and mucus glands were seen (Fig. 12.5d). The liver of control fish showed typical paranchymatous appearance with glycogen vacuole, kuppfer cells, and central and round nucleus, whereas in experimentally infected fish, irregular vacuolations were observed and kuppfer cells were absent (Fig. 12.5e). In the fish treated with the NPs glycogen vacuole was seen (Fig. 12.5f). Similar histopathological changes were reported in catfish infected with *Aeromonas* and tilapia infected with *P. aeruginosa*.[1,17,18]

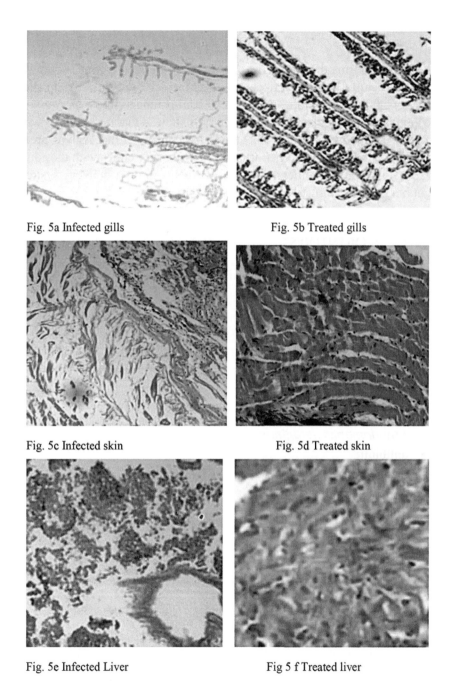

Fig. 5a Infected gills   Fig. 5b Treated gills

Fig. 5c Infected skin   Fig. 5d Treated skin

Fig. 5e Infected Liver   Fig 5 f Treated liver

FIGURE 12.5   (a) Infected gills; (b) treated gills; (c) infected skin; (d) treated skin; (e) infected liver; (f) treated liver

12.5   CONCLUSION

This study confirms that silver nanopartilcle synthesized from *C. scalpelli-formis* can be used as good antibacterial agent which can be used as an alternative to synthetic antibiotics to treat bacterial pathogen like *P. aeruginosa* infection in *C. catla*. Characterization of NPs showed that the particles were of standard nanodimensions. This study also confirmed that AgNps are as efficient as antibiotics and other drugs used in pharmaceutics with the added advantage of eco friendliness. The use of AgNps in drug delivery systems might be the future thrust in the field of nanomedicine.

CONFLICTS OF INTEREST

Authors do not have any known conflicts of interest and copyrights violations regarding this publications.

KEYWORDS

- silver nanoparticles
- baceterial pathogens
- *Pseudomonas aeruginosa*
- antibiotic
- fish pathogen

REFERENCES

1. Thomas, J.; Thanigaivel, S.; Vijayakumar, S.; Kuntal, A.; Dhairyasheel, S.; Seelan, T. S.; Amitava, M.; Natarajan, C. Pathogenecity of *Pseudomonas aeruginosa* in *Oreochromis mossambicus* and Treatment Using Lime Oil Nanoemulsion. *Colloids Surf B Biointerfaces* **2014,** *116,* 372–377.
2. Park, E. J.; Lee, S. W.; Bang, I. C.; Park, H. W. Optimal Synthesis and Characterization of Ag Nanofluids by Electrical Explosion of Wires in Liquids. *Nanoscale Res. Lett.* **2011,** *6* (1), 1–10.
3. Schultz, S.; Smith, D. R.; Mock, J. J.; Schultz, D. A. Single-Target Molecule Detection with Nonbleaching Multicolor Optical Immunolabels. *Proc. Natl. Acad. Sci.* **2000,** *97* (3), 996–1001.

4. Bhainsa, K. C.; D'souza, S. F. Extracellular Biosynthesis of Silver Nanoparticles Using the Fungus *Aspergillus fumigatus*. *Colloids Surf B Biointerfaces*. **2006,** *47* (2), 160–164.

5. Parashar, V.; Parashar, R.; Sharma, B.; Pandey, A. C. Parthenium Leaf Extract Mediated Synthesis of Silver Nanoparticles: A Novel Approach towards Weed Utilization. *Dig. J. Nanomater. Biostruct.* **2009,** *4* (1), 45–50.

6. Saifuddin, N.; Wong, C. W.; Yasumira, A. A. Rapid iosyBnthesis of Silver Nanoparticles Using Culture Supernatant of Bacteria with Microwave Irradiation. *J. Chem.* **2009,** *6* (1), 61–70.

7. Willner, I.; Basnar, B.; Willner, B. Nanoparticle–Enzyme Hybrid Systems for Nanobiotechnology. *FEBS J.* **2007,** *274* (2), 302–309.

8. Bhati-Kushwaha, H.; Malik, C. P. Biopotential of Verbesina Encelioides (Stem and Leaf Powders) in Silver Nanoparticle Fabrication. *Turk. J. Biol.* **2013,** *37* (6), 645–654.

9. Geethalakshmi, R.; Sarada, D. V. L. Synthesis of Plant-Mediated Silver Nanoparticles Using *Trianthema decandra* Extract and Evaluation of Their Anti Microbial Activities. *Int. J. Eng. Sci. Technol.* **2010,** *2* (5), 970–975.

10. Jacob, J. A.; Murugaiyan, N.; Durairaj, S.; Ganesapandy, P.; Palanivel, K.; Seenivasan, K.; Arunachalam, S.; Athmanathan, B.; Vivekanandan, M.; Shanmugam, A. Antimicrobial Activity of *Leucas aspera* Engineered Silver Nanoparticles against *Aeromonas hydrophila* in Infected *Catla catla*. *Colloids Surf B Biointerfaces*. **2013,** *109,* 20–24.

11. Thakkar, K. N.; Mhatre, S. S.; Parikh, R. Y. Biological Synthesis of Metallic Nanoparticles. *Nanomedicine.* **2010,** *6* (2), 257–262.

12. Malabadi, R. B.; Mulgund, G. S.; Meti, N. T.; Nataraja, K.; Vijayakumar, S. Antibacterial Activity of Silver Nanoparticles Synthesized by Using Whole Plant Extracts of *Clitoria ternatea*. *Res. Pharm.* **2012,** *2* (4), 10–21.

13. Jeong, S. H.; Yeo, S. Y.; Yi, S. C. The Effect of Filler Particle Size on the Antibacterial Properties of Compounded Polymer/Silver Fibers. *J. Mater. Sci.* **2005,** *40* (20), 5407–5411.

14. Raffi, M.; Hussain, F.; Bhatti, T. M.; Akhter, J. I.; Hameed, A.; Hasan, M. M. Antibacterial Characterization of Silver Nanoparticles against *E. coli* ATCC-15224. *J. Mater. Sci. Technol.* **2008,** *24* (2), 192–196.

15. Percival, S. L.; Bowler, P. G.; Russell, D. Bacterial Resistance to Silver in Wound Care. *J. Hosp. Infect.* **2005,** *60* (1), 1–7.

16. Taylor, P. L.; Ussher, A. L.; Burrell, R. E. Impact of Heat on Nanocrystalline Silver Dressings: Part I: Chemical and Biological Properties. *Biomaterials* **2005,** *26* (35), 7221–7229.

17. Thanigaivel, S.; Vijayakumar, S.; Amitava, M.; Natarajan, C.; John, T. Antioxidant and Antibacterial Activity of *Chaetomorpha antennina* against Shrimp Pathogen *Vibrio parahaemolyticus*. *Aquaculture* **2014,** *433,* 467–475.

18. Thanigaivel, S.; Vijayakumar, S.; Gopinath, S.; Amitava, M.; Natarajan, C.; John, T. Invivo and Invitro Antimicrobial Activity of *Azadirachta indica* (Lin) against *Citrobacter freundii* Isolated from Naturally Infected Tilapia Fish. *Aquaculture*. **2014,** *437,* 252–255.

19. Wiley, B. J.; Im, S. H.; Li, Z. Y.; McLellan, J.; Siekkinen, A.; Xia, Y. Maneuvering the Surface Plasmon Resonance of Silver Nanostructures through Shape-Controlled Synthesis. *J. Phys. Chem B.* **2006,** *110* (32), 15666–15675.

20. Panneerselvam, C.; Ponarulselvam, S.; Murugan, K. Potential Anti-Plasmodial Activity of Synthesized Silver Nanoparticles Using *Andrographis paniculata* Nees (Acanthaceae). *J. Ecobiotechnol.* **2011,** *3,* 24–28.

21. Muthukannan, R.; Karuppiah, B. Rapid Synthesis and Characterization of Silver Nanoparticles by Novel *Pseudomonas* sp. "ram bt-1." *J. Ecobiotechnol.* **2011,** *3,* 24–28.

22. Bharathi, R. S.; Suriya, J.; Sekar, V.; Rajasekaran, R. Biomimetic of Silver Nanoparticles by *Ulva lactuca* Seaweed and Evaluation of Its Antibacterial Activity. *Int. J. Pharm. Pharm. Sci.* **2012,** *4* (3), 139–143.

23. Geoprincy, G.; Saravanan, P.; Nagendra, G. N.; Renganathan, S. A Novel Approach of Studying the Combined Antimicrobial Effects of Silver Nanoparticles and Antibiotics through Agar Over Layer Method and Dicc Diffusion Method. *Dig. J. Nanomater. Biostruct.* **2011,** *6* (4), 1557–1565.

24. Srivastava, A. A.; Kulkarni, A. P.; Harpale, P. M.; Zunjarrao, R. S. Plant Mediated Synthesis of Silver Nanoparticles Using a Bryophyte: Fissidens Minutus and Its Antimicrobial Activity. *Int. J. Eng. Sci. Technol.* **2011,** *3* (12), 8342–8347.

25. Ramya, M.; Sylvia, S. M. Green Synthesis of Silver Nanoparticles. *Int. J. Pharm. Med. Biol. Sci.* **2012,** *1* (1), 54–61.

26. Aditi, P. K.; Ankita, A. S.; PSravin, M. H.; Rajendra, S. Z. Plant Mediated Synthesis of Silver Nnanoparticles-Tapping the Unexploited Sources. *J. Nat. Prod. Plant Resour.* **2011,** *1* (4), 100–107.

27. Shirley, A. D.; Sreedhar, B.; Syed, G. D. Antimicrobial Activity of Silver Nanopaarticles Synthesized from Novel Streptomyces Species. *Dig. J. Nanomater. Biostruct.* **2010,** *5* (2), 447–451.

28. Karthick, R. N. S.; Ganesh, S.; Avimanyu. Evaluation of Anti Bacterial Activity of Silver Nanoparticles Synthesized from *Candida glabrata* and *Fusarium oxysporum*. *Int. J. Med. Res.* **2011;** *1* (3), 131–136.

29. Kasture, M. B.; Patel, P.; Prabhune, A. A.; Ramana, C. V.; Kulkarni, A. A.; Prasad, B. L. V. Synthesis of Silver Nanoparticles by Sophorolipids: Effect of Temperature and Sophorolipid Structure on the Size of Particles. *J. Chem. Sci.* **2008,** *120* (6), 515–520.

CHAPTER 13

ASSESSMENT OF ANTIBACTERIAL AND ANTIFUNGAL ACTIVITY OF ZERO VALENT IRON NANOPARTICLES

CHANDRAPRABHA M. N*, SAMRAT K., and SHRADDHA SHAH

Department of Biotechnology, M. S. Ramaiah Institute of Technology, Bangalore 560054, Karnataka, India

*Corresponding author. E-mail: chandraprabhamn@yahoo.com

CONTENTS

ABSTRACT

Antimicrobial activity was observed for zero valent iron nanoparticles (nZVI) against Gram-negative (*Escherichia coli*, *Klebsiella pneumonia*, and *Pseudomonas aeruginosa*) bacteria and fungal (*Candida albicans* and *Trichophyton rubrum*) strains. The nZVI showed significant anti-fungal and anti-bacterial activity against tested strains. These results suggest the possibility of using nZVI for effective development of antimicrobial drugs with high efficacy toward the pathogenic microorganisms.

13.1 INTRODUCTION

Resistance of the bacteria to current antibiotics is a serious clinical problem, so it is important to develop new antimicrobials from various sources as alternative agents. The advancement in the field of nanotechnology has provided various nanomaterials as alternative antibacterial agents.[1-3] In recent years, synthesis of nanometer-size metallic nanoparticles has become an important area of research and is attracting a variety of scientific and technological applications, such as biosensor, antimicrobial activity, food preservation, magnetic resonance imaging, and targeted drug delivery. It is found that metal-based nanoparticles due to their biological and physico-chemical properties are used as antimicrobials and therapeutic agents.[4] The main mechanism by which these particles showed antibacterial activity might be via oxidative stress generated by reactive oxygen species (ROS).[4] ROS, including superoxide radicals (O^{2-}), hydroxyl radicals ($-OH$), hydrogen peroxide (H_2O_2), and singlet oxygen (O_2), can cause damage to proteins and DNA in bacteria. A similar process was also described by Kim et al.[5] in which Fe^{2+} reacted with oxygen to create H_2O_2.

There are various methods of synthesizing nanoparticles, such as chemical reduction method, combustion method, etc. The synthesis of nanoparticles has become a particularly important area of research and is attracting a growing interest because of the potential applications such advanced magnetic materials, catalysts, high-density magnetic recording media, and medical diagnostics. Zero valent iron nanoparticles (nZVI) and their dispersions in various media have long been of scientific and technological interest.[6]

In this study, nZVI are synthesized by chemical reduction method and it was found that, these nanoparticles were uniformly dispersed in dimethyl sulfoxide (DMSO). nZVI were tested against bacteria and fungi to check its antibacterial and antifungal activity. Absorption spectrophotometer

(UV–Vis), X-ray diffraction (XRD), Fourier transform infrared spectroscopy (FTIR), and scanning electron microscope (SEM) were used for character-ization of nZVI. Antibacterial and antifungal activities are evaluated against human pathogenic Gram-negative (*Escherichia coli*, *Klebsiella pneumonia*, and *Pseudomonas aeruginosa*) bacteria and fungal (*Candida albicans* and *Trichophyton rubrum*) strains.

nZVI synthesized by chemical reduction was validated by UV–Vis spec-troscopic analysis. The characteristics peaks of nZVI were observed at 295.0 nm. The XRD analysis of nZVI shows the peak at 2θ of 44.2° indicating the presence of nZVI.[7] FTIR techniques provide information about vibrational state of adsorbed molecule and hence the nature of surface complexes. The band at 3362.70 cm^{-1} indicates the OH stretching vibration.[7]

The spherical shape of nZVI was observed in the analysis of SEM image. The size of most of the nanoparticles ranges from 71.00 to 109.5 nm and mean particle size of 90.25 nm.

13.2   REVIEW OF LITERATURE

13.2.1   SYNTHESIS OF ZERO VALENT IRON NANOPARTICLES

nZVI were synthesized by the reduction of ferric chloride (FeCl$_3$) with sodium borohydride (NaBH$_4$) using previously described method. In this, 1:1 volume ratio of NaBH$_4$ (0.2 M) and FeCl$_3$ solution (0.05 M) was vigor-ously mixed in a flask reactor for 30 min. Excessive borohydride (0.2 M) was applied to accelerate the synthesis. A black precipitate of nZVI was obtained soon after the addition of NaBH$_4$ which was later separated and stored in ethanol. The reaction was carried out in an inert atmosphere. For the preparation of fully dispersed and stabilized nZVI, 0.8% carboxymethyl cellulose (CMC) was added to nZVI and sonicated for homogenization.[7]

$$4Fe^{3+} + 3BH_4^- + 9H_2O \rightarrow 4Fe^0 + 3H_2BO_3^- + 12H^+ + 6H_2 \qquad (13.1)$$

Synthesis of nZVI involves a reduction method using two main chemicals which were FeSO$_4$.7H$_2$O and NaBH$_4$. The NaBH$_4$ functions as a reducing agent in order to reduce the iron sulfate (FeSO$_4$.7H$_2$O) in form of solution to produce zero valent iron. The method comprised of four stages which were mixing, separating, washing, and drying. These two chemicals were mixed by dropping 0.75 M of NaBH$_4$ solution into 100 mL of 0.1 M FeSO$_4$.7H$_2$O continuously while stirring the reaction mixture well. Excess NaBH$_4$ was typically applied in order to accelerate the reaction and ensured uniform growth of iron particles. The resulted reaction occurs as;

$$Fe^{2+} + 2BH_4^- + 6H_2O \rightarrow Fe^0 + 2B(OH)_3 + 7H_2\uparrow \qquad (13.2)$$

Immediately after the first drop of reducing agent into iron solution, black particles were formed. Then further mixing for 15–20 min resulted in maximum yield of black iron particles. Black iron particles were then separated from the solution by vacuum filtration using Whatmann cellulose nitrate membrane filter (0.45 μm).[8]

The reverse micelle reaction is carried out using cetyl trimethyl ammonium bromide (CTAB) as the surfactant, octane as the oil phase, and aqueous reactants as the water phase. Varying the water to surfactant ratio () can form micelles ranging in size from 5 to 30 m thus leading to careful control over the particle size. Use of n-butanol a co-surfactant helped to decrease and neutralize the fraction of the micellar head group thereby increasing micellar stability. Without the addition of the co-surfactant, the amount of free water available to carry on the reactions is greatly reduced, as most of the water is locked in the head group of the CTAB.

The metal particles are formed inside the reverse micelle by the reduction of a metal salt using $NaBH_4$. The sequential synthesis offered by reverse micelles is utilized to first prepare an iron core by the reduction of ferrous sulfate by $NaBH_4$. After the reaction has been allowed to go to completion, the micelles within their action mixture are expanded to accommodate the shell using a larger micelle containing additional $NaBH_4$. The shell is formed using an aqueous hydrogen tetra chloroaurate solution. The particles are then washed, collected in a magnetic field, and dried under vacuum.[9]

nZVI used for the study were synthesized in the laboratory by the reduction of $FeSO_4.7H_2O$ with $NaBH_4$. The iron nanoparticles were synthesized by drop wise addition of stoichiometric amounts of $NaBH_4$ solution containing $FeSO_4 \cdot 7H_2O$ aqueous solution simultaneously with electrical stirring at ambient temperature. The ferrous iron was reduced to zero valent iron according to the following reaction:

$$4Fe^{3+}_{(aq)} + 3BH_4^- + 9H_2O \rightarrow 4Fe^0_{(s)}\downarrow + 3H_2BO_3 + 12H^+_{(aq)} + 6H_{2(g)}\uparrow \quad (13.3)$$

The Fe^0 nanoparticles were then rinsed several times with deionized water.

In the preparation of critical micelle concentration (CMC)-stabilized nanoparticles, FeSO4.7H2O stock solution was added to CMC solution to yield a solution with desired concentration. Then, the solution was allowed to form a complex with CMC. Here the addition of CMC serves as a dispersant and prevents the agglomeration of nanoparticles, thereby extends their reactivity. The FeSO4 concentration used in this study was 1 g/L and CMC concentration was 1% (w/w) of Fe. In the next step, Fe^{+2} is reduced to Fe^0 using stoichiometric amount of NaBH4 solution at ambient temperature with vigorous stirring. NaBH4 solids were dissolved in 0.1 M NaOH because NaBH4 is unstable in water and can quickly result in a loss of reducing capacity, addition of NaBH4 to the FeSO4 solution resulted in the rapid formation of fine black precipitate of CMC-stabilized Fe^0.[10]

13.2.2 CHARACTERIZATION OF ZERO VALENT IRON NANOPARTICLES

13.2.2.1 UV–VIS SPECTROSCOPY

A spectrum is a graphical representation of the amount of light absorbed or transmitted by matter as a function of the wavelength. A UV–Vis spectrophotometer measures absorbance or transmittance from the UV range to which the human eye is not sensitive to the visible wavelength range to which the human eye is sensitive.

Because only small numbers of absorbing molecules are required, it is convenient to have the sample in solution (ideally the solvent should not absorb in the ultraviolet/visible range, however, this is rarely the case). In conventional spectrometers electromagnetic radiation is passed through the sample which is held in a small square-section cell (usually 1 cm wide internally). Radiation across the whole of the ultraviolet/visible range is scanned over a period of ~30 s, and radiation of the same frequency and intensity is simultaneously passed through a reference cell containing only the solvent. Photocells then detect the radiation transmitted and the spectrometer records the absorption by comparing the difference between the intensity of the radiation passing through the sample and the reference cells. In the latest spectrometers' radiation across the whole range is monitored simultaneously.

No single lamp provides radiation across the whole of the range required, so two are used. A hydrogen or deuterium discharge lamp covers the ultraviolet range, and a tungsten filament (usually a tungsten/halogen lamp) covers the visible range. The radiation is separated according to its frequency/wavelength by a diffraction grating followed by a narrow slit. The slit ensures that the radiation is of a very narrow waveband—it is monochromatic. The cells in the spectrometer must be made of pure silica for ultraviolet spectra because soda glass absorbs below 365 nm, and pyrex glass below 320 nm. Detection of the radiation passing through the sample or reference cell can be achieved by either a photomultiplier or a photodiode, that converts photons of radiation into tiny electrical currents; or a semiconducting cell (that emits electrons when radiation is incident on it) followed by an electron multiplier similar to those used in mass spectrometers. The spectrum is produced by comparing the currents generated by the sample and the reference beams.

Modern instruments are self-calibrating, though the accuracy of the calibration can be checked if necessary. Wavelength checks are made by passing the sample beam through glass samples (containing holmium oxide) that have precise absorption peaks, and the absorption is calibrated by passing the sample beam through either a series of filters, each with a specific, and known absorption, or a series of standard solutions.[11]

13.2.2.2   X-RAY DIFFRACTION (XRD)

XRD is a non-destructive technique widely applied for the characterization of crystalline materials. The method is normally applied to data collected under ambient conditions, but in situ diffraction as a function of external constraints (temperature, pressure, stress, electric field, atmosphere, etc.) is important for the interpretation of solid-state transformations and materials behavior. Various kinds of micro- and nano-crystalline materials can be characterized from X-ray powder diffraction, including inorganics, organics, drugs, minerals, zeolites, catalysts, metals, and ceramics. The physical states of materials can be loose powders, thin films, polycrystalline, and bulk materials. For most applications, the amount of information which is possible to extract depends on the nature of the sample microstructure (crystallinity, structure imperfections, crystalline size, and texture), the complexity of the crystal structure (the number of atoms in the asymmetric unit cell, unit cell volume), and the quality of the experimental data (instrument performances, counting statistics).

13.2.2.3   FUNDAMENTAL PRINCIPLES XRD

XRD is based on constructive interference of monochromatic X-ray and a crystalline sample. These X-ray are generated by a cathode ray tube, filtered to produce monochromatic radiation, collimated to concentrate, and directed toward the sample. The interaction of the incident rays with the sample produces constructive interference (and a diffracted ray) when conditions satisfy Bragg's law ($\lambda = 2d \sin\theta$). This law relates the wavelength of electromagnetic radiation to the diffraction angle and lattice spacing in a crystalline sample. These diffracted X-ray are then detected, processed, and counted. By scanning the sample through a range of 2θ angles, all possible diffraction directions of the lattice should be attained due to the random orientation of the powdered materials. Conversion of the diffraction peaks to d-spacings allows identification of the mineral because each mineral has a set of unique d-spacings. Typically, this is achieved by comparison of d-spacings with standard reference patterns.

All diffraction methods are based on generation of X-ray in an X-ray tube. These X-rays are directed at the sample, and the diffracted rays are collected. A key component of all diffraction is the angle between the incident and diffracted rays. Powder and single crystal diffractions vary in instrumentation beyond this.[11]

13.2.2.4   FOURIER TRANSFORM INFRARED SPECTROMETRY (FTIR)

FTIR is preferred method of infrared spectroscopy. In infrared spectroscopy, IR radiation is passed through a sample. Some of the infrared radiation is absorbed by the sample and some of it is passed through (transmitted). The resulting spectrum represents the molecular absorption and transmission, creating a molecular fingerprint of the sample. Like a fingerprint, no two unique molecular structures produce the same infrared spectrum. This makes infrared spectroscopy useful for several types of analysis.

An infrared spectrum represents a fingerprint of a sample with absorption peaks which correspond to the frequencies of vibrations between the bonds of the atoms making up the material. Because each different material is a unique combination of atoms, no two compounds produce the exact same infrared spectrum. Therefore, infrared spectroscopy can result in a positive identification (qualitative analysis) of every different kind of material. In addition, the size of the peaks in the spectrum is a direct indication of the amount of material present.[11]

13.2.2.5   SCANNING ELECTRON MICROSCOPY (SEM)

The SEM uses a focused beam of high-energy electron to generate a variety of signals at the surface of solid specimens. The signals that derive from electron-sample interaction reveal information about the sample including external morphology (texture), chemical composition and crystalline structure, and orientation of materials making up the sample. In most applications, data are collected over a selected area of the surface of the sample, and a two-dimensional image is generated that displays spatial variations in these properties. Areas ranging from ~1 cm to 5 μm in width can be imaged in a scanning mode using conventional SEM techniques (magnification ranging from 20× to ~30,000×, spatial resolution of 50–100 nm). The SEM is also capable of performing analyses of selection point location on the sample.

The main signals which are generated by the interaction of the primary electrons (PE) of the electron beam and the specimen's bulk are secondary electrons (SE) and backscattered electrons (BSE) and furthermore X rays. They come from an interaction volume in the specimen which differs in diameter according to different energies of the PE (typically between 200 eV and 30 keV). The SE come from a small layer on the surface and yields the best resolution, which can be realized with a SEM. The well-known topographical contrast delivers micrographs which resemble on conventional light optical images.[11]

13.2.3 ANTIMICROBIAL ACTIVITY

Resistance to antibiotics is a ubiquitous and relentless clinical problem that is compounded by a dearth of new therapeutic agents. Therefore, there is an immediate need to develop new approaches to handle this problem. The emergence of nanoscience and nanotechnology in the last decade presents opportunities for exploring the bactericidal effect of metal nanoparticles. In recent scenario, much attention has been paid to metal nanoparticles which exhibit novel chemical and physical properties owing to their extremely small size and high surface area to volume ratio. It is evident that metal-based nanoparticles due to their biological and physicochemical properties are promising gas antimicrobials and therapeutic agents.[1–3]

Antimicrobial activity of the nanoparticles is known to be a function of the surface area in contact with the microorganisms. Fe^0 nanoparticles have several advantages, such as low cost, easy preparation, and high reactivity compared to other metal nanoparticles. Nanoscale zero valent iron has been used increasingly over the last decade to clean up polluted waters, soils, and sediments but little is known about the antimicrobial activity of nano-Fe^0. Typically, Fe(0)-based nanoparticles are prepared by reducing Fe(II) or Fe(III) in an aqueous phase using $NaBH_4$ appears most suitable for environmental applications because of its minimal use of environmentally harmful solvents or chemicals. Nano-Fe^0 has shown good antimicrobial activity against bacteria and fungi.[13]

The antibacterial effect of Fe^0 has been revealed to involve the generation of intracellular oxidants (e.g., HO^0 and Fe^{IV}) produced by the reaction with H_2O_2 or other species, as well as a direct interaction of Fe^0 with cell membrane components. The ions released by the nanoparticles may attach to the negatively charged bacterial cell wall and rupture it, thereby leading to protein denaturation and cell death. Increasing concentration of Fe^0 nanoparticles substantially inhibited the growth of *E. coli, K. pneumonia, P. aeruginosa, C. albicans*, and *T. rubrum*. Therefore, this study has been focused to synthesize and assess the antibacterial activity of nZVI and to evaluate the interaction of these nanoparticles and antibiotics on bacterial and fungal strains.[1–3]

13.3   METHODOLOGY

13.3.1   SYNTHESIS OF ZERO VALENT IRON NANOPARTICLES (nZVI)

Chemical method is used for synthesis of nZVI. In this method, chemical reduction of ferrous sulfate solution (0.1 M) with $NaBH_4$ (0.75 M) in the presence of ethylenediaminetetraacetic acid (EDTA) as a stabilizing agent is used to prepare the nanoparticles. In this process, 150 mL of nitrogen purged distilled water is used to dissolve 0.1 M of $FeSO_4$ and the solution was heated upto 60°C with constant stirring. Then, 100 mL of 0.1 M EDTA solution was added to the above solution and allowed to stir for 5 min. The aqueous solution of $NaBH_4$ (0.75 M) was then added drop wise to it under constant stirring until color changes to dark black, indicating the formation of nZVI. The particles were then immediately filtered from the solution and oven dried. The dried particles were then scraped and used for characterization.[1–3,8] nZVI were then dispersed in DMSO. Zero valent iron (1 mg) was added to 1 mL of DMSO and sonicated for 10 min (20 s on period and 40 s off period).

13.3.2   CHARACTERIZATION

Ultraviolet–visible spectroscopy (UV–Vis spec): The UV–Vis spectroscopy instrument ELICO model no. SL 159 was used. The analysis was performed at Chemistry laboratory, MSRIT.

X-ray diffractometer (XRD): The Zero valent iron samples were characterized at Poornaprajna Institute of Scientific Research (PPISR), Bidlur, Bangalore (BRUKER-D2-PHASER).

Fourier transform infrared spectroscopy (FTIR): The zero valent iron samples were characterized at PPISR, Bidlur, Bangalore (BRUKER-ALPHA FTIR Spectrometer).

Scanning electron microscope (SEM): The SEM analysis was performed at Centre for Nano Science and Engineering (CeNSE), Indian Institute of Science (IISC), Bangalore (ZEISS-GEMINI-ULTRA 55).

13.3.3   ANTIMICROBIAL ACTIVITY STUDIES

Three pathogenic bacteria, namely, Gram-negative (*E. coli, K. pneumonia,* and *P. aeruginosa*) (clinical isolates, Chrompet medical college, Chennai) and two fungal (*C. albicans*-MTCC NO. 4748, *T. rubrum*-MTCC No. 7859) (MTCC Chandigarh, India) strains were used in this study. The organisms were maintained on agar (Hi Media, India) slopes at 4°C and sub cultured before use.

Well diffusion method: Well diffusion method was carried out to establish the antibacterial activity of nZVI against the pathogens (three bacterial and one fungal (*C. albicans*) strains). The pathogens were swabbed on the agar plates and four wells were punched using sterile gel puncture. nZVI (1 mg) were added to 1 mL of DMSO and sonicated for 10 min so that the nanoparticles get uniformly dispersed in DMSO. Then, 25, 50, 75, and 100 µL (1 mg/1 mL) of the dispersed solution were dispensed into each well. The plates were incubated at 37°C for 24 h. The zone of the inhibition around each well after the incubation period, confirms the antimicrobial activity of the nZVI extract.

Food poisoning method: Food poisoning method is used for *T. rubrum*. In this method, the nanoparticles were directly added to the molten media while it is about to solidify and mixed thoroughly.

The media was poured in the plate and allowed to solidify. A loop of inoculum of fungi was then transferred to the center of the plate. In control plate, no nanoparticles were added to the media. *T. rubrum* takes 7 days to grow, so the plates were incubated for 7 days at 37°C.[12]

13.4   RESULTS AND DISCUSSION

13.4.1   SYNTHESIS OF ZERO VALENT IRON NANOPARTICLES

Chemical reduction method was used to synthesize nZVI. The color change of the reaction mixture from pale yellow to black indicates the formation of nZVI. Black color formation was observed 10 min after drop wise addition of $NaBH_4$.

13.4.2   CHARACTERIZATION OF ZERO VALENT IRON NANOPARTICLES

UV–Vis spectroscopy: UV–Vis spectroscopic analysis was used to validate nZVI synthesized by chemical reduction method and the absorbance vs. wave length (λ) was established. The characteristics peaks of nZVI were observed at 295.0 nm (Fig. 13.1a).

XRD: The XRD analysis of nZVI is shown in Figure 13.1b. The peak at 2θ of 44.2° indicates the presence of nZVI.

FTIR: FTIR techniques provide information about vibrational state of adsorbed molecule and hence the nature of surface complexes. The band at 3362.70 cm^{-1} indicates the OH stretching vibration. The various bands showed in Figure 13.1c indicate the presence of nZVI.

SEM: The spherical shape of nZVI was absorbed in the analysis of SEM image (Fig. 13.1d). The size of most of the nanoparticles ranges from 71.00 to 109.5 nm and mean particle size of 90.25 nm.

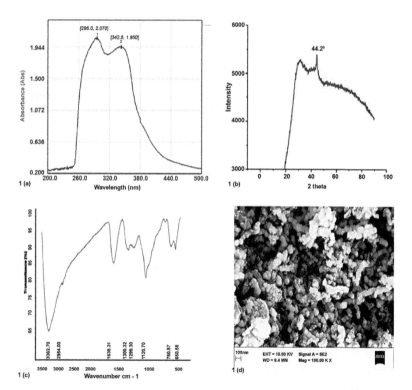

FIGURE 13.1   (a) UV–Vis spectrum (b) XRD pattern (c) FTIR spectrum, and (d) scanning electron microscopy of zero valent iron nanoparticles (nZVI).

13.4.3   EVALUATION OF ANTIMICROBIAL ACTIVITY OF ZERO VALENT IRON

nZVI showed antibacterial activity against three bacterial strains and one fungal strain (*C. albicans*) (Fig. 13.2(i)). The concentration used was 1 mg/1 mL (1 mg of nZVI in 1 mL of DMSO) and 25, 50, 75, and 100 µL were dispensed in four wells, respectively. The DMSO was used as positive control for all pathogens which showed no zone of inhibition (Fig. 13.2(ii)). The present results are comparable with that of the standard antibiotic ciprofloxacin (50 µg/disc) (Fig. 13.2(iii)). The zone of inhibition for all the bacteria and *C. albicans* is shown in Table 13.1.

TABLE 13.1   Antimicrobial Activity of Zero Valent Iron Nanoparticles (nZVI).

Organism	DMSO	Zone of Inhibition of Zero Valent Iron Nanoparticles (nZVI) mm				Standard Drug
		25 µL	50 µL	75 µL	100 µL	
Escherichia coli	0	14	10	12	15	28
Klebsiella pneumonia	0	12	10	13	10	27
Pseudomonas aeruginosa	0	14	15	15	15	28
Candida albicans	0	14	15	15	15	

For *T. rubrum*, food poisoning technique was used and it showed growth of 12 mm in diameter for nZVI and control showed 18 mm of growth (Fig. 13.2(iv)).

13.5   DISCUSSION

Pathogenic microbes that have become resistant to drug therapy are an increasing public health problem. One of the measures to combat this increasing rate of resistance is to have continuous investigations into new, safe, and effective antimicrobials as alternative agents to substitute with less effective ones. Owing to their high antibacterial properties, nanoparticles of silver, oxides of Zinc, titanium, copper, and iron are the most commonly used nanoparticles in antimicrobial studies. Furthermore, these nanoparticles have been used to deliver other antimicrobial drugs to the site of pathological process. In general, a variety of preparation routes have been reported for the preparation of metallic nanoparticles.

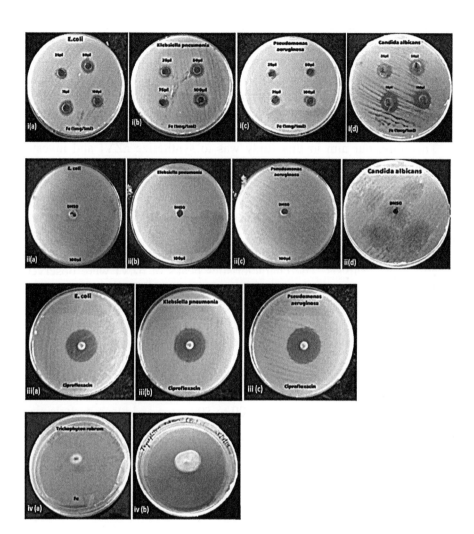

FIGURE 13.2 Study of antibacterial and antifungal activity (zone of inhibition) of (i) zero valent iron nanoparticles (ii) control (DMSO), (iii) standard (ciprofloxacin), against (a) *Escherichia coli*, (b) *Klebsiella pneumonia*, (c) *Pseudomonas aeruginosa*, and (d) *Candida albicans*, and (iv) (a) shows antifungal activity of *Trichophyton rubrum* against zero valent iron nanoparticles (b) control.

Although metals and metal oxides are known to be toxic at relatively high concentrations, they are not expected to be toxic at low concentrations. Several resistance mechanism to metals have been described, the most common, which is enhanced efflux of metal ions from the cell, is a high-level, single-step, and target-based mutation. This mutation enhances efflux of metal ions from cell and makes metal resistance less probable owing to its multifaceted mode of action.

Therefore, our aim in this study was to synthesize nZVI and investigate the antibacterial and antifungal effect against Gram-negative (*E. coli*, *K. pneumonia*, and *P. aeruginosa*) bacteria and fungal (*C. albicans* and *T. rubrum*) strains.

The synthesized nZVI were characterized by UV–Vis spectroscopy, XRD, FTIR, and SEM.

nZVI synthesized by chemical reduction was validated by UV–Vis spectroscopic analysis. The characteristics peaks of nZVI were observed at 295.0 nm. These peaks were nearly matching to that of the peaks given in the literature.[8] The XRD analysis of nZVI shows the peak at 2θ of 44.2° indicating the presence of nZVI.[7] FTIR techniques provide information about vibrational state of adsorbed molecule and hence the nature of surface complexes. The band at 3362.70 cm^{-1} indicates the OH stretching vibration.[7]

SEM analysis revealed spherical morphology of nZVI. The size of most of the nanoparticles ranges from 71.00 to 109.5 nm and mean particle size of 90.25 nm.

13.6 CONCLUSION

Antimicrobial activity was observed for nZVI against Gram-negative (*E. coli*, *K. pneumonia*, and *P. aeruginosa*) bacteria and fungal (*C. albicans* and *T. rubrum*) strains. The nZVI showed significant anti-fungal and anti-bacterial activity against tested strains. These results suggest the possibility of using nZVI for effective development of antimicrobial drugs with high efficacy toward the pathogenic microorganisms. Further studies will be useful to evaluate improved antimicrobial activity against the pathogenic microorganisms and to study the toxic effect on animal models.

KEYWORDS

- zero valent iron nanoparticles
- antifungal activity
- antibacterial activity
- X-ray diffraction
- Fourier transform infrared spectroscopy
- scanning electron microscope

REFERENCES

1. Selvarani, M.; Prema, P. Synergistic Antibacterial Evaluation of Commercial Antibiotics Combined with Nano Iron against Human Pathogens. *Int. J. Pharm. Sci. Rev. Res.* **2013,** *18,* 183–190.
2. Selvarani, M.; Prema, P. Evaluation of Antibacterial Efficacy of Chemically Synthesized Copper and Zero Valent Iron Nanoparticles. *Asian J. Pharm. Clin. Res.* **2013,** *6,* 223–227.
3. Selvarani, M.; Prema, P. Synergistic Antibacterial Evaluation of Commercial Antibiotics Combined with Nano Iron against Human Pathogens. *Int. J. Pharm. Sci. Rev. Res.* **2013,** *18* (1), 183–190.
4. Mahdy, S. A.; Raheed, Q. J.; Kalaichelvan, P. T. Antimicrobial Activity of Zero Valent Iron Nanoparticles. *Int. J. Modern Eng. Res.* **2012,** *2,* 578–581.
5. Kim, J. S.; Kuk, E.; Yu, K. N. Antimicrobial Effects of Silver Nanoparticles. *Nanomedicine.* **2007,** *3,* 95–101.
6. Saha, S.; Bhunia, A. K. Synthesis of Fe_2O_3 Nanoparticles and Study of Its Structural, Optical Properties. *J. Phys. Sci.* **2013,** *17,* 191–195.
7. Ritu, S.; Virendra, M.; Rana, P. S. Synthesis, Characterization and Role of Zero-Valent Iron Nanoparticle in Removal of Hexavalent Chromium from Chromium-Spiked Soil. *J. Nanopart. Res.* **2011,** *13* (9), 4063–4073.
8. Vemula, M.; Prasad, T. N. V. K.; Gajulapalle, M. Synthesis and Spectral Characterization of Iron Based Micro and Nanoparticles. *Int. J. Nanomater. Biostruct.* **2013,** *3,* 31–34.
9. Everett, C. E. Iron Nanoparticles as Potential Magnetic Carriers. *J. Magn. Magn. Mater.* **2001,** *225,* 17–20.
10. Vemula, M.; Ambavaram, V. R.; Kalluru, G. R.; Gajulapalli, M. A Simple Method for the Determination of Efficiency of Stabilized Fe^0 Nanoparticles for Detoxification of Chromium (VI) in Water. *J. Chem. Pharm. Res.* **2012,** *4* (3), 1539–1545.
11. He, Y. P.; Wang, S. Q.; Li, C. R.; Miao, Y. M.; Wu, Z. Y.; Zou, B. S. Synthesis and Characterization of Functionalized Silica-Coated Fe_3O_4 Superparamagnetic Nanocrystals for Biological Applications. *J. Phys. D Appl. Phys.* **2005,** *38,* 1342–1350.

12. Mohana, D. C.; Raveesha, K. A. Anti-Fungal Evaluation of Some Plant Extracts against Some Plant Pathogenic Field and Storage Fungi. *J. Agric. Tech.* **2007,** *4* (1), 119–137.
13. Sudhanshu, S. B.; Jayanta, K. P.; Krishna, P.; Niladri, P.; Hrudayanath, T. Characterization and Evaluation of Antibacterial Activities of Chemically Synthesized Iron Oxide Nanoparticles. *World J. Nano. Sci. Eng.* **2012,** *2,* 196–200.

CHAPTER 14

SPINEL FERRITES—A FUTURE BOON TO NANOTECHNOLOGY-BASED THERAPIES

SHARATH R.[1*], NAGARAJU KOTTAM[2*], MUKTHA H.[1], SAMRAT K.[1], CHANDRAPRABHA M. N.[1] HARIKRISHNA R.[2], and BINCY ROSE VERGIS[2,3]

[1]*Department of Biotechnology, M.S. Ramaiah Institute of Technology, Bangalore 560054, Karnataka, India*

[2]*Department of Chemistry, M.S. Ramaiah Institute of Technology, Bangalore 560054, Karnataka, India*

[3]*Department of chemistry, BMS Institute of technology, Bangalore, India.*

Corresponding author. E-mail: sharathsarathi@gmail.com; knrmsr@gmail.com

CONTENTS

ABSTRACT

Spinel ferrite nanoparticles (NPs) exhibit unique optical, electronic, and magnetic properties which have made them NPs of greater interest. The spinel ferrite NPs have high permeability, magnetization, good saturation, and they have no preferred direction of magnetization. Magnetic NPs have a unique property that makes them to be applied in nanomedicine—they can address targets, such as cellular therapy, tissue repair, drug delivery, magnetic resonance imaging (MRI), and nanobiosensors. These applications require high values of magnetization of NPs and the size must be <100 nm with uniform chemical and physical properties. Over the years, iron oxide, especially magnetite (Fe_3O_4), was the most investigated magnetic NP. The small magnetic NPs allow delivery of an antibiotic when targeting certain organs such as brain and kidney. The Fe_3O_4 NPs coated with silver improves bacterial activity. Cobalt–ferrite (Co–Fe) NPs which belong to the crystal family of spinel ferrites (MFe_2O_4), possess larger magnetic anisotropy than other ferrites (e.g., magnetite) making them more attractive for nanotechnology-based therapies. A significant decrease in cytokinesis-blocked proliferation index and increase in the frequency of micro nucleated binucleated lymphocytes are shown with 5.6 nm Co–Fe NPs. Coating the surface of these NPs leads to a fourfold reduced level of toxicity. The substitution of zinc and copper in Co–Fe NPs significantly improves antibacterial activity. A copper-substituted Co–Fe NP exhibits the most effective contact biocidal property among all of the NPs. Zinc and copper-substituted Co–Fe NPs have high potential to be used in drug delivery systems as well as in other biomedical and biotechnology applications.

14.1   INTRODUCTION

Nanotechnology incorporates the control and understanding of matter at dimensions of 1–100 nm, where a unique physical phenomenon enables novel applications. The properties of nanomaterials both physical and chemical are governed largely by their size and shape or morphology.[1] For this reason, scientists are focusing on developing simple but effective methods for fabricating nanomaterials with controlled size and morphology and hence with a view to tailoring their properties. Magnetic nanoparticles (NPs) possess unique properties which can be applied in nanomedicine: they address targets, such as cellular therapy, tissue repair, nanobiosensors, drug delivery,

MRI, and magnetic fluid hyperthermia. All these applications require high magnetization values of NPs and size of <100 nm with uniform physical and chemical properties. Additionally, it is essential to understand the biological fate and potential toxicity of magnetic NPs for their successful application in nanomedicine. Over the years, iron oxide, especially magnetite (Fe_3O_4), was the most investigated magnetic NP. In the last decade, it became easier to synthesize new and more effective types of magnetic NPs.[2] Co–Fe NPs which belong to the crystal family of spinel ferrites (MFe_2O_4), possess larger magnetic anisotropy than other ferrites (e.g., magnetite) making them more attractive for nanotechnology-based therapies.[3]

14.2 FERRITES

Ferrites are chemical compounds with the formula of AB_2O_4, where A and B represent various metal cations; usually including iron. Ferrites are considered a class of spinels that consist of cubic closed pack oxides with a cation occupying one-eighth of the octahedral voids and B cations occupying half of the octahedral voids. For the inverse spinel structure, half the B cations occupy tetrahedral sites and both the A and B cations occupy the octahedral sites. Meanwhile, divalent, trivalent, and quadrivalent cations can occupy the A and B sites and they include Mg, Zn, Fe, Mn, Al, Cr, Ti, and Si. Ferrites are ferromagnetic ceramic compounds usually non-conductive derived from iron oxides as well as oxides of other metals. Like most other ceramics, ferrites are hard and brittle. Ferrites are manufactured by homogeneously mixing iron oxides and then calcinating the mixture which causes the partial decomposition of carbonates and oxides. This mixture is then sintered which is done by gradually raising the temperature to 1500°C in a klin. Typically, the cores will shrink by 10–20% of its original size after sintering. The mixture crystallizes into cubic structure; the resulting ferrite has a black non-porous ceramic appearance. Ferrites are often classified as soft ferrites and hard ferrites, which is based on their ability to be magnetized or demagnetized and can also be classified based on their coercive field strength. Soft ferrites can be easily magnetized and demagnetized thus making them to be used as electromagnets they have low coercive field strength, while hard ferrites are used as permanent magnets and have high coercive field strength. Ferrites are widely used in high-frequency applications because an alternating current (AC) field does not induce undesirable eddy currents in an insulating material.[4]

14.2.1 SPINEL FERRITES

Spinel ferrites possess the crystal structure of the natural spinel $MgAl_2O_4$, first determined by Bragg. This structure is particularly stable, since there are an extremely large variety of oxides which adopt it, fulfilling the conditions of overall cation-to-anion ratio of 3/4, a total cation valency of 8, and relatively small cation radii. The spinel structure has two cation sites for metal cation occupancy. Spinel structure is shown in Figure 14.1. In ferrites with applications as magnetic materials, Al^{3+} has usually been substituted by Fe^{3+}. An important ferrite is magnetite, $Fe^{2+}Fe^{3+}_2O_4$ (typically referred as Fe_3O_4), probably the oldest magnetic solid with practical applications and currently a very active research field, due to the interesting properties associated with the coexistence of ferrous and ferric cations. Another important material by its structure, as well as by its applications in audio recording media, is maghemite or γ-Fe_2O_3, which can be considered as a defective spinel O_4, where represents vacancies on cation sites.

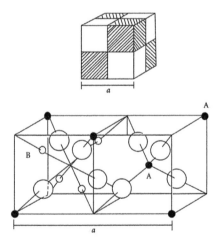

FIGURE 14.1 The unit cell of the spinel structure, divided into octants to show the tetrahedral (small, black spheres A) and octahedral (small white spheres B) sites. Oxygen atoms are the large white spheres.[4] (Reprinted from Raul, V. Novel Applications of Ferrites. *Phys. Res. Int.* **2012,** *2012,* 1–9. Copyright © 2012 Raúl Valenzuela. https://creativecommons. org/licenses/by/3.0/)

14.2.1.1 SYNTHESIS OF SPINEL FERRITE NANOPARTICLES

Top-down and bottom-up approaches have been used to synthesize spinel ferrites. Methods like high energy ball milling, microemulsion, sol–gel, co-precipitation, and combustion have been used extensively.

14.2.1.1.1   Co-Precipitation Method

The precursors are of metal nitrates and chlorides with reducing agents such as, sodium borohydride, hydrazine hydrate, etc. In a typical synthesis, desired concentrations of precursor solutions are prepared and vigorously mixed under stirring for 1 h at 80°C. Capping agent is added to the solution through which particle size can be controlled. Subsequently, a reducing agent is added drop by drop into the solutions and brown/black color precipitates are formed. Finally, the precipitates are separated by centrifugation and dried in hot air oven.

14.2.1.1.2   Solution Combustion Method

The principle of solution combustion synthesis is based on heating which initiates a chemical reaction which is exothermic and eventually becomes self-sustaining within a time interval, resulting in a powder as final product. The exothermic reaction starts at the ignition temperature and a certain amount of heat is generated which is in turn manifested as the maximum temperature or temperature of combustion. Solution combustion synthesis has the advantage of rapidly producing fine and homogeneous powders, as it is an exothermic and auto-propagated process.

The solution combustion method is used to synthesize spinel ferrites where in nitrates as oxidant and constituents of a suitable fuel or complexing agent (e.g., citric acid, urea, and glycine) in water are mixed in stoichiometric ratios and an exothermic redox reaction between the fuel and the oxidant takes place.

14.2.1.1.3   Microemulsion Method

Spinel ferrites have been prepared by reverse micelle micro emulsion method. In this method, NPs synthesis involves preparation of aqueous solution of precursor material (nitrate salts) with addition of a surfactant (sodium dodecyl sulfate (SDS), cetrimonium bromide (CTAB), Triton-X 100, Tween 80, or polysorbate 80), and toluene to form a reverse micelles. This solution is then refluxed following removal of excess toluene through distillation; the resulting brown particles are then washed with water and ethanol to ensure that any excess surfactant is removed. The solution mixture is subjected to centrifugation and particles are collected. The sample is annealed at a ramping rate which results in a fine black powder. The size of the particles can be varied by adjusting the water to toluene ratios.

14.2.1.1.4 Sol–Gel Method

Sol–gel processing is based on chemical method, a technique used to manufacture ceramic powders, especially oxides. It has been down-selected for the synthesis of spinel ferrite NPs due to relative simplicity in combination with novelty concerning the intended application.[1] The term sol refers to the initial solution of the chemical components for the final powder whereas the term gel describes the final consolidation stage that forms the ceramic product. Sol–gel procedures have been successful in the preparation of bulk metal oxides, for example, ceramics, glasses, films, and fibers and, therefore, they have been applied for NP synthesis (Fig. 14.2).

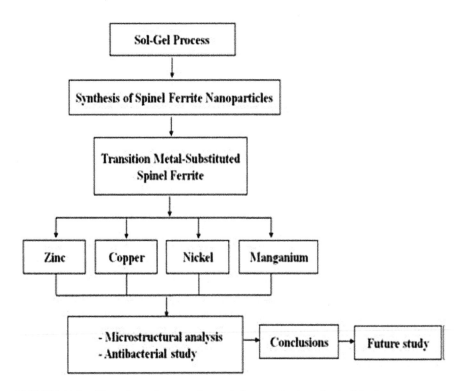

FIGURE 14.2   Flow chart of preparation and characterization of spinel ferrite.

The sol–gel process consists of the following five prime steps:

i) Preparation of a homogeneous solution either by dissolution of metal organic precursors in an organic solvent that is miscible with water, or by dissolution of inorganic salts in water.
ii) Conversion of the homogeneous solution into a sol by treatment with a suitable reagent that is generally water with or without an acid base.
iii) Aging of the solution.
iv) Shaping of the gel.
v) Thermal treatment or sintering of the final product.

14.2.1.2 ANTIBACTERIAL PROPERTIES OF SPINEL FERRITES

The need for highly effective antibiotics is due to the growing resistance of bacteria to conventional antibiotics and increasing incidence of bacterial infections. Accordingly, new methods for reducing bacterial activity are immensely needed. NPs have greater antibacterial properties as the particle size is in the nanometer range they can easily penetrate into bacteria which appears to provide a unique bactericidal mechanism. Various methods have been employed to check the antibacterial activity and they are disc diffusion, well diffusion, agar dilution, and broth dilution. The diffusion and dilution methods have been popularly used for many years to measure the antibacterial activity. The most commonly and widely used methods are well or disc diffusion method with different media formulations, different incubation times, and different volumes of test samples.

14.2.1.2.1 Disc Diffusion Method

1. Take sterile Petri plates.
2. Pour the molten nutrient agar into Petri plates and allow it to set.
3. Spread an overnight culture of bacteria over the surface of the agar plate using a sterile glass spreader and incubated at 37°C for a 30 min.
4. Add calculated volume of sample solution on sterile blank antimicrobial susceptibility discs.
5. Keep the sample impregnated discs onto the inoculated surface of the agar plate (maximum of five discs per plate).
6. The agar plates incubated overnight at 37°C and the zones of bacterial inhibition recorded.

14.2.1.2.2   *Well Diffusion Method*

1. Take sterile Petri plates.
2. Pour the molten nutrient agar into Petri plates and allow it to set.
3. Spread an overnight culture of bacteria over the surface of the agar plate using a sterile glass spreader and incubated at 37°C for a 30 min
3. Create four wells using a 6 mm cork borer.
4. Load the test sample into these wells, incubate the plates at 37°C overnight and record the zones of inhibition.

Buteica et al.[5] prepared magnetic, core-shelled Fe_3O_4 NPs to improve colloidal dispersion and to control particle sizes. It was observed that, for the same time interval, the inhibition zone diameters for cephalosporins were higher than those for the cephalosporin-nanofluid. An advantage of this technology was that the nanofluid acted only as a carrier for the antibiotic. In addition, small magnetic NPs allowed delivery of an antibiotic when targeting certain organs such as brain and kidney. Sun et al.[6] developed a thermal decomposition method that used a mixture of an iron salt, 1,2-hexadecanediol, oleic acid, oleylamine, and biphenyl ether to obtain Fe_3O_4 NPs. The Fe_3O_4 NPs were then coated with silver to improve bacterial activity. The distinctive properties of magnetic NPs required for biomedical applications require precise control of particle size, shape, dispersion, and any external factors that influence these properties. In principle, it is necessary to stabilize the magnetic NP dispersion in the aqueous environment. Thus, coating the magnetic NPs with a polymer shell, including organic (e.g., polyethylene glycol, dextran, chitosan, polyethyleneimine, and phospholipids) or inorganic (e.g., silica) materials, leads to highly dispersed and high quality NPs with good biocompatibility. Recently, Sanpo et al.[7] suggested a new approach to enhance the antibacterial property of ferrite NPs for biomedical applications. They found that substitution of spinel ferrite with a transition metal can improve the antibacterial capability of NPs. However, the biomedical application of these newly synthesized NPs requires further study concerning their biocompatibility and antibacterial properties. Techniques reported for synthesizing spinel ferrite NPs of chemical formula, MFe_2O_4 (where M = Co, Mg, Mn, Zn, etc.) include solid state reaction, microemulsion, combustion, the redox process, chemical co-precipitation, the hydrothermal method, and microwave synthesis. However, sol–gel techniques offer enhanced control over homogeneity, elemental composition, and powder morphology. In addition, uniform nano-sized metal clusters can be achieved, which are crucial for enhancing the properties of the NPs.

These advantages favor the sol–gel route over other conventional preparation methods of ceramic oxide composites.[8]

The contact biocidal property of all ferrite NPs was investigated using a modified Kirby–Bauer technique. Filter papers are partially covered with and without ferrite NPs and placed on a lawn of *Escherichia coli* in an agar plate. The contact antibacterial property can be measured by the clear zone of inhibition around the filter papers after 24 h incubation. The diameter of the zone of inhibition for the copper-substituted Co–Fe NPs is the largest, followed by zinc-substituted decreasing order. The cobalt ferrite, nickel-substituted cobalt ferrite, pure cobalt ferrite, and manganese-substituted cobalt ferrite, in results agree well with the quantitative bacterial tests that indicated the copper-substituted cobalt ferrite NPs have the most effective antibacterial property against *E. coli* among all of the NPs (Fig. 14.3).[1,9]

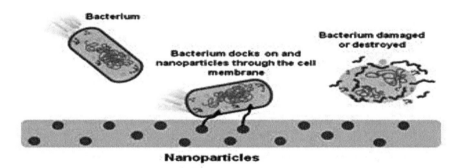

FIGURE 14.3   Action of nanoparticles on bacterium.

14.2.1.3   TOXICITY OF SPINEL FERRITES

NPs possess tremendous capability to revolutionize medical imaging, diagonostics, and therapeutics but with all these promising features of NPs, they could pose health risk due to the characteristics like size, shape, chemical composition, and surface characteristics. The toxicity of Co–Fe NPs was explored in a number of studies. A significant decrease in cytokinesis-blocked proliferation index and increase in the frequency of micronucleated-binucleated lymphocytes were shown when employing 5.6 nm Co–Fe NPs. However, coating the surface of these NPs led to a four-fold reduced level of toxicity. Similarly, toxicological study of silica-coated Co–Fe possessing a silica shell of 50 nm thickness revealed that although the particles were found in the mice's brain, no significant changes in the hematological and clinical biochemistry tests were found. Another study investigated the embryo toxicity of Co–Fe NPs (17 ± 3 nm) through an embryonic stem-cell test which shows differentiation into cardiomyocytes. The obtained ID50 for the inhibition of differentiation classified Co–Fe NPs coated with gold and silanes as non-embryotoxic. However, Co–Fe NPs coated only with silanes were found to be weakly embryotoxic, but less embryotoxic than the cobalt ferrite salt ($CoFe_2O_4$). Obviously, further toxicological work should proceed with the aim to achieve a larger toxicological database to enable the prediction of toxicology by in silico approaches.[1]

14.2.1.4 PHOTOCALYTIC ACTIVITY OF FERRITES

Photocatalysts (PCs) are those materials which can bring about a chemical change by utilizing the solar energy. Photocatalysis is applied in many areas including the elimination of contaminants from water and air, odor control, bacterial inactivation, water splitting to produce H_2, the inactivation of cancer cells, and many others applications.

Solar energy active PCs have been a promising material for an environmental purification process. Utilizing this technology for the removal of potentially toxic and dangerous compounds from the environment has generated a great interest in the last decade. Efforts have been made to synthesize materials capable of utilizing solar spectrum for the photo degradation of industrial waste pollutants and dyes.

The textile industries are becoming a major source of environmental contamination because an alarming amount of dye pollutants are generated during the dyeing processes. A wide use of dyes as redox indicators, biological stains, and pharmaceutical industries had adverse effects on gastrointestinal, genitourinary, and cardiovascular system of human body. Conventionally, photocatalysis involved the use of semiconductor materials, which mainly included oxides like TiO_2, WO_3, ZnO, and certain sulfides like ZnS and CdS which are used for the removal of large dye molecules by degradation. Thus, photocatalysis of nano metal oxides involve the removal of contaminants by degradation, ideally to CO_2 and water by oxidation and reduction reaction of these large dye molecules. But then the removal of these nano metal oxides from the water bodies is a major concern. Solution to this problem is the use of magnetically active PC, giving them added advantage of easy recovery after photocatalytic reaction. Thus, ferrites can be used as magnetic PCs as these are magnetic in nature.[10,11]

14.2.1.5   MECHANISM OF PHOTOCATALYSIS

PCs utilize light energy ($h\acute{v}$) to carry out oxidation and reduction reactions. When irradiated with light energy, an electron (e^-) is excited from the valence band (VB) to the conduction band (CB) of the PC, leaving a photo-generated hole (h^+). Figure 14.4 describes the process of a PC absorbing light energy in order to produce e^-/h^+ pairs. The produced e^- and h^+ enable oxidation and reduction processes to occur. When photocatalytic processes take place in aqueous solutions, water and hydroxide ions react with photo generated h^+ to form hydroxyl radicals (OH^-), which is the primary oxidant in the photocatalytic oxidation of organic compounds. OH^- ($E^0 = 2.80$ V vs. normal hydrogen electrode (NHE)) has a higher oxidation potential than other common oxidants such as ozone (O_3) ($E^0 = 2.07$ V) and hydrogen peroxide (H_2O_2) ($E^0 = 1.77$ V), and it has been shown that repeated attack of organic pollutant (OP) by OH^- eventually leads to complete oxidation. The process of generating OH^- can occur by two pathways; first O_2 present in water is reduced to form O_2^-, which then reacts with H^+ to form OH^- followed by rapid decomposition to OH^-.[12] The second pathway involves the oxidation of OH^-.

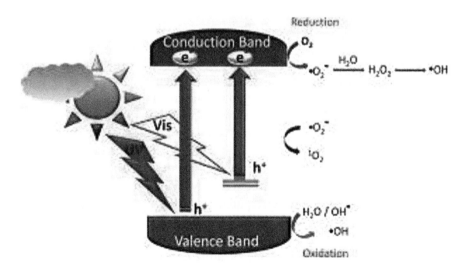

FIGURE 14.4   The change taking place with the absorption of light energy.

The possible reactions involved in photocatalysis are as follows:[13]

1. Photo generation of electrons and holes
    - $PC + hv \rightarrow e^- + h^+$

2. Adsorption of dissolved O_2
    - $O_2 \text{(dissolved)} + e^- \rightarrow O_2^-$

3. Photo splitting of water
    - $H_2O + h^+ \rightarrow H^+ + OH^-$

4. Formation of hydroxyl free radical by holes
    - $H^+ \text{(aq)} + O_2^- \rightarrow HO_2\cdot$
    - $OH^- + h^+ \rightarrow OH\cdot$

5. Dissociation of OP
    - $OP \text{(aq)} \rightarrow OP^+ \text{(aq)} + OP^- \text{(aq)}$

6. Decolorization of dye
    - $OP\cdot + OH\cdot/HO_2\cdot \rightarrow Mn\cdot + \text{neutral site}$
    - $Mn\cdot + OH\cdot/HO_2\cdot \rightarrow CO_2 + H_2O + \text{mineral acids}$

When selecting a PC, the band gap of the material determines the wavelength of light that can be absorbed. As can be seen in Figure 14.5, the difference between the VB and the CB is inversely related to a wavelength of light ($E = hc/k = 1240$ eVnm/k), where E is the band gap energy (eV).

Also the amount of visible light falling on Earth's surface accounts for 46% while UV only accounts for 5% of the total energy from the sun, with the remaining portion corresponding to the infrared region.

Many of the PCs that are commonly used have wide band gaps (> 3.1 eV) and are capable of utilizing only a small portion of sunlight. It is ideal to use visible solar energy due to the large amount that reaches the Earth's surface annually, which is ~10,000 times more than the current yearly energy consumption. In order for a PC to effectively absorb visible solar energy a maximum band gap of 3.1 eV is required.[14]

TiO_2, which is a good catalyst, has a wide band gap (3.03 eV rutile and 3.18 eV anatase) and can, therefore, absorb only a small portion of sunlight. Therefore, developing PCs capable of utilizing safe and sustainable solar energy effectively and efficiently is important. There are two approaches in which visible light irradiation can be utilized by PCs. One approach is to dope a UV active PC with elements in an effort to make them visible light active. The second method is the development of materials that have a narrow band gap, which allows for photocatalytic activity under visible light irradiation. Of these PCs, it can be seen that some ferrites such $CaFe_2O_4$, Mg Fe_2O_4, $ZnFe_2O_4$, etc. can effectively utilize visible light due to smaller band gaps. It is important to use narrow band gap semiconductors rather than those with a wide band gap to utilize solar visible energy.[15]

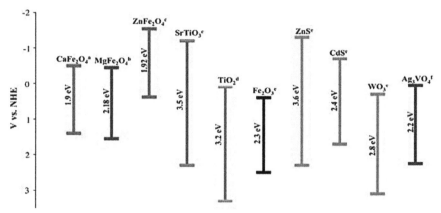

FIGURE 14.5 The band gaps of selected photocatalysts.

Spinel ferrites, with a general formula of MFe_2O_4, where M represents a metal cation, are chemically and thermally stable magnetic materials that have been used for many applications. Their magnetic properties make them useful in MRI, electronic devices, information storage and drug delivery, and other applications. Ferrites possess important photocatalytic properties for many industrial processes, including the oxidative dehydrogenation of hydrocarbons, the decomposition of alcohols and hydrogen peroxide, the treatment of exhaust gases, the oxidation of compounds such as CO, H_2, CH_4, and chlorobenzene, the hydroxylation of phenol, alkylation reactions, the hydrodesulfurization of crude petroleum, and the catalytic combustion of methane. The presence of different metals in the lattice structure can modify the redox properties of the ferrites show that the replacement of Fe^{2+} occurs at the octahedral sites in the crystal structure. For example, Co^{2+} and Mn^{2+} form redox pairs with Co^{3+} and Mn^{3+}, respectively, and, therefore, can play the same role as Fe^{2+}/Fe^{3+} in the system.[16]

14.2.1.6 DYE DEGRADATION USING FERRITES

The photocatalytic degradation of organic molecules is of great importance in water treatment. In most cases, dyes, such as methyl orange (MO), methylene blue (MB), rhodamine B (RhB), and a wide range of other organic dye molecules are studied as model compounds. These dyes have large and complex structure.[17]

Various Ferrites with Zn, Co, Cu, Mn, Ca, etc., are used alone or in combination with other PC successfully. In most cases, ferrites alone show a much lower degradation percentage, but are significantly enhanced when combined with other PCs, including TiO_2. For example, $ZnFe_2O_4$ alone does not effectively degrade MO dye (4%), but when irradiated with visible light and combined with TiO_2, degradation as high as 84% is accomplished (Table 14.1).[18] Another common way to increase the degradation efficiency is by heat-treating the PC. By doing so, the crystal structure of the PC is altered and, therefore, the surface defects are changed, which decreases reaction of recombination of e^-/h^+ pairs, thereby leading to higher photocatalytic activity. Enhanced photocatalytic degradation of MB using $Zn\ Fe_2O_4$/multi-walled carbon nanotubes (MWCNT) composite synthesized by hydro-thermal method was reported by Sonal Singhal et al. (Table 14.2).[19]

Magnetic PCs cobalt ferrite ($CoFe_2O_4$) coated with titania–silica (TiO_2–SiO_2) were successfully prepared by coating TiO_2–SiO_2 onto $CoFe_2O_4$ by the sol–gel technique. Controlling the modification process of TiO_2–SiO_2 with $CoFe_2O_4$ NPs is a key factor for obtaining appropriate catalytic performance. Under optimized conditions, a core-shell structure could be obtained in which $CoFe_2O_4$ is a core while TiO_2–SiO_2 forms a shell. This PC exhibits remarkable catalytic activity in the degradation of MB dye in water which could remove as high as 98.3% of the organic dye in just 40 min. $CoFe_2O_4$@SiO_2@TiO_2 core-shell magnetic nanostructures can be used as potential magnetically retrievable and re-usable photocatalysts.[20]

Doping of $CoFe_2O_4$ with Zn, Mn, etc., also found to show enhanced photocatalytic activity.

For example, $Zn_{1-x}Co_xFe_2O_4$ samples ($x = 0.03, 0.1$, and 0.2) exhibited enhanced photocatalytic activity in the degradation of MB under visible light irradiation as shown by Guoli Fan et al. (Table 14.2).[21]

Usually coating of ferrites with SiO_2 helps in protecting the nanostructure against degradation and oxidation of the magnetic core. For example, in SiO_2-coated Fe_3O_4, the silica coating on γ-Fe_2O_3 can prevent the thermal transition of γ-Fe_2O_3 to the less magnetic α-Fe_2O_3. As a result, the silica-coated magnetic NPs are stable against degradation which improves their biocompatibility and facilitates their utilization. Furthermore, the ease of silica surface modification allows for further specific functionalization to perform catalysis, biolabeling, and drug delivery. The surface of silica has also been shown to be easily modified with cationic groups which can improve its electrostatic binding to DNA enhancing its use as a drug delivery agent and improving its cellular transport.

Ternary titania–cobalt ferrite–polyaniline nanocomposite prepared by Pan et al.[22] is a good magnetically recyclable hybrid for adsorption and photodegradation of dyes under visible light irradiation. It follows pseudo-second-order and Langmuir models for the adsorption of MO onto the PCs (Fig. 14.6).

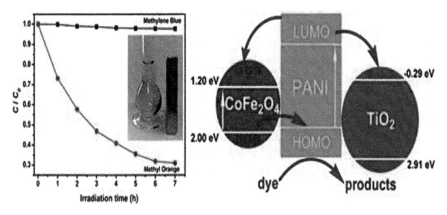

FIGURE 14.6   The dye degradation process.

Yao-Jen et al.[23] recommended that the nano-copper ferrite as a good adsorbent for As (V). Magnetic copper ferrite was successfully fabricated from industrial sludge and was used as an effective adsorbent for As (V). The experimental results show that this adsorbent has a great potential for treating As-containing groundwater and adsorption capacity of the copper ferrite was found to be very good at pH 3.7.

Again cubic copper ferrite $CuFe_2O_4$ nanopowders synthesized via hydrothermal route using industrial waste is found to be a good PC. A good catalytic efficiency of 95.9% in the degradation of the MB was obtained for the copper ferrite which is prepared at hydrothermal temperature of 200°C for hydrothermal time 24 h at pH 12.

PCs can also be deposited as a film onto a substrate, which is placed into the dye solution. Depending on the number of layers that are deposited, the photocatalytic activity varies, leading to changes in the degradation efficiency. For example, a single layer of $ZnFe_2O_4$, which is ~75 nm thick, degrades 35% of the MO dye. However, when the amount of $ZnFe_2O_4$ is increased to three layers (135 nm), degradation increases to 53%. It is also notable that as the layers are further increased to four (165 nm) and five (207 nm), the degradation drops to 47 and 37%, respectively.[24]

Other examples for the ferrite photocatalysis for MO, MB, and RhB are listed below (Tables 14.1–14.3).[25]

TABLE 14.1   Ferrites and Mixed Ferrites Reacted with Methyl Orange Dye under Visible Light Irradiation.

Photocatalyst	Dye (mg/L)	Catalyst (g/L)	Irradiation Source	Irradiation Time (min)	Degradation (%)
$ZnFe_2O_4$	25	5	400 nm	240	4
$ZnFe_2O_4$	10	4	400 nm	60	75
$TiO_2/ZnFe_2O_4$ (0.3%)	25	5	400 nm	240	75
$TiO_2/ZnFe_2O_4$ (1.5%)	25	5	400 nm	240	84
$TiO_2/ZnFe_2O_4$ (3.0%)	25	5	400 nm	240	73
$TiO_2/ZnFe_2O_4$ (6.0%)	25	5	400 nm	240	55
$SiO_2/NiFe_2O_4$	10	1.0	400 nm	60	5
$TiO_{2-x}N_x/NiFe_2O_4$	10	1.1	400 nm	60	50
$CoFe_2O_4/ZnO$	50	30	UV	300	93.9
$TiO_{2-x}N_x/SiO_2/NiFe_2O_4$	10	1.1	400 nm	60	95
$Bi_{12}TiO_2 0SiO_2/NiFe_2O_4$	10	1.1	400 nm	30	85

TABLE 14.2   Methylene Blue Degradation under Visible Light Irradiation.

Photocatalyst	Dye (mg/L)	Catalyst (g/L)	Irradiation Source	Irradiation Time (min)	Degradation (%)
$ZnFe_2O_4$	10	0.6	180	400–700 nm	32
$MnFe_2O_4$	7	0.3	1200	Visible	15.17
$MnFe_2O_4$	7	0.3	1200	Visible	67.18
$ZnFe_2O_4 + H_2O_2$	10	1	360	Dark	45
$ZnFe_2O_4 + H_2O_2$	10	1	360	420 nm	52
$ZnFe_2O_4 + MWCNT\ H_2O_2$	10	1	360	420 nm	99
$Fe_{2.25}Co_{0.75}O_4 + H_2O_2$	100	30	50	None	10
$Fe_{2.46}Co_{0.54}O_4 + H_2O_2$	100	30	50	None	99
$Fe_{2.47}Mn_{0.53}O_4 + H_2O_2$	100	30	50	None	99

TABLE 14.3 Rhodamine B Degradation under Visible Light Irradiation.

Photocatalyst	Dye (mg/L)	Catalyst (g/L)	Irradiation Source	Irradiation Time (min)	Degradation (%)
$ZnFe_2O_4$	10	2.0	150	200–700 nm	60
$ZnFe_2O_4$/Ag	10	2.0	150	200–700 nm	95
$ZnFe_2O_4$	20	0	300	200–700 nm	100
Ag/Fe_3O_4/SiO_2	25	0.5	75	420 nm	10
AgI/Fe_3O_4/SiO_2	25	0.5	75	420 nm	20
Ag-AgI/Fe_3O_4/SiO_2	25	0.5	75	420 nm	99
Ag-AgI/Fe_3O_4/SiO_2	25	0.5	75	420 nm	99

Another study on the degradation of dyes involved the addition H_2O_2 to the reaction mixture. This is due to the fact that H_2O_2 creates a fenton-type system, which causes a larger number of reactive oxygen species formations, and, therefore, results in much high reactivity. This fact is evident as seen with $ZnFe_2O_4$ and H_2O_2 without light irradiation, which shows a 45% degradation of the dye. When irradiated with light, degradation is increased to 52%, showing that H_2O_2 can play an important role in the photocatalytic process. Composite PCs also show increased degradation efficiency with the addition of H_2O_2.[17]

14.3 CONCLUSION

Spinel ferrite NPs are of interest because of their well-known unique optical, electronic, and magnetic properties. These NPs have high permeability, good saturation magnetization, and no preferred direction of magnetization. They are magnetically "soft," being easily magnetized and demagnetized, and electrically insulating. For these reasons, ferrites have been used as magnetic materials as well as refractory materials and catalysts. However, using such magnetic NPs for biological and medical purposes remains a challenge due to lack of much research in this field.

The effectiveness of the ferrite on dye degradation depends on both the ferrite and the dye. Ferrites are shown to be effective visible light PCs to perform oxidation processes. Different preparation methods affect the size, shape, and overall structure, all of which can alter the photocatalytic activity. The activity changes by heat-treatment of the PC. Heating changes the crystal structure leading to change in surface defects, hence may affect the activity. Mechanical treatment of the PC also results in similar effects on photocatalytic activity.

The combination of ferrites with other PCs at different ratios or stoichiometric amounts showed a synergistic effect that produces enhanced photocatalytic activity. Also, with the addition of ferrites, PCs that previously are effective only under UV irradiation can become effective under visible light irradiation. Furthermore, combining two PCs with different band gap positions effectively causes a greater separation of e^-/h^+ pairs, allowing more of the species to be available for reactions to degrade contaminants. The ferrites have been shown that they can be recycled for the degradation of a number of different contaminants with little or no loss of the photocatalytic activity. Significantly, ferrites react to degrade organic dyes, however, in most cases complete degradation occurs only in the presence of another PC or oxidizing agent such as H_2O_2.

KEYWORDS

- **magnetic nanoparticles**
- **toxicity**
- **cytokinesis**
- **antibacterial activity**

REFERENCES

1. Sanpo, N.; Wen, C.; Berndt, C, C.; Wang, J. Antibacterial Properties of Spinel Ferrite Nanoparticles Microbial Pathogens and Strategies for Combating Them: Science, Technology and Education. *Formatex* **2013**, 1, 239–250.
2. Jiang, K.; Li, K.; Peng, C.; Zhu, Y. Effect of Multi-Additives on the Microstructure and Magnetic Properties of High Permeability Mn-Zn Ferrite. *J. Alloys Compd.* **2012**, *541*, 472–476.
3. Sanpo, N.; Wang, J.; Berndt, C. C. Effect of Zinc Substitution on Microstructure and Antibacterial Properties of Cobalt Ferrite Nanopowders Synthesized by Sol-Gel Methods. *Adv. Mater. Res.* **2012**, *535–537*, 436–439.
4. Raul, V. Novel Applications of Ferrites. *Phys. Res. Int.* **2012**, *2012*, 1–9.
5. Buteica, A. S.; Mihaiescu, D. E.; Grumezescu, A. M.; Vasile, B. S.; Popescu, A.; Mihaiescu, O. M. The Anti-Bacterial Activity of Magnetic Nanofluid: Fe$_3$O$_4$/Oleic Acid/Cephalosporins Core/Shell/Adsorption Shell Proved on *S. aureus* and *E. coli* and

Possible Applications as Drug Delivery Systems. *Dig. J. Nanomater. Biostruct.* **2010,** *5,* 927–932.

6. Sun Shouheng.; Hao Zeng. Size-Controlled Synthesis of Magnetite Nanoparticles. *J. Am. Chem. Soc.* **2002,** *124* (28), 8204–8205.

7. Sanpo, N.; Berndt, C. C.; Wen, C.; Wang, J. Transition Metal-Substituted Cobalt Ferrite Nanoparticles for Biomedical Applications. *Acta Biomater.* **2013,** *9,* 5830–5835.

8. Sanpo, N.; Berndt, C. C.; Wang, J. Microstructural and Antibacterial Properties of Zinc-Substituted Cobalt Ferrite Nanopowders Synthesized by Sol-Gel Methods. *J. Appl. Phys.* **2012,** *112,* 1–6.

9. Raffi, M.; Mehrwan, S.; Bhatti, T. M.; Akhter, J. I.; Hameed, A.; Yawar, W. Investigations into the Antibacterial Behavior of Copper Nanoparticles against *Escherichia coli.* *Annal. Microbiol.* **2010,** *60,* 75–80.

10. Kostedt, IV W. L.; Drwiega, J.; Mazyck, D. W.; Lee, S.; Sigmund, W.; Wu, C.; Chadik, P. Magnetically Agitated Photocatalytic Reactor for Photocatalytic Oxidation of Aqueous Phase Organic Pollutants. *Environ. Sci. Technol.* **2005,** *39,* 8052–8056.

11. Valenzuela, M. A.; Bosch, P.; Jimenez-Becerrill, J.; Quiroz, O.; Paez, A. I. Preparation, Characterization and Photocatalytic Activity of ZnO, Fe_2O_3 and $ZnFe_2O_4$. *J. Photochem. Photobiol. A Chem.* **2002,** *148,* 177–182.

12. Wikins, T. D.; Holdeman, L. V.; Abramson, I. J.; Moore, W. E. Standardized Single-Disc Method for Antibiotic Susceptibility Testing of Anaerobic Bacteria. *Antimicrob. Agents Chemother.* **1972,** *1,* 451–459.

13. Dutta, P. K.; Pehkonen, S. O.; Sharma, V. K.; Ray, A. K. Photocatalytic Oxidation of Arsenic (III): Evidence of Hydroxyl Radicals. *Environ. Sci. Technol.* **2005,** *39,* 1827–1834.

14. Gratzel, M. Photoelectrochemical Cells. *Nature.* **2001,** *414,* 338–344.

15. Cai-Hong, C.; Yan-Hui, L.; Wei-De, Z. $ZnFe_2O_4$/MWCNTs Composite with Enhanced Photocatalytic Activity under Visible-Light Irradiation. *J. Alloys Compd.* **2010,** *501,* 168–172.

16. Rashad, M. M. Magnetic and Catalytic Properties of Cubic Copper Ferrite Nanopowders Synthesized from Secondary Resources. *Adv. Powder Technol.* **2012,** *23,* 315–323.

17. Kornmuller, A.; Karcher, S.; Jekel, M. Adsorption of Reactive Dyes to Granulated Iron Hydroxide and Its Oxidative Regeneration. *Water Sci. Technol.* **2002,** *46,* 43–50.

18. Cheng P.; Deng C.; Gu M.; Shangguan W. Visible-Light Responsive Zinc Ferrite Doped Titania Photocatalyst for Methyl Orange Degradation. *J. Mater. Sci.* **2007,** *42,* 9239–9244.

19. Sonal, S.; Rimi, S.; Charanjit, S.; Bansal, S. Enhanced Photocatalytic Degradation of Methylene Blue Using $ZnFe_2O_4$/MWCNT Composite Synthesized by Hydrothermal Method. *Indian J. Mater. Sci.* **2013,** *2013,* 1–6.

20. David, G.; Raquel, S. G.; Joseph, G.; Yurii, K.; Gun'ko. Synthesis and Characterisation and Photocatalytic Studies of Cobalt Ferrite-Silica- Titania Nanocomposites. *Nanomaterials* **2014,** *4,* 331–334.

21. Guoli, F.; Ji, T.; Feng, L. Visible-Light-Induced Photocatalyst Based on Cobalt-Doped Zinc Ferrite Nanocrystals. *Ind. Eng. Chem. Res.* **2010,** *2,* 1995–1996.

22. Pan, X.; Lianjun, W.; Xiaoqiang, S.; Binhai, X.; Xin, W. Ternary Titania–Cobalt Ferrite–Polyaniline Nanocomposite: A Magnetically Recyclable Hybrid for Adsorption and Photodegradation of Dyes under Visible Light. *Ind. Eng. Chem. Res.* **2013,** *52* (30), 10105–10113.

23. Yao-Jen, T.; Chen-Feng, Y.; Chien-Kuei, C.; Shan-Li, W.; Ting-Shan, C. Arsenate Adsorption from Water Using a Novel Fabricated Copper Ferrite. *Chem. Eng. J.* **2012,** *198–199,* 440–448.

24. Qiu, J.; Wang, C.; Gu, M. Photocatalytic Properties and Optical Absorption of Zinc Ferrite Nanometer Films. *Mater. Sci. Eng. B.* **2004,** *112,* 1–4.

25. Erik, C.; Virender, K. S.; Xiang-Zhong, L. Synthesis and Photocatalytic Activity of Ferrites under Visible Light: A Review. *Sep. Purif. Technol.* **2012,** *87,* 1–14.

PART IV

Electronics Application of Nanomaterials

CHAPTER 15

COPPER/NiZn FERRITE NANOCOMPOSITE FOR MICROINDUCTOR APPLICATIONS

S. R. MURTHY[1,2*]

[1]*Department of Mechanical Engineering, Geethanjali College of Engineering and Technology, Cheeryal (V), Keesara (M), RR Dist., Hyderabad 501301, Telangana, India*

[2]*Department of Physics, Osmania University, Hyderabad 500007, Telangana, India*

Corresponding author. E-mail: ramanasarabu@gmail.com

CONTENTS

ABSTRACT

The nanoparticles of copper metal and $Ni_{0.5}Zn_{0.5}Fe_2O_4$ were synthesized by using electroplating and microwave-hydrothermal (M-H) methods, respectively, and they were further used to prepare $Cu/Ni_{0.5}Zn_{0.5}Fe_2O_4$ nanocomposites by mechanical milling method. The phase identification, crystallinity, and morphology of the prepared nanoparticles were characterized by using X-ray diffraction (XRD), transmission electron microscopy (TEM), and Fourier transforms infrared spectroscopy (FTIR). The XRD analysis was carried out on prepared nanocomposite to confirm the phases formed. The microstructure of nanocomposites has been examined by scanning electron microscopy (SEM). The magnetic hysteresis measurements were carried out to obtain important magnetic properties such as, saturation magnetization (Ms) and coercivity (Hc) of the samples. The performance of the copper/ NiZn ferrite composites has been estimated from the studies of dependence of permittivity spectra on the frequency. The higher frequency response characteristics (>1 GHz) of the nanocomposite has been obtained by the measurement of S-parameters. The applicability of the present nanocomposites for the use of electromagnetic microwave absorption was examined in terms of their dielectric, magnetic properties, and shielding effectiveness (SE). The microinductors were fabricated with the help of composite material by using screen-printing method. The electrical properties, such as inductance (L) and quality factor (Q) of the fabricated inductors were measured over a wide frequency range.

15.1 INTRODUCTION

Composite materials are one of the main types of the engineering materials, next to metals and alloys, ceramics, and polymeric materials. They are made at least of two separate types of substances each with its own characteristics, one of them is called matrix while the second is filler. In composite materials, these two phases are immiscible and are separated by boundary interface layer. Such a wide range of constituent materials gives materials scientists' very big field of possibilities to create new innovative material selection and design; it is possible to change properties of products depending on requirements and future applications. Composite materials find practical application in many domains of industry for instance in civil engineering, machine building, sport and leisure industry, automotive industry, aerospace industry, etc. Nowadays some kinds of composite materials demonstrate gradual distribution of structure and/or composition, for example, gradual distribution of fillers in the matrix. Such continuous structure or composition changes can eliminate one of basic drawbacks of other composites, namely sharp boundaries between joined substances and stresses concentrations at the same time. In this way scientists try to increase strength properties of ready products and to differentiate properties across them. Scientific interest is primarily motivated by the possibility of exploring totally new phenomena in novel class of materials with properties mainly determined by the tremendously high surface/volume ratio, which offers the possibility of applications ranging from the field of new catalysts to the fabrication of ceramic nanocomposites with significant property improvement.

Inductive components are extensively used in high frequency (>1 MHz) electronic devices from radar, satellite, and telecommunication systems to home stereo radios. Conventional inductive components use metallic alloys and ferrites as core materials. The major problem for metallic materials is their low resistivity (~10–6 Wcm). Since it is impossible to dramatically increase their resistivity, metallic materials were excluded in high frequency applications and ferrites have been the only choice for six decades since World War II. Although efforts have been made extensively to improve the performance of the ferrites, very limited progress was obtained. Magnetic materials have been a key impediment for the miniaturization of electronic equipment. To overcome the difficulties of both metallic alloys and ferrites, one can think of metal/ferrite nanocomposites for the next generation of high frequency magnetic applications. Nanocomposite processing has provided a new approach for fabricating soft magnetic materials. In a metal/ferrite nanocomposite, the resistivity can be drastically increased, leading to significantly reduced eddy current loss. In addition, the exchange coupling between neighboring magnetic nanoparticles can overcome the anisotropy and demagnetizing effect, resulting in much better soft magnetic properties than conventional bulk form materials.

A completely new phenomenon has been observed when reducing the particle size and the separation between neighboring particles at a nanometer scale in composite materials. For example, it has been found that a Co- or Fe-based nanocomposite can possess permeability much higher than that obtainable from the bulk Co or Fe metal. This large enhancement in permeability is due to the exchange coupling effect. The exchange interaction which leads to magnetic ordering within a grain also extends out to neighboring environments within a characteristic distance, the so-called exchange length. Thus, neighboring grains separated by distances shorter than a nanometer can be magnetically coupled by exchange interaction. For a traditional powder material of large particle sizes, exchange coupling effect is negligibly small in determining magnetic properties. However, when the particle size plus the separation between particles is reduced, intergrain exchange coupling plays a dominant role and the material will possess a variety of properties different from the bulk size material. One important effect is the cancellation of the magnetic anisotropy of individual nanoparticles: when the particle–particle separation is significantly less, the intergrain exchange interaction makes all the neighboring particles coupled. This coupling averages out the magnetic anisotropy of individual nanoparticles. As a consequence, the permeability of an exchange-coupled nanocomposite can be even much higher than the permeability of its bulk counterpart. There has been a great deal of interest in recent years in artificially engineered nanomaterials with novel physical properties. Among the researches, perhaps the effort in magnetic nanostructures attained the biggest rewards to date. These achievements imply a bright future for nanomagnetics.

The electronics industry is directed toward high frequency of operation, which in turn requires bulk sized high frequency soft magnetic materials. Conventionally used ferrites possess poor magnetic properties at elevated frequencies. The metal/insulator nanocomposite design, the metal nanograins are insulated by insulating layers, thus the resistivity of the system will be dramatically increased, leading to a significantly reduced eddy current loss, while the exchange coupling between neighboring magnetic nanoparticles can overcome the anisotropy and demagnetizing effect, resulting in much better soft magnetic properties than conventional ferrites. This design provides more degrees of freedom (phase constituents, their ratio, and grain size) to tailor magnetic as well as electric properties. We will be studying metal/insulator system to meet various requirements.

All electronic devices generate and emit radiofrequency waves that can interfere with the operation of electronic components within the same device as well as other electronic devices. In present day, all electronic devices are moving toward miniaturization and in this process the electronic components are to be packed very close to each other, which increase the problem of electromagnetic interference (EMI). When an electromagnetic wave is incident upon a conductive surface, energy is reflected and absorbed. The ability of a material to shield electromagnetic energy, whether it be unwanted energy entering a system or escaping a system, is called as its shielding effectiveness (SE). It is consists of losses due to absorption, reflection, and multifull reflections. EMI suppression over a wideband frequency range requires tunability of the impedance (Z), which depends on the tunability of the complex permeability and complex dielectric constant. The conductivity plays an important role in a material's ability to shield electromagnetic energy.

Metals and ferrites are commonly used in EMI suppression and these materials have many disadvantages in terms of their weight, corrosion, and physical rigidity. These properties could be overcome with the development of new composite materials. Out of all composite materials, conducting and insulating materials with magnetic nanoparticles are lightweight, flexible alternatives to the micron-sized metal components. Hence, metal/ferrite nanocomposites are best choice. Another importance of nanocomposites is that critical parameters, such as loss tangents and impedance matching, which are important in microwave devices, may be controlled in these materials. The loss tangent is a measure of the inefficiency of a magnetic system. The loss tangents are primarily determined by magnetic and eddy current losses and which depend on the resistivity of the material. The resistivity of conducting metal is increases with an addition of magnetic nanoparticles, thereby decreasing the eddy current losses. Generally, magnetic losses in composite materials are controlled by the material grain structure; domain wall resonances, etc., and these parameters can be manipulated effectively in nanocomposite materials by the size distribution of nanoparticles and their dispersion in the host matrix.

During the transmission, impedance mismatch between source and load in circuits is the main source of signal attenuation. By manipulating the properties of nanocomposites desired impedance values at specific frequencies are attained and the attenuation can be minimized. An impedance of nanocomposite material can be adjusted by the type and amount of magnetic nanoparticles dispersed in the metal. The combination of the ferrimagnetic nanoparticles with conducting metal leads to formation of an important nanocomposite possessing with unique combination of electrical and magnetic properties. This property of the nanocomposite materials can be used for utilizing them as electromagnetic shielding material. This is because, the electromagnetic wave consists of an electric (E) and the magnetic field (H) right angle to each other. The ratio between E and H factor (impedance) has been subjugated in the shielding application. In metal/ferrite type of nanocomposites, the conducting copper metal can effectively shield electromagnetic waves generated from an electric source, whereas electromagnetic waves from a magnetic source can be effectively shielded only by magnetic materials. Thus, having both conducting and magnetic components in a single system could be used as an EMI shielding material. The applicability of the present nanocomposites for the use of electromagnetic microwave absorption was examined in terms of their dielectric, magnetic properties, and SE.

Composites made of micronic or nanometric metallic particles dispersed in an oxide matrix, have received great attention because of their specific or improved mechanical, optical, electrical, thermal, or magnetic properties.[1-8] Metal–ceramic composites are widely studied for their mechanical, magnetic properties, and catalysts.[9] However, no studies were reported on Cu/ferrite nanocomposites used for high frequency devices such as microinductor applications.

With the development of integrated wireless communication systems, on-chip inductors with satisfactory performance (enough quality factor and self resonant frequency) are required. In recent years, on-chip spiral inductor is a key component to the success and of high performance radio frequency integrated the circuitry (RFIC) design. In general, hundreds of inductors have been implemented in a RF transceiver. Much effort has been made to improve the performance of on-chip inductors by using copper metal with higher conductivity. The coil function ranges from impedance matching, electromagnetic energy transforming, to signal oscillating due to its low quality characteristics (>1 GHz) of the superparamagnetic nanoparticles and lower electrical resistivity of copper. In an order to resolve these problems, several techniques have been developed and proposed to improve the performance of the spiral inductor. For example, silicon micromachining technique is utilized to remove the substrate loss mechanism for high Q performance.[10] Similarly, the use of spiral coil made with soft ferrimagnetic materials increases an areaeffective of integrated inductor.[11] However, these approaches cannot fully solved problems occur in the on-chip spiral inductor. The micromachined inductors possess a high Q-factor, but still have smaller area. Due to this, it is difficult to achieve a high inductance value using this type of inductor. But, using ferrimagnetic materials as a core may reduce the size of inductor. But, the low self-resonant frequency and eddy current loss of the ferrimagnetic inductor still hinder its application for high frequency RFIC.

The properties of the nanocomposites depend on the microstructure which is related to the method of preparation. The later plays a very important role with regard to the chemical and structural properties of the composite. Many methods to prepare magnetic material and CuO nanocomposites have been published.[12-14] Therefore, in order to solve all the problems, we have undertaken a detailed study on metal (Cu)/ferrite ($Ni_{0.5}Zn_{0.5}Fe_2O_4$) nanocomposite materials for the fabrication of on-chip spiral inductor. The nanocomposites were prepared using mechanical milling method. This method is simple, considers an environment friendly method that offers scalability for large scale production, uses low cost starting materials and low synthesis temperature.

15.2 EXPERIMENTAL

15.2.1 SYNTHESIS OF NANOPOWDERS OF FERRITE AND COPPER

For the synthesis of nanopowders of NiZn ferrite, high pure (99.98%) nickel nitrate [$Ni(NO_3)2.6H_2O$] (99.97%) zinc nitrate [$Zn(NO_3)2.6H_2O$], and (99.98%) iron nitrate [$Fe(NO_3)2.9H_2O$] were dissolved in 50 mL of deionized water. The molar ratio of powders was stoichiometrically adjusted in such a way to obtain the composition $Ni_{0.5}Zn_{0.5}Fe_2O_4$. The powders were dissolved in 50 mL of deionized water and aqueous NaOH solution was added until to obtain the desired pH (~9.5). The precipitate was transferred into a teflon lined vessel and treated using a microwave digestion system (Model MDS-2000, CEM Corp., Mathews, NC) at 160°C/30 min. This system uses 2.45 GHz microwaves and can operate at 0–100% full power (1200 ± 50 W). The products obtained were filtered and washed repeatedly with deionized water and finally with ethanol to remove the residual nitrates present. This final slurry was dried up for about 12 h at 65°C. The obtained powders were weighed and the percentage yields were calculated from the total expected based on the solution concentration and volume and the amount that was actually crystallized. The percentage of yield obtained was ~98%. As synthesized powder was characterized using X-ray diffraction (XRD) (PhilipsPW-1730 X-ray diffractometer) with Cu-Kα radiation (λ = 1.5406 A°) and transmission electron microscopy (TEM) (Model JEM-2010, JEOL, Tokyo, Japan). Fourier transforms infrared spectroscopy (FTIR, Brucker tensor 27 DTGS TEC) studies were carried out in the wavenumber range of 375–4000 cm^{-1}.

Nanocrystalline copper has been synthesized using electrolytic cathode method. A homogenous aqueous sulfate solution was kept in a cleaned glass vessel, surface-cleaned anode and cathode copper plates were kept inside the glass vessel. Electrolysis processes have been carried out by passing a current of 2 A inside the solution through anode and cathode. The different experimental conditions were adopted with constant current and with constant concentration of the copper sulfate solution for various time periods at room temperature. A layer of copper deposition on the cathode surface was observed at the end of electrolyzing process. Then copper powder was collected from the cathode surface. The powder was characterized by using XRD, TEM, and FTIR.

15.2.2 PREPARATION OF NANOCOMPOSITE COPPER + NiZn FERRITES

The methods which have been commonly employed to synthesize nanoscale materials include inert gas condensation, spray conversion processing, and controlled crystallization of amorphous phases, vapor deposition, coprecipitation, sol–gel process, plasma processing, laser ablation, hydrothermal pyrolysis, and sonochemical processing. Although significant progresses have been achieved in the preparation of nanomaterials, there are problems. For example, sol–gel process uses metal alkoxides as the starting materials, which are very expensive and extremely sensitive to the environmental conditions, such as moisture, light, and heat. Moisture sensitivity makes it necessary to conduct the experiment in dry boxes or clean rooms. Coprecipitation processes involve repeated washing in order to eliminate the anions coming from the precursor salts used, making the process complicated and very time consuming. Furthermore, it is difficult to produce large batches by using most of the chemical solution processing routes. Therefore, exploring alternative methods for the preparation of nanomaterials is still of technological as well as scientific significances.

Mechanochemical synthesis, which is also known as mechanical alloying, high-energy mechanical milling, high-energy milling, and high-energy activation, was initially invented to prepare oxide-dispersed metallic alloys for structural applications and subsequently applied to extensions of metallic solid solubility, synthesis of intermetallics, disordering of intermetallics, solid-state amorphization, nanostructured materials, and mechanochemical synthesis of nanosized oxides or metal powders.[15-22] Recently, this novel technique has been successfully employed to synthesize a wide range of nanosized ceramic powders,[23,24] superconductor[25] ferrites,[26,27] ferroelectrics,[28,29] as well as composites.[30,31] The most significant characteristic of this technique is that the formation of the designed compounds is due to the reactions of oxide precursors which are activated by mechanical energy, instead of the heat energy required in the conventional solid-state reaction process. The novel mechanical technique is superior to both the conventional solid-state reaction and the wet-chemistry-based processing routes for several reasons. First, it uses cost-effective and widely available oxides as the starting materials and skips the intermediate temperature calcination step, leading to a simpler process. Second, it takes place at room temperature in well-sealed containers, thus effectively controlling the loss of the volatile components. Furthermore, due to their nanometer scale size and very high homogeneity, the mechanochemically derived ceramic powders demonstrate much better sinterability than those synthesized by the conventional solid-state reaction and wet-chemical processes. Also, the high-energy milling can greatly improve the reactivity of precursors by reducing the phase formation temperatures of some ferroelectric materials which cannot be directly synthesized. In this chapter, we will report the enhanced formation of copper and ferrite nanocomposite prepared by a high-energy ball milling process.

The synthesized nanopowders of $Ni_{0.5}Zn_{0.5}Fe_2O_4$ (NZ) and pure copper (Cu) were mixed at different weight percentage to obtain the composites of Cu (x) + $(1-x)$ $Ni_{0.5}Zn_{0.5}Fe_2O_4$ and named as copper (Cu), 25 wt% Cu + 75 wt% NiZn ferrite (NZC1), 50 wt% Cu + 50 wt% NiZn ferrite (NZC2), 75 wt% Cu + 25 wt% of NiZn ferrite (NZC3), and $Ni_{0.5}Zn_{0.5}Fe_2O_4$ (NZ). The required amounts were weighed and mixed accordingly in a vial to achieve the stoichiometry. Mechanical milling was carried out in the hardened WC vial together with ten 12 mm WC balls for 20 h using a Retsch Co. high energy planetary ball mill. A ball to powder mass charge ratio of 14:1 was chosen. The speed of the mill was set at 400 rpm with interval at 40 min. Finally, at the end of 20 h grinding time the powder was annealed at 600°C/1 h. Then, the samples were characterized by XRD and scanning electron microscopy (SEM, LEICA, S440i, UK).

The magnetic properties, such as saturation magnetization (Ms) and coercive field (Hc) for the samples were obtained from the recorded hysteresis loops obtained with the help of vibrating sample magnetometer (VSM, Model DMS 1660 model) at room temperature. The complex permittivity (ε^*) was measured in the frequency range of 1 MHz–1.8 GHz using impedance analyzer (HP 4291B) at room temperature. The shielding efficiency was measured according to the transmission/reflection method though a two-port vector network analyzer (VNA, Agilent, ENA, E5071) in the range of 1–18 GHz. This apparatus allows for the characterization of microwave devices, circuits, and the properties of materials at the specified frequencies. Samples were characterized with a 7 mm outer and 3.04 mm inner diameter coaxial transmission line adapter. Samples were precisely machined into toroidal shapes and inserted within the coaxial sample holder. Using the built-in software, a geometry correction was applied for correcting small deviations from nominal geometry. The microinductors were fabricated using the screen-printing and coding method with the help of nanocomposite material paste. The inductance (L) and quality factor (Q) of the fabricated inductors were also measured. The SE is determined as a ratio of the power received without shielding material (P_0) to that of the power received with shielding material (P_1): SE = $-10 \log(P_0/P_1)$.

15.3 RESULTS AND DISCUSSION

Figure 15.1a gives the XRD pattern for nanopowder of $Ni_{0.5}Zn_{0.5}Fe_2O_4$ (NZ). It can be seen from the figure that the powder possesses spinel phase (JCPDS card no. 03-0864) and no other phases were detected. Particle size was estimated by Scherrer's formula: $D = 0.9\ \lambda/\beta\ cos\theta$, where D is the diameter of the crystal particle, λ is the wavelength of the target used (Cu-Kα = 1.5406 Å), and β is the full width at half of diffracted line in radian. The average particle size obtained from the above equation is 20 nm. Particle size and morphology were determined using TEM. Figure 15.1b shows the XRD pattern for pure copper powder. It can be seen from the figure that the obtained particles are (FCC) copper nanopowder. The presently obtained XRD pattern has been compared with the standard powder diffraction card of JCPDS, copper file NO. 04-0836. The crystalline size has been estimated using Scherer's formula and it is 48 nm. Lattice constant has been estimated from XRD data and it is 3.6140 ± 0.0002 Å and which is fair agreement with reported value of 3.6150 Å (Powder Diffraction File ~Joint Committee on Powder Diffraction Standards, Swarthmore, PA, 1990, No. 40836). It can be seen from the TEM picture (Fig. 15.2a) for the ferrite, the particles are evenly distributed, aggregated, and exhibited rounded irregular shapes. The selected-area diffraction (SAD) patterns of these specimens reveal the formation of single-phase ferrite. The diffraction rings, characteristic of nanocrystalline aggregates, have been indexed. Indeed the bright and dark field images reveal nanometric particles with a broad (15–20 nm)

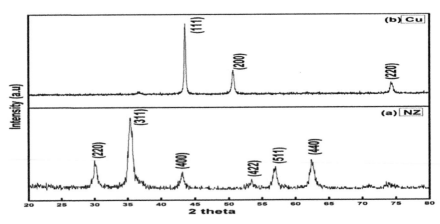

FIGURE 15.1 The XRD patterns for as synthesized nanopowders of (a) $Ni_{0.5}Zn_{0.5}Fe_2O_4$ (NZ) and (b) pure copper (Cu).

size distribution. Occasional spots in the SAD pattern may arise from coarse crystallites or agglomerates. The mean particle sizes obtained from the TEM picture is 18 nm. Thus, the particle sizes calculated from TEM and XRD are almost comparable with each other.

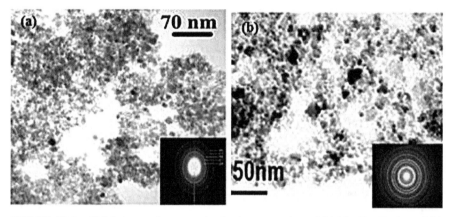

FIGURE 15.2 TEM pictures for as synthesized nanopowders of (a) $Ni_{0.5}Zn_{0.5}Fe_2O_4$ and (b) pure copper.

Copper has a finest microstructure (Fig. 15.2b) along with the most narrow particle size distribution. Assuming that each particle is circular, the area of each particle was then determined. The average largest particle diameter was calculated based on averaging the five largest particles found in the size distributions. Twin boundaries were ignored in particle size measurements, that is, a particle that contained many twin boundaries was considered as only one particle. The average particle size was 21 nm.

Figure 15.3 gives the FTIR spectra of the $Ni_{0.5}Zn_{0.5}Fe_2O_4$ (NZ), pure copper (Cu), and nanocomposites (NZC1, NZC2, and NZC3). It can be seen from the figure that the formation of cubic spinel ferrite (NZ) structure is confirmed for 20 h milled powders. It can be observed from the figure, in the case of NZ and NZC1, NZC2, and NZC3 samples, vibrational frequencies of IR bands of ferrite are observed in the range $v_1 = 584$ cm^{-1} and $v_2 = 440$ cm^{-1} and these are attributed to the intrinsic vibration of the tetrahedral and octahedral sites, respectively. It can be seen from the figure that a small band of absorption is observed at 3470 cm^{-1}, which characterizes stretching vibrations of O—H bonds in H_2O molecules adsorbed on the nanoparticles surface. The band observed at 1610 cm^{-1} is characteristic for bending vibrations of O—H bonds in OH groups.[32,33] A FTIR spectrum of the Cu nanoparticles is also shown in the figure. The bands observed in the region 1000–1600 cm^{-1} are gives an account of antisymmetrical and symmetrical stretching vibrations of B—O bonds in BO$_3$ and BO$_4$ groups formed on the nanoparticle surface.[34-36] The band at 615 cm^{-1} can be referred to symmetrical stretching vibrations of B—O bonds in BO$_4$ groups involving the oxygen atoms connecting the groups.[36] The bands of absorption in the 750–500 cm^{-1} region can be assigned to bending vibrations of B—O bonds in BO3 and BO4 groups. The weakly expressed bands at < 500 cm^{-1} give information about the Cu—O bond vibrations in copper oxides (Cu$_2$O and CuO).

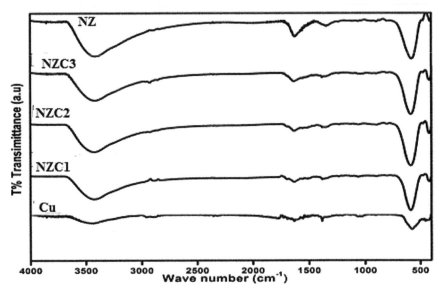

FIGURE 15.3   FTIR spectra for NiCuZn ferrite (NZ), nanocomposites (NZC1, NZC2 and NZC3), and pure copper (Cu).

Figure 15.4a–c shows the XRD patterns of the annealed samples of composites. It can be seen from the figure that only pure Cu metal and ferrite peaks are present in the XRD patterns. Three peaks at 2θ values at 43.640°, 50.800°, and 74.420° corresponding to (111), (200), and (220) planes of copper were observed. The characteristic peak (311) of ferrite can be observed at $2\theta = \sim35°$. No impurity phases were detected in the XRD patterns. The intensity of the major peaks, such as (311) for ferrite and (111) for Cu, depends on the amount of their individual phase fraction. The number of copper metal peaks increases with an increase of copper content (x), and the peaks are intensified while the intensity and numbers of ferrite-phase peaks decrease. The crystallite size of the ferrite and copper has been calculated from the X-ray peak broadening of the (311) and (111) diffraction peak using Scherer's formula and presented in Table 15.1. It can be seen from the table that the crystallite size decreases with an increase of copper content. The lattice parameters of the composites were calculated using XRD patterns and results are given in Table 15.1. It can be seen from the table that the lattice parameter is found to decrease with an increase of Cu content. The shift of XRD peaks to lower angle side with an increase of Cu content implies change in the lattice parameter due to Cu substitution. This is because of the ionic radius difference of Cu^{2+} (0.87 A°) and Ni^{2+} (0.83 A°). But, the non-linear variation of lattice constant with Cu content indicates that the distribution of divalent copper ions in the different ionic states in the A- and B-sites.

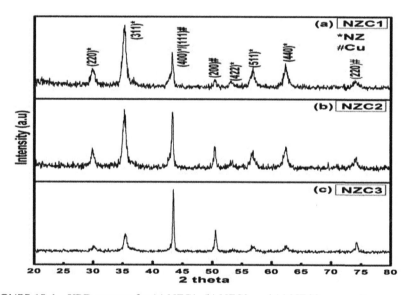

FIGURE 15.4 XRD patterns for (a) NZC1, (b) NZC2, and (c) NZC3 composites.

Figure 15.5 gives the TGA-DTA plots for the nanocomposites. It is evident from the figure that rather rapid weight loss (0.4%) with small exothermic peak on DTA is observed in the temperature range of 25–100°C, due to vaporization of the residual water content. No significant weight loss was observed in the temperature range of 100–400°C, indicating no quantitative oxidation behavior of Cu nanoparticles. These results show that a thermal stability of the Cu nanoparticles. The decomposition of particles begins at 410°C (a regular decrease of weight loss on TGA curve).

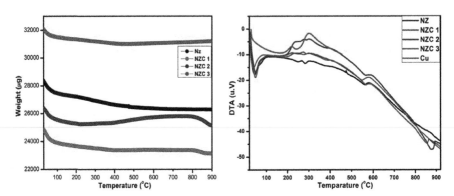

FIGURE 15.5 TGA-DTA plots for the nanocomposites.

Figure 15.6 shows the DSC curves of as synthesized $NiZnFe_2O_4$, Cu, and composites powders. In all the samples an endothermic peak is observed at 100°C and which is associated with the removal of water retained in the partially dried powder. The endothermic peak observed around 320°C is due to the formation of ferrite phase. The exothermic peak around 220°C indicates crystallization process begins in the Cu powder.

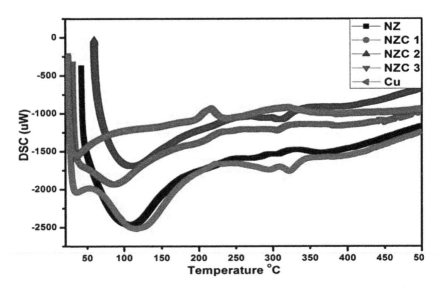

FIGURE 15.6 DSC curves of as synthesized $NiZnFe_2O_4$, Cu, and composites powders.

Figure 15.7 gives the SEM photographs of all the nanocomposites. It can be seen from figure that the microstructure of ferrite confirming that the particles are in nanoscale. SEM micrographs of nanocomposites exhibit a two-phase system with copper grains (white) and NiZn ferrite grains (black). Energy-dispersive X-ray spectroscopy has been used to confirm the grains of these two phases. The connectivity of NiZn ferrite grains is dispersed by the distribution of metal grains, which leads to the variation of magnetic properties of the composites. It can be observed from the images that the average grain sizes of the samples are in the range 50–80 nm. The grain size of the composites is somewhat smaller than the crystallite size calculated by the XRD method, which is due to the existence of the partial particle aggregation. The theoretical density of the composites is calculated from equation: $\rho = m1 + m2/(m1\rho1 + m2\rho2 + \rho1\rho2)$, where $m1$ and $m2$ are the masses of NiZn ferrite and copper, respectively, $\rho1$, $\rho2$, and ρ are the theoretical densities of NiZn ferrite, copper, and composites, respectively. The bulk density of the composites is measured by Archemidics principles and they are in the range of 94–98% of the theoretical density. The density of the nanocomposites are found to increase (Table 15.1) with increasing of copper content, because ferrite, and metal were densified by the interaction of the two phases that coexist in one material. From the SEM pictures and appearance of the sintered samples, no mismatch has occurred in the composites, which indicates that the two phases have better cofiring properties. The average value of %P (porosity) for the present nanocomposites is 4.

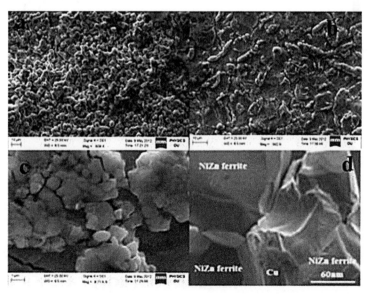

FIGURE 15.7 SEM pictures for (a) NZC1, (b) NZC2, and (c) NZC3 composites and (d) enlarger view of nanocomposites.

TABLE 15.1 Experimental Data for Nanocomposites.

Sr. No	Sample Name	Ferrite Crystallite Size D (nm)	Cu Metal Crystallite Size D (nm)	Lattice Parameter (a) A°	Theoretical Density g/cm^3	Dielectric Constant at 1 MHz	M_s emu/g	H_c Oe
1	NZ	35		8.408	5.313	24	38	55
2	NZC1	17	54	8.440	5.432	20	29	30
3	NZC2	13	72	8.431	5.763	16	12	15
4	NZC3	18	62	8.390	6.521	8	5	4
5	Cu		48	3.6	9.045			

Figure 15.8 shows magnetic hysteresis loops for all composites at room temperature. The samples exhibit typical magnetic hysteresis, indicating that the composites are magnetically ordered. The values of saturation magnetization (*Ms*) and coercivity (*Hc*) are obtained from hysteresis loops and are presented in Table 15.1. It can be seen from the table that the coercivity for the NZC1 nanocomposite is lower than that of NZ. This indicates that the magnetization ability becomes weak because of the existence of

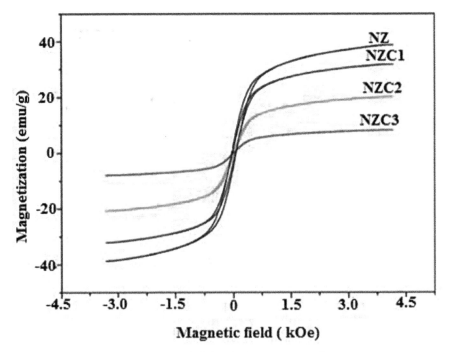

FIGURE 15.8 Hysteresis loops for Cu/NiZn ferrite nanocomposites.

nonmagnetic Cu phase, in which domain wall pinning can occur, which leads to the decrease of coercivity. The coercivity further decreases for NZC2 and NZC3 nanocomposites and which indicates that no domain wall pinning occurs. The saturation magnetization of the composites linearly decreases with the increase of Cu component. The saturation magnetization of Cu is inherent nonmagnetic nature. Therefore, the saturation magnetization of composites decreases linearly with the increasing of Cu phase. The distribution of cations in A and B sites influences the magnetic properties. Owing to the presence of diamagnetic zinc ions, the magnetization of the A-sites will be smaller than the B sites although the Fe^{3+} ions have the largest moment (5 μ_B). The magnetization of the A sublattice is expected to be constant due to unchanged Zn-content and the B sublattice will be decreased with increase of lower magnetic moment of Cu^{2+} (1 μ_B) that will make a net decrease in the magnetization (M_S) of the composite. The magnetization decreases with an increase of Cu-phase.

Figure 15.9 gives the frequency dependence of the complex permittivity (ε^*) for all nanocomposites at room temperature. It can be seen from Figure 15.7 that the value of dielectric constant (ε') increases from 8 to 24 with an increase of ferrite phase at 1 MHz. It is evident from the figure that the ε' decreases with an increase of the frequency range of 1–10 MHz and nearly remain constant upto 500 MHz. The value of ε' increases with further increase of frequency and shows a resonance and anti-resonance above 1 GHz. The dispersion occurring in the lower frequency region is attributed to interfacial polarization, since the electronic and atomic polarizations remain by and large unchanged at these frequencies. In the high frequency region the decrease in the value of dielectric constant is small and constant. The polarization mechanism contributing to the polarizability is observed to show lagging with the applied field at these frequencies. This result can be reduced polarization leading to diminished ε' value at higher frequencies. The variation in dielectric constant can be explained in a different perspective by considering the ferrite system as a heterogeneous system with grains and grain boundaries possessing different conducting properties. Previously, the dielectric dispersion observed in a number of ferrite systems was explained satisfactorily on the basis of the Maxwell–Wagner theory of interfacial polarization[37] in consonance with the Koop's phenomenological theory.[38] According to this model, it is the conductivity of grain boundaries that contributes more to the dielectric value at lower frequencies. In ferrites, it is observed that the mechanism of dielectric polarization is similar to the mechanism of electrical conduction. The variation of dielectric constant can be related to the collective behavior of both types of electric charge carriers, electrons, and holes. Hence, the electrons exchanging between Fe^{2+} and Fe^{3+} ions and the holes that transfer between Ni^{3+} and Ni^{2+} ions are responsible for electric conduction and dielectric polarization in these composites. At higher frequencies, the frequency of electron/hole exchange will not be able to follow the applied electric field, thus resulting in a decrease in polarization. Consequently, the dielectric constant remains small and constant.

The frequency variation of imaginary part of permittivity (ε'') for all samples is shown in Figure 15.9. It can be seen from the figure that the ε'' in all samples has been found to low and remains constant upto 500 MHz and shows a resonance and anti-resonance behavior around 1 GHz. The reason for obtaining low values of ε'' is due to the curtailing of the Fe^{2+} ions on account of the resulting in better stoichiometry of crystal structure. The resonance peak observed in the frequency dependence of ε'' around 1 GHz may be due to the hoping frequency of electrons is equal to the applied field frequency, maximum electrical energy is transferred to the oscillating ions and power loss shoots up, thereby resulting in resonance. Thus, dielectric constant and loss remains nearly constant for all samples, over a wide range of frequencies (10–500 MHz). Therefore, the Cu/NiZn ferrite composites can be a good candidate material for high frequency EMI device applications.

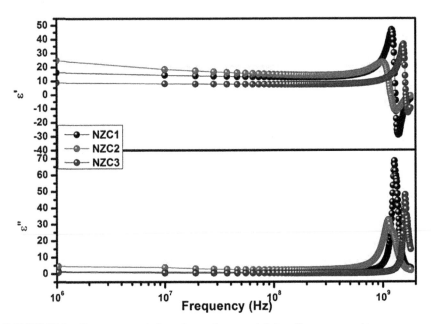

FIGURE 15.9 Frequency variation of complex permittivity of nanocomposites.

Figure 15.10 gives the plots of SE versus frequency for all composite samples. It can be seen from the figure that the total SE decreases steadily with frequency. The most important feature in this graph is that above 2–10 GHz, the main contribution to SE is absorption while reflection remains very low. Concerning the typical multiple maxima related to reflections from the output interface of the sample are observed at low frequencies and these maximum and minimum dips appear when samples are electromagnetically thin or, in other words, when samples are thinner than their skin depth. In these composites, since conductivity increases with frequency[39] and, therefore, the slabs become electromagnetically thick at some point, these dips are only observed at low frequencies. It can also observe from the figure that the SE increases with copper content in composite. The maximum value of SE = 33 dB has been observed for the sample NZC2. These results show that the present composites are useful for EMI applications.

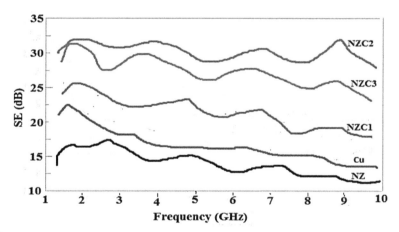

FIGURE 15.10 A plot of shielding effectiveness (SE) vs. frequency for NiCuZn ferrite, copper, and nanocomposites.

The higher frequency response characteristics (>1 GHz) of the nanocomposite (NZC2) and lower electrical resistivity of copper for the application for RFIC technology have been obtained by the measurement of S-parameters (Fig. 15.11a,b. The two-pod S-parameters measurement of the pure Cu and composite is performed in a frequency range of 0.1–20 GHz using the HP8510C Network Analyzer. S-parameters are transformed into Y-parameters from which inductance and quality factors for the samples are calculated.[40] $L = Im\,(1/Y)\,2\pi f$ and $Q = Im\,(1/Y)\,Re(1/Y)$, where f is the signal frequency. The L and Q values are measured on pure Cu and nanocomposite

(NZC2) and results are given in Figure 15.11a,b, respectively. It can be observed from the figure that the inductance of Cu and composite is almost same but their quality factors are quite different. The quality factor of the nanocomposite degrades 20% at 2 GHz in comparison with that of the pure Cu. It may be resulted by the increase of electrical resistivity.

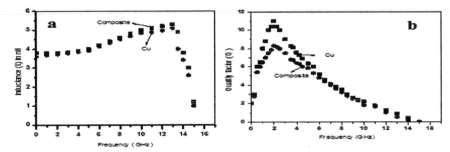

FIGURE 15.11   Frequency variation of (a) inductance (L) and (b) quality factor for copper and nanocomposite.

Figure 15.12a shows a photograph of a fabricated inductor with bottom nanocomposite (NZC2) core. The inductor consisted of six low temperature co-fired ceramic (LTCC) layers and a silver (Ag) coil created on the top ferrite layers. The conductive coil has three turns and dimensions of 0.20 mm in width, 12 mm in thickness, and 0.5 mm in pitch. The sandwiched inductor was fabricated of six ferrite layers in the top and bottom. The conductive coils screen printed on the middle nanocomposite layer. The multilayer inductor has been fabricated using the composite materials powder forming the core sheet was mixed with binder solutions and solvent to make slurry for tape casting.

The inductance (L) and quality factor (Q) of the fabricated inductor has been measured with the help of impedance analyzer (HP 4291B) in the frequency range (1 MHz–1.8 GHz) at room temperature. Figure 15.12b gives plots of frequency variation of L and Q for fabricated inductor. It can be seen from the figure that the L of fabricated inductor is 75 nH at 1 MHz and decreases with an increase of frequency. The increase of L in inductor is due to close of magnetic flux path and enhance internal mutual inductance and decrease of leak inductance and coil with more turns in fixed area. It can be also seen from the figure that the fabricated inductor has higher quality factor more than 12 in the MHz range. Thus high quality factor in MHz range will increase efficiency in the high frequency energy conversion and improve the operating frequency of inductor.

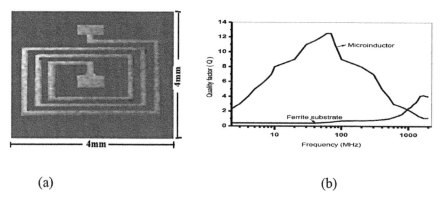

(a)   (b)

FIGURE 15.12 (a) Photograph and (b) frequency variation of inductance (L) and quality factor (Q) for a fabricated microinductor.

15.4 CONCLUSION

Cu/NiZn ferrite nanocomposites were successfully prepared using the mechanical-milling process. With the ferrite content, the permittivity of the all composites increased. The real part of permittivity of all the samples has shown good frequency stability. The magnetic properties such as saturation magnetization and coercivity were studied. It is also observed that the present composite shows a low dielectric loss within the measurement range. Such magnetic composites are candidates for the EMI applications. The higher frequency response characteristics (>1 GHz) of the nanocomposite for its application for RFIC technology has been obtained by the measurement of S-parameters. The microinductors were fabricated with the help of microwave sintered sample by using screen-printing method and cofiring. The electrical properties, such as inductance and quality factor of the prepared inductors were measured over a wide frequency range.

KEYWORDS

- **nanocomposites**
- **ferrites**
- **microwave-hydrothermal method**

- **mechanical milling**
- **magnetic properties**
- **shielding effectiveness**
- **microinductor**

REFERENCES

1. Niihara, K. A.; Nakahira, T. S. *Mater. Res. Soc. Symp. Proc.* **1993,** *286,* 405.

2. Narayan, J.; Chen, Y. *Philos. Mag. A.* **1948,** *49,* 475.

3. Abeles, B.; Ping, S.; Coutts, M. D.; Arie, Y. *Adv. Phys.* **1975,** *24,* 407.

4. Chakravorty, D. *Bull. Mater. Sci.***1992,** *15,* 411.

5. Dormann, J. L.; Sella, C.; Renaudin, P.; Kaba, A.; Gibart, P. *Thin Solid Films.* **1979,** *58,* 265.

6. Dormann, J. L.; Djega-Mariadassou, C.; Jove, J. *J. Magn. Magn. Mater.* **1992,** *104/107,* 1567.

7. Marchand, A.; Devaux, X.; Barbara, B.; Mollard, P.; Brieu, M.; Rousset, A. *J. Mater. Sci.* **1993,** *28,* 2217.

8. Malats, A.; Riera, I.; Pourroy, G.; Poix, P. *J. Magn. Magn. Mater.* **1993,** *125,* 125.

9. Tihay, F.; Roger, A. C.; Pourroy, G.; Kiennemann, A. *Energy Fuels.* **2002,** *16,* 1271.

10. Srivastava, M.; Ojha, A. K.; Chaubey, S.; Singh, J. *J. Solid State Chem.* **2010,** *183,* 2669.

11. Liu, K. L.; Yuan, S. L.; Duan, H. N.; Yin, S. Y.; Tian, Z. M.; Zheng, X. F.; Huo, S. X.; Wang, C. H. *Mater. Lett.* **2010,** *64,* 192.

12. Stewart,S. J.; Goya, G. F.; Punte, G.; Mercader, R. C. *J. Phy. Chem. Solids.* **1997,** *58,* 73.

13. Nakamoto, K. *Infrared and Raman Spectra in Inorganic and Coordination Compound,* 3rd ed.; John Wiley and Sons: New York, 1966.

14. Stuart, B. *Modern Infrared Spectroscopy ACOL series;* Wiley: Chichester, UK, 1996.

15. Suryanarayana, C. *Prog. Mater. Sci.* **2001,** *46,* 1–184.

16. Weeber, A. W.; Bakker, H. *Phys. B.* **1988,** *153,* 93–135.

17. Zhang, D. L. *Prog. Mater. Sci.* **2004,** *49,* 537–560.

18. Harris, J. R.; Wattis, J. A. D.; Wood, J. V. *Acta Mater.* **2001,** *49,* 3991–4003.

19. Froes, F. H.; Senkov, O. N.; Baburaj, E. G. *Mater. Sci. Eng. A.* **2001,** *301,* 44–53.

20. Koch, C. C. *Mater. Sci. Eng. A.* **1998,** *244,* 39–48.

21. Hong, L. B.; Fultz, B. *J. Appl. Phys.* **1996,** *79* (8), 3946–3955.

22. Suryanarayana, C.; Ivanov, E.; Boldyrev, V. V. *Mater. Sci. Eng. A.* **2001,** *304–306,* 151–158.

23. Jiang, J. Z.; Poulsen, F. W.; Mørup, S. *J. Mater. Res.* **1999,** *14,* 1343–1352.

24. Kong, L. B.; Ma, J.; Zhu, W.; Tan, O. K. *J. Alloy Comp.* **2002,** *335* (1–2), 290–296.

25. Simoneau, M.; L'Esperance, G.; Trudeau, J. L.; Schulz, R. *Struct. J. Mater. Res.* **1994,** *9,* 535–540.

26. Wang, S.; Ding, J.; Shi, Y.; Chen, Y. J. *J. Magn. Magn. Mater.* **2000,** *219,* 206–212.

27. Fatemi, D. J.; Harris, V. G.; Browning, V. M.; Kirkland, J. P. *J. Appl. Phys.* **1998,** *83,* 6767–6769.

28. Kong, L. B.; Ma, J.; Zhu, W.; Tan, O. K. *J. Alloys Comp.* **2001,** *322,* 290–297.

29. Wang, J; Xue, J. M.; Wan, D. M.; Gan, B. K. *J. Solid State Chem.* **2000,** *154,* 321–328.

30. Kanakadurga, M.; Raju, P.; Murthy, S. R. *J. Mag. Mag. Mate.* **2013,** *341,* 112–117.

31. Khamkongkaeo, A.; Jantaratana, P.; Sirisathitkul, C.; Yamwong, T.; Maensiri, S. *Trans. Nonferrous. Met. Soc. China.* **2011,** *21,* 2438–2442.

32. Griffiths, P. R.; de Haseth, J. A. *Fourier Transform Infrared Spectroscopy,* 2nd ed.; Wiley-InterScience: New York, 2007; ISBN 978-0-471-19404-0.

33. Zeng, C.; Hossienty, N.; Zhang, C.; Wang, B. *Polymer.* **2010,** *51,* 655.

34. Sakavati-Niasari, M.; Dava, F.; Mir, N. *Polyhedron.* **2008,** *27,* 3514.

35. Ram, S.; Ram, K.; Raman, I. R. *J. Mater. Sci.* **1989,** *23,* 4541.

36. Macedo, P. B.; Moynihan, C. T.; Bose, R. *Phys. Chem. Glasses.* **1972,** *13* (6), 171.

37. Maxwell, J. C. *Electricity and Magnetism;* Oxford University Press: London, 1973.

38. Wagner, K. W. *Ann. Phys.* **1993,** *40,* 818–826.

39. Raju, P. "Nanocomposites for EMI Applications." Ph.D Thesis, Osmania University, 2015.

40. Jr-Wei, L.; Chen, C. C; Yu-Ting, C. *IEEE Trans. Electron Devices* **2005,** *52* (7), 1489.

CHAPTER 16

THE ROLE OF NANOSIZED MATERIALS IN LITHIUM ION BATTERIES

BIBIN JOHN*, SANDHYA C. P., and GOURI C.

Lithium Ion and Fuel Cell Division, Energy Systems Group, Vikram Sarabhai Space Centre, Thiruvananthapuram 695022, India

Corresponding author. E-mail:bbnjohn@yahoo.com

CONTENTS

ABSTRACT

Nanosized materials are gaining attention in lithium ion batteries. The use of nanosized materials can significantly improve the energy density and power density of lithium ion batteries. This review deals with the role of nanosized materials on the performance of lithium ion batteries. The potential advantages of nanomaterials with a focus on the recent advances in cathode, anode, and electrolyte are reviewed. The challenges in the area of nanomaterials for lithium ion batteries are also summarized.

16.1 INTRODUCTION

Lithium ion (Li-ion) battery has become the most popular power source for portable electronics due to its high energy density, low self-discharge, and absence of memory effect. It is also used in high tech areas like military and space applications. With the advent of electric vehicles (EV) and plug-in hybrid electric vehicles (PHEV), there has been a growing need to build Li-ion batteries that provide high energy and power densities in order to be considered as a potential replacement for conventional gasoline engines.[1] The replacement of conventional vehicles with EV or PHEV can address at least partially, the problem of air pollution, climate change, and oil shortage.[2,3] Li-ion batteries with high energy and power densities and long cycle life are required for meeting the ever increasing energy demands for various other applications also.

A Li-ion cell, like any other electrochemical cell is made up of three basic components: cathode (or positive electrode), anode (or negative electrode) and electrolyte (which serves as a medium for the movement of ions). Typical cathode materials used in Li-ion cells are $LiCoO_2$, $LiFePO_4$, etc. Conventionally used anode materials are graphite or carbon. The cathode and anode active materials are coated on aluminium and copper foil respectively, to get the corresponding electrodes. The electrolyte consists of a Li salt ($LiPF_6$, $LiBF_4$, etc.) dissolved in organic solvents (ethylene carbonate, diethyl carbonate, etc.). The working principle of a typical Li-ion cell is depicted in Figure 16.1 with $LiCoO_2$ as the cathode and graphite as the anode.[4]

During the charging of the cell, $LiMO_2$ deintercalates Li^+ ions which move from cathode to anode and get intercalated in between the layers of graphite anode. Thus, electrochemical oxidation takes place at cathode (positive electrode) and reduction at anode (negative electrode) during the charging of the cell. When the cell is discharged, the above processes are

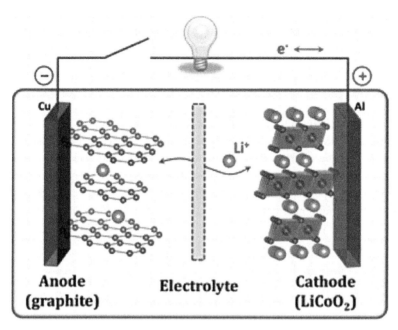

FIGURE 16.1 Working principle of a Li-ion cell. (Reprinted with permission from Goodenough, J. B.; Park, K. S. The Li-ion rechargeable battery: A Perspective. *J. Am. Chem. Soc.* **2013**, *135* (4), 1167–1176. © 2013 American Chemical Society.)

reversed, oxidation occurs at anode and reduction at cathode and Li^+ ions move from anode to cathode. The reactions that take place at the electrodes as well as the overall reaction within the cell during the discharging step are depicted in Scheme 16.1. The reactions take place in the reverse direction during the charging step.

Anode: $Li_xC_6 \rightarrow C_6 + xLi^+ + e^-$

Cathode: $MO_2 + Li^+ + e^- \rightarrow LiMO_2$

Cell (Overall): $Li_xC_6 + xMO_2 \rightarrow C_6 + xLiMO_2$

SCHEME 16.1   Reactions that take place during the discharge of the Li-ion cell.

The performance of a Li-ion cell is decided by the individual components that constitute the cell and their complex interactions. The improvement in performance of Li-ion cells can be achieved by using materials with new chemistry. Most of the research works in Li-ion battery area focus on development of materials mainly cathode, anode, and electrolyte.[1] Since the energy density of a cell is the product of voltage and capacity, the studies are focused on developing cathode materials which deintercalate Li at a higher potential and also which can exhibit higher capacity. In the case of anode, most of the research is on the development of high capacity materials. Studies are also being carried out to improve the rate capability, cycle stability, and safety.

The improvement in performance of Li-ion batteries has also been achieved with the use of nanomaterials. Conventional Li-ion batteries use micro-meter sized particles with the exception of the conductive carbon additive used in cathode. The use of nanomaterials can improve the electrochemical properties such as specific energy, rate capability, and cycle stability over their bulk counterparts.[2] With the development of new nanomaterials, there is an interest in the replacement of conventional materials with nanomaterials. This article gives an overview of the nanomaterials that are being developed/studied for Li-ion batteries and presents the advances in nanomaterials with a focus on the progress in cathode, anode, and electrolyte. The challenges in nanomaterials have also been discussed. Finally, the application of nanomaterials in commercial Li-ion batteries is also presented.

16.2 ROLE OF NANOMATERIALS

Nanomaterials have been widely applied in almost all areas of science. Recently, they have attracted attention in energy storage devices, particularly, Li-ion batteries. The use of nanomaterials is very promising in improving the performance of Li-ion batteries due to their high surface area, porosity, etc.[2] A plethora of nanomaterials have been developed in the past few years; some of them are now being used in commercial Li-ion batteries.

The potential advantages by the use of nanomaterials in Li-ion batteries are summarized below:

a. **New reaction pathways:** Certain materials which are electrochemically inactive in the bulk form could exhibit excellent electrochemical performance at the nanoscale.[2] When the dimensions of materials, grains, or domains becomes comparable to (or less than) the characteristic length scale (such as the mean free path) of phonons, photons, electrons, ions, and molecules, many physical phenomena involving them are strongly influenced. This leads to new modes for the transport of charge, mass, and energy and for chemical and energy transformation processes. The length scales of mobile species in batteries (electrons, ions, and molecules) fall generally in the order of 0.1–100 nm in typical electrochemical systems. Therefore, the use of materials in nanoscale regime can lead to unique physiochemical properties and novel reaction pathways.[5]

b. **Short diffusion length:** The characteristic time constant for diffusion (equilibration time), τ for solid-state diffusion of Li in electrode materials, is determined by the diffusion constant, D, and the diffusion length, L, according to the equation:

$$\tau = \frac{L^2}{D}$$

The time for Li intercalation, τ, decreases with the square of the diffusion length, L, illustrating the remarkable effect of nanomaterials: fast Li storage in nanomaterials and high rate capability (high power). For example, the diffusion length would be reduced to the diameter of nanowires, tubes, and rods and the thickness of thin films if they are properly dispersed in the electrolyte (ionic conductor) and the current collector (electronic conductor).[5] In short, the small size of the nanoparticles increases the rate of Li insertion/removal considerably, thereby increasing the rate capability of the cell.

c. **High surface area:** Nanomaterials have high surface area. This leads to a large area of contact with the electrolyte, thereby increasing the active sites for electrode reactions, which in turn reduces electrode polarization loss and improves power density (or rate capability), energy efficiency, and usable energy density.[5]

d. **Enhanced mechanical properties:** Nanostructured materials exhibit significantly enhanced mechanical strength, toughness, and structural integrity. Nanomaterials like nanowires, nanotubes, etc., are more resistant to mechanical damage and can be engineered to allow volume change only in certain directions or dimensions.[5] For example, alloy anodes have a fatal problem of pulverization and cracking due to volume change during charging and discharging, resulting in poor cycle performance. This problem can be solved by utilizing 1D nanowire materials or alloy-encapsulated hollow structures due to their stable mechanical structure.[6]

e. **Better accommodation of volume changes:** Electrode active materials undergo volume changes during cycling. Nanosized materials can accommodate the volume change of alloy anode material like Sn and Si, which undergoes drastic volumetric change during cycling.[6]

The small absolute volume changes occurring for nanomaterials due to their small particle size and uniform particle distribution can mitigate the pulverization of the particles and reduce capacity fading.[7]

f. **Interfacial charge storage mechanism**: The high Li storage capacity in nanometer-sized transition metal oxides at low potential has recently been explained by an interfacial charge storage mechanism.[8,9] According to this model, Li-ions are stored on the oxide side of the interface, while electrons are localized on the metallic side, resulting in a charge separation. On this basis, the electrochemically driven size confinement of the metal particles is believed to enhance their electrochemical activity toward the formation/decomposition of Li_2O. With decreasing particle size, an increasing proportion of the total number of atoms lies near or on the surface, making the electrochemical reactivity of the particles more and more important.[2,8,9]

g. **Modification of redox potential:** The redox potentials of electrode materials can also be modified by nanostructures, resulting in a change of cell voltage or energy density.[10]

From the ongoing discussion, it is evident that nanomaterials offer the potential to create unique Li-ion batteries with both high energy and power density. A large number of cathode and anode materials are developed and evaluated in the past few years which could find application in Li-ion batteries. Different morphologies of nanomaterials like nanoparticles, nanotubes, nanospheres, nanoflakes, etc., have been synthesized. The developments in composite nanostructured materials designed to include electronic conductive paths could decrease the inner resistance of Li-ion batteries, leading to higher specific capacities, even at high charge/discharge current rates.[2] New synthesis routes have been evolved to improve the yield and reduce the cost of manufacturing nanomaterials. Bionanotechnology has also been now applied to synthesize nanomaterials for Li-ion batteries.[11]

Based on their dimension, nanomaterials can be classified into zero dimensional (0D), one dimensional (1D), two dimensional (2D), and three dimensional (3D). 0D nanomaterials are materials wherein all the dimensions are within the nanoscale, that is, < 100 nm. The most common representation of 0D nanomaterials are nanoparticles (e.g., solid nanoparticles, hollow nanoparticles, core-shell nanoparticles, etc.). On the other hand, 1D nanomaterials differ from 0D nanomaterials in that the former have one dimension that is outside the nanoscale. This difference in dimension leads to needle-shaped materials like nanotubes, nanorods, nanowires, etc. 2D nanomaterials are materials in which two of the dimensions are not confined to the nanoscale. 2D nanomaterials include nanofilms, nanolayers, and nano-coatings. 3D nanomaterials or bulk nanomaterials are relatively difficult to classify. These materials are not confined to the nanoscale in any dimension and are thus characterized by having three arbitrary dimensions above 100 nm. Despite their nanoscale dimensions, these materials possess a nano-crystalline structure or involve the presence of features at the nanoscale.[12] 0D, 1D, 2D, and 3D nanomaterials and their composites have been widely studied for Li-ion batteries. The following sections give an overview of these nanomaterials used as cathode, anode, and electrolyte additives in Li-ion batteries.

16.2.1 CATHODE

The most commonly used cathode materials in Li-ion batteries are lithiated transition metal oxides (e.g., $LiCoO_2$, $LiNi_{0.8}Co_{0.15}Al_{0.05}O_2$, $LiMn_2O_4$, etc.) and phospho-olivines (e.g., $LiFePO_4$.). Each material is having its own advantages and disadvantages. Synthesizing these materials in nanoscale improves the performance of the material. The properties are also improved by forming nanocomposites with carbon/CNT, etc. and by providing coating on the material. The nano effect in various cathode materials is discussed below:

16.2.1.1 LiCoO₂

$LiCoO_2$ has been the most widely used cathode material in Li-ion batteries. It was used in the first Li-ion cell commercialized by Sony in 1991. This material offers significant advantages with respect to ease of synthesis, high working potential, and excellent cycle ability at room temperature.[1] However, there are certain limitations with $LiCoO_2$. The practical capacity of $LiCoO_2$ is 140 mAhg⁻¹ corresponding to a charging voltage of 4.2 V, and reversible delithiation of 0.5Li. This is only 50% of the theoretical capacity for $LiCoO_2$ (274 mAhg⁻¹). Cycling at voltages above 4.2 V can lead to a rapid degradation in capacity, due to structural changes and dissolution of cobalt in the electrolyte.[13] Also, the diffusion coefficient of $LiCoO_2$ is between 10^{-12} cm²s⁻¹ and 10^{-9} cm²s⁻¹, which limit its use for high power applications.[1]

Nanosized $LiCoO_2$ showed improved performance compared to micro-sized ones. Chen et al. reported the high rate capabilities of $LiCoO_2$ nano particles having grain sizes less than 100 nm. At 10 C, the cathode exhibited a first capacity of ~130 mAhg⁻¹, comparable to the performance of conventional $LiCoO_2$ cathodes operating at much lower cycling rates. The capacity at 50 C rate was 100 mAhg^{-1}. The capacity for conventional $LiCoO_2$ at 10 C rate was only 20 mAhg^{-1}.[14] $LiCoO_2$ nano crystals having 200 nm size synthesized by Qian et al. also exhibited high rate capabilities. At a higher charge/ discharge rate of 15 C, it showed a rapid deterioration in capacity over the first 10 cycles after which it stabilized to give capacities in excess of 100 mAhg^{-1} over the next 40 cycles.[15] Li et al. synthesized nano-structured $LiCoO_2$ thin film by coupling a sol–gel process with a spin-coating method using polyacrylic acid (PAA) as chelating agent.[16]

Luo et al. developed binder-free electrodes using super-aligned carbon nanotubes (SACNTs) and $LiCoO_2$. SACNTs form a three-dimensional conductive and flexible structural network in which active material can be embedded. In the binder-free $LiCoO_2$-SACNT cathode, $LiCoO_2$ particles are uniformly distributed in the continuous SACNT network. Both electron and ion transfer are greatly improved without the hinder of insulating binder. The $LiCoO_2$-SACNT cathode exhibited high specific capacities (144–151 mAhg^{-1} at 0.1 C) and excellent cycling performance with capacity retention more than 95% after 50 cycles. The $LiCoO_2$-5 wt% SACNT composite cathode exhibited the best cycling stability (151.4 mAhg^{-1} at 0.1 C with retention of 98.4% after 50 cycles) and rate capability (137.4 mAhg^{-1} at 2 C, corresponding to capacity retention of 90.8% compared to 0.1 C).[17]

$LiCoO_2$ nanotubes were synthesized by Li et al. using porous alumina templates and were reported to show good promise as cathode materials for Li-ion batteries. The intercalation/deintercalation of Li$^+$ could be more efficiently accomplished with nanotubes than other nanocrystalline structures or morphologies.[18]

Surface coating on $LiCoO_2$ also improved the electrochemical performance. Carbon coatings on $LiCoO_2$ improved the capacity by reducing charge transfer resistance and facilitating Li-ion diffusion. Metal phosphate (e.g., $AlPO_4$ or $FePO_4$) coatings and metal oxide (e.g., ZrO_2, Al_2O_3, SnO_2, MgO, or ZnO) coatings on $LiCoO_2$ have been used to suppress cobalt dissolution and improve uniform stress/strain distribution.[5] For example, a coating of CuO nanoparticles on $LiCoO_2$ prevented dissolution of Co into the electrolyte when charged to a voltage above the conventional 4.2 V limit. At 5 C rate, CuO coated $LiCoO_2$ displayed a significant improvement in performance as compared to pristine $LiCoO_2$. Even at much higher rates of 50 C, an average capacity of ~100 mAhg^{-1} was obtained with capacity retention of 76.9% after 10 cycles.[19] A nanoscale surface coating of $AlPO_4$ was applied on $LiCoO_2$ for improving the thermal stability.[20]

16.2.1.2 $LiNi_{0.8}Co_{0.15}Al_{0.05}O_2$

$LiNi_{0.8}Co_{0.15}Al_{0.05}O_2$ showed improved electrochemical and thermal stabilities by employing a stabilizing layered structure through the substitution of Co and Al for Ni sites in $LiNiO_2$ which is a high energy cathode material for Li-ion batteries.[21] $LiNi_{0.8}Co_{0.15}Al_{0.05}O_2$ nanotubes prepared by alumina template route provides a promising cathode for Li-ion batteries.[18] $Co_3(PO_4)_2$ nanoparticle coating on $LiNi_{0.8}Co_{0.16}Al_{0.04}O_2$ has resulted in a 30% enhancement of the cycle life, compared to a bare sample without a loss in the first discharge capacity.[22]

16.2.1.3 $LiMn_2O_4$

Spinel $LiMn_2O_4$ and its substituted derivatives have gained much attention as alternatives to $LiCoO_2$ in Li-ion batteries for electric vehicle applications because they are less expensive and environmentally benign.[5] Kim et al. reported the hydrothermal synthesis of single-crystalline β-MnO_2 nanorods and their chemical conversion into free-standing single-crystalline $LiMn_2O_4$ nanorods using a simple solid-state reaction. $LiMn_2O_4$ nanorods have a high charge storage capacity at high rates compared with commercially available $LiMn_2O_4$ powders. More than 85% of the initial charge storage capacity was maintained for over 100 cycles.[23] Figure 16.2a,b shows the scanning electron microscope (SEM) image of $LiMn_2O_4$ nanorods and comparison of the cycling performance with commercial powders, respectively.

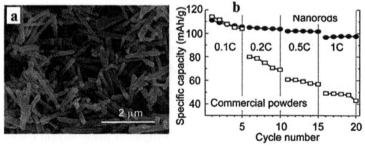

FIGURE 16.2 (a) SEM image of $LiMn_2O_4$ nanorods, (b) comparison of electrochemical performance of $LiMn_2O_4$ nanorods with that of commercial powders. (Reprinted with permission from Kim, D. K.; Muralidharan, P.; Lee, H. W.; Ruffo, R.; Yang, Y.; Chan, C. K.; Peng, H.; Huggins, R. A.; Cui, Y. Spinel LiMn2O4 Nanorods as Lithium Ion Battery Cathodes. *Nano Lett.* **2008**, *8* (11), 3948–3952. © 2008 American Chemical Society.)

A nanochain $LiMn_2O_4$ with beads of 100 nm was synthesized by Tang et al. using a starch-assisted sol–gel method. The starch chain structure played a decisive role in the formation of the $LiMn_2O_4$ nanochains. This type of interconnected nanocrystalline morphology helps the transferring process of Li^+ ions and allow a better rate performance while the commercial $LiMn_2O_4$ materials are bulk particles that consist of aggregated submicron-sized particles (on the average about 200–300 nm). The nano chain $LiMn_2O_4$ cathode exhibited good rate capability with a reversible capacity of 100 mAhg^{-1} at 100 mAg^{-1} (about 1 C) and 58 mAhg^{-1} even at a charge rate of 20 C. In contrast, the commercial $LiMn_2O_4$ exhibited a reversible capacity of 92 and 36 mAhg^{-1} at 1 C and 5 C, respectively[3].

Hosono et al. synthesized novel single crystalline spinel $LiMn_2O_4$ nanowires using $Na_{0.44}MnO_2$ nanowires as a self-template. The as-prepared single crystalline $LiMn_2O_4$ nanowires have a diameter of 50–100 nm. The $LiMn_2O_4$ nanowires exhibited specific capacity of 118, 108, 102, and 88 mAhg^{-1} at a rate of 0.1, 5, 10, and 20 Ag^{-1} respectively, and demonstrated good cycling performance up to 100 cycles at a high rate of 5 Ag^{-1}. The unique morphology and the high quality single crystalline structure of $LiMn_2O_4$ nanowires effectively reduced Li-ion diffusion length and improved the electronic transport, demonstrating large capacities at higher cycling rates.[24]

$LiMn_2O_4$ nanoparticles synthesized by Shaju et al. demonstrated high rate capabilities and good cycle life. At cycling rates of 40and 60 C, the capacity retention was as high as 90 and 85%, respectively. A comparison of the performance of $LiMn_2O_4$ nanoparticles and conventionally synthesized $LiMn_2O_4$, at different rates is shown in Figure 16.3. The fading was very high in the case of conventional $LiMn_2O_4$ compared to $LiMn_2O_4$ nanoparticles at high rates. Even at 10 C, $LiMn_2O_4$ nanoparticles provided near 100% power retention after as many as 1000 cycles. This excellent performance of $LiMn_2O_4$ nanoparticles is ascribed to better structural stability, lower Mn dissolution, and a more stable electrode interface than conventionally synthesized $LiMn_2O_4$.[25]

Single-crystalline nanotubes of spinel $LiMn_2O_4$, with a diameter of about 600 nm, a wall thickness of about 200 nm and a length of 1–4 μm have been synthesized via a template-engaged reaction using β-MnO_2 nanotubes as a self-sacrifice template. The $LiMn_2O_4$ nanotubes exhibited superior high-rate capabilities and good cycling stability. About 70% of its initial capacity was retained after 1500 cycles at 5 C rate. Importantly, the tubular nanostructure and the single-crystalline nature of $LiMn_2O_4$ nanotubes are also well preserved after prolonged charge/discharge cycling at a relatively high current density, indicating good structural stability of the single-crystalline nanotubes during Li intercalation/deintercalation process.[26]

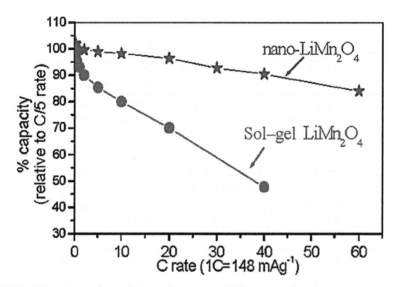

FIGURE 16.3 Comparison of the performance at different rates for $LiMn_2O_4$ nanoparticles and conventional $LiMn_2O_4$ synthesized by a sol–gel method. (Reprinted with permission from Shaju, K. M.; Bruce, P. G. A Stoichiometric Nano-$LiMn_2O_4$ Spinel Electrode Exhibiting High Power and Stable Cycling. *Chem. Mater.* **2008**, *20* (17), 5557–5562.© 2008 American Chemical Society.)

It is reported that the electrochemical performance of $LiMn_2O_4$ can be improved by nano-sized ZnO coating. The electrochemical characteristics of nano-sized ZnO-coated $LiMn_2O_4$ were investigated at 55°C. After 50 cycles, the coated $LiMn_2O_4$ exhibited a capacity retention 97%, higher than that of the pristine $LiMn_2O_4$ electrode material. ZnO coating decreases the Mn dissolution in the electrolyte and consequently reduces the interfacial resistance.[27]

16.2.1.4   LiFePO₄

$LiFePO_4$ has received considerable interest as cathode material in Li-ion batteries owing to long cycle life, low cost, low toxicity, and safety. The excellent reversibility during cycling has been attributed to similarities in the structures of $LiFePO_4$ and its delithiated state, $FePO_4$. However, $LiFePO_4$ cathode suffers from poor conductivity that could hinder its application in high power batteries. Several studies involving doping, nanostructuring LiFePO$_4$ cathodes, and carbon coating have been effectively explored to overcome the conductivity limitation, enabling the use of LiFePO$_4$ cathodes in high power Li-ion batteries.[1]

Reducing the particle size of LiFePO$_4$ to a nanometer scale could help overcome the poor Li-ion transport that is associated with the two-phase (LiFePO$_4$/FePO$_4$) boundary with a miscibility gap, improving capacity, and capacity retention upon cycling.[10] Lim et al. reported the synthesis of LiFePO$_4$ nanowires and hollow LiFePO$_4$ cathodes for high rate applications. Both the nanowires and hollow LiFePO$_4$ cathodes showed superior capacity retention at high C-rates as illustrated in Figure 16.4. At 15 C, nanowires retained 89% of the capacity compared to discharge at 0.2 C while hollow cathodes displayed 95% capacity retention. Hollow cathodes exhibited a capacity of 165mAhg^{-1} when cycled at a low C-rate (0.2 C). The high rate capability of hollow cathode is due to its high Brunauer−Emmett−Teller surface area.[28]

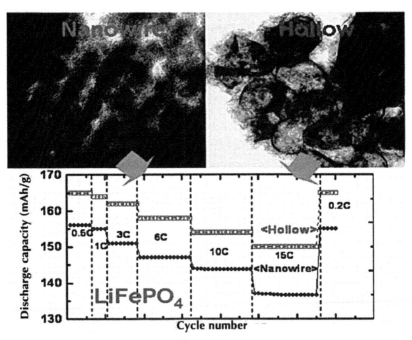

FIGURE 16.4 The performance at different rates for nanowire and hollow LiFePO$_4$. (Reprinted with permission from Lim, S.; Yoon, C. S.; Cho, J. Synthesis of Nanowire and Hollow LiFePO4 Cathodes for High-Performance Lithium Batteries. *Chem. Mater.* **2008,** *20* (14), 4560–4564.© 2008 American Chemical Society.)

Olivine $LiMPO_4$ (M = Mn, Fe, Co, and Ni) with nanorod morphologies were developed by a novel microwave assisted solvothermal (MW-ST) process. A reaction of the metal acetates with H_3PO_4 and LiOH under MW-ST conditions produced highly crystalline $LiMPO_4$ nanorods. These morphologies are particularly attractive to achieve fast Li diffusion and high rate capability.[29,30]

Nanoscale $LiFePO_4/C$ composite cathodes provide an effective solution for high power applications owing to the shorter distance for Li-ion diffusion and the improved electrical conductivity.[31] Kang and Ceder, fabricated a nanoscale $LiFePO_4/C$ cathode which could deliver power densities as high as 170 kWkg^{-1}, almost two orders of magnitude higher than conventional Li-ion batteries. The cathodes were composed of pyrophosphate coatings on $LiFePO_4$ nanoparticles with an average diameter ~50 nm. Carbon black (15 wt%) was added in order to improve the conductivity. At 60 C, the cathodes could deliver capacities of ~100 mAhg^{-1}, continuously over 50 cycles. When the carbon loading was increased to 65 wt%, a further improvement in the rate capabilities was observed. With a much higher carbon loading, rates as high as 400 C could be reached, delivering a capacity of 60 mAhg^{-1} and a power density of 170 kWkg^{-1}. Higher loading of conductive carbon reduced the volumetric energy densities but were effective in establishing ultrahigh rate capabilities of the active cathode material.[32]

Nanostructured mesoporous $LiFePO_4$ embedded in conductive interconnected carbon networks resulted in high electronic conductivity to facile mass transfer and charge transfer, which delivered an initial discharge capacity of 115 mAhg^{-1} and retaining 91% of initial capacity till 1000 cycles at 10 C current rate.[33] Jin et al. synthesized $LiFePO_4$-MWCNT nanocomposites by a combination of hydrothermal process, ball-milling, and heating reaction. The electronic conductivity significantly improved for the nanocomposites compared to pure $LiFePO_4$.[34] Wang et al. synthesized a $LiFePO_4/$ carbon composite consisting of a highly crystalline $LiFePO_4$ core (20–40 nm in size) and a highly graphitic carbon shell (1–2 nm in thickness) via in situ polymerization. The complete coating of carbon layer ensures that electrons pass through the surface of each $LiFePO_4$ particle, effectively shortening the path length for electronic transport. Moreover, Li-ions can easily intercalate into the $LiFePO_4$ through the graphitic carbon coating layer. As-prepared, $LiFePO_4/$carbon composite exhibited a discharge capacity of 168 mAhg^{-1} at a rate of 0.6 C and 90 mAhg^{-1} at a very high rate of 60 C.[35]

$LiFePO_4$ nanorods encapsulated within a mixed ionic-electronic conducting p-toluene sulfonic acid (p-TSA) doped poly (3,4- ethylenedioxy thiophene) (PEDOT) has been reported. The composite nanorod sample exhibited high capacity (166 mAhg⁻¹) and excellent cycling performance after coating due to the enhancement in electronic conductivity and the additional effect provided by ion-conducting doped PEDOT.[29]

Tobacco mosaic virus (TMV) as "bottom-up" templates was used for fabricating $LiFePO_4$ based nanoforest cathode arrays with a multilayered hierarchy. Both ionic and electronic migration lengths are significantly reduced in the TMV-enabled dense nanoforest cathodes. Also, the mechanical and electrochemical stresses between the cathode ($LiFePO_4$) and the main current collector are minimized. The resultant multilayered C/$LiFePO_4$/Ti/Ni composites with a shell–core–shell structure show excellent electrochemical performance; the detailed recombination process is illustrated in Figure 16.5a. The carbon coated cathode, C/$LiFePO_4$/Ti/Ni delivered a discharge capacity of 165 mAhg⁻¹ at 0.1 C which is better compared to $LiFePO_4$/Ti/Ni and $LiFePO_4$/Ti. Also, the C/$LiFePO_4$/Ti/Ni electrode exhibited good cyclability and coulombic efficiency as shown in Figure 16.5b. The capacity fading is only 0.014% per cycle during cycling at 1 C for 450 cycles, while the corresponding coulombic efficiency quickly rises to ~100% after the first 5 cycles.[36]

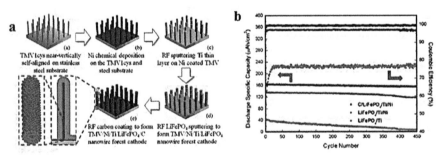

FIGURE 16.5 (a) Schematic description of TMV-templated near-vertical assembly of a $LiFePO_4$ nanoforest on a current collector with multilayered nanohierarchical arrangement of active materials and electron conducting pathway. (b) Cycling performance and coulombic efficiency of $LiFePO_4$ cathodes at 1 C. (Reprinted with permission from Liu, Y.; Zhang, W.; Zhu, Y.; Luo, Y.; Xu, Y.; Brown, A.; Culver, J. N.; Lundgren, C. A.; Xu, K.; Wang, Y.; Wang, C. Architecturing Hierarchical Function Layers on Self-Assembled Viral Templates as 3D Nano-Array Electrodes for Integrated Li-Ion Microbatteries. *Nano Lett.* **2013**, *13*, 293–300.© 2013 American Chemical Society.)

Su et al. employed phytic acid (PA) as both a template and phosphorus source for fabricating self-assembled $LiFePO_4$/C nano/microsphere by a hydrothermal method (Fig.16.6a). Figure 16.6b shows the high-magnification SEM image of a single $LiFePO_4$ microsphere, which is made from many ordered nanoparticles with grain size less than 100 nm. The HRTEM image

(Fig.16.6c) shows that there is an amorphous carbon coating layer with a thickness of 4–6 nm, uniformly covering the surface of the $LiFePO_4$ grains, which will enhance the electrical conductivity of the $LiFePO_4$ particles. $LiFePO_4$/C nano/microspheres, exhibited a high reversible specific capacity of 155 mAhg^{-1} at 0.1 C, as well as excellent rate capability and cycling performance.[37]

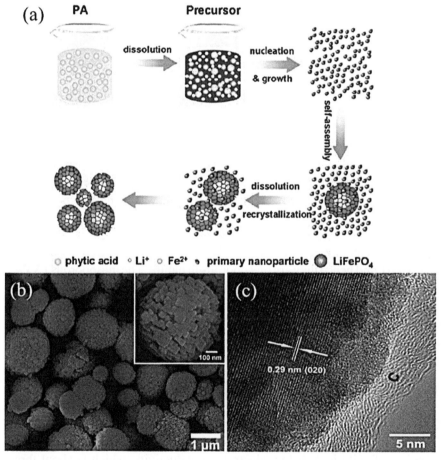

FIGURE 16.6 (a) Schematic illustration of the self-assembly of $LiFePO_4$ nano/microspheres templated by biomass PA. (b) SEM and (c) TEM images of the as-synthesized $LiFePO_4$ nano/microspheres. (Reprinted with permission from Su, J.; Wu, X. L.; Yang, C. P.; Lee, J. S.; Kim, J.; Guo, Y. G.; Self-Assembled $LiFePO_4$/C Nano/Microspheres by Using Phytic Acid as Phosphorus Source. *J. Phys. Chem. C.* **2012,** *116,* 5019–5024. © 2012 American Chemical Society.)

Another strategy to improve the performance of LiFePO$_4$ is to combine carbon coating and electronically conducting oxides such as RuO$_2$ with LiFePO$_4$. RuO$_2$ with a particle size of ~5 nm can be deposited on a carbon-coated LiFePO$_4$. During RuO$_2$ deposition, the original morphology was maintained and in good contact with carbon layer. Co-existing carbon and RuO$_2$ layers improve the kinetics and rate capability of the composite. Carbon coated LiFePO$_4$ electrode deteriorates the performance at a high rate, while carbon-LiFePO$_4$/RuO$_2$ showed improved electrochemical performance at high rates.[38]

16.2.1.5 $Li_3V_2(PO_4)_3$

Monoclinic Li$_3$V$_2$(PO$_4$)$_3$ (LVP) offers an optimal combination of high operating voltage, high Li capacity, good ion mobility and excellent thermal stability.[11] Du et al. synthesized LVP nanoparticles covered with amorphous carbon by biomimetic mineralization process. Mesoporous microspheres are formed on heat treatment which is composed of these nanoparticles. Here, low-cost microbe yeast cells are used as templates. During the biomimetic mineralization process, Li$_3$V$_2$(PO$_4$)$_3$ replicates the hierarchical nanostructures of a Baker's yeast cell. The microspheres are composed of densely aggregated nanoparticles (20–40 nm) as well as interconnected nanopores (2–15 nm), and hence they are mesoporous in nature. The mesoporous LVP-C microspheres have a high discharge capacity (about 126.7 mAhg^{-1}), only 2% capacity loss after 50 cycles at a current density of 0.2 C, and a high rate capacity of 100.5 mAhg^{-1} at 5 C in the region of 3.0–4.3 V. The microspheres could be an ideal cathode-active material for high power applications.[39]

16.2.1.6   ORTHOSILICATES BASED Li_2MSiO_4

Orthosilicate is a class of polyanionic cathode material having high capacity for application in next generation Li-ion batteries.[40] The main disadvantage of silicate family is their low electronic conductivity, which is about three orders of magnitude lower than that of LiFePO$_4$.[41] Murliganth et al. synthesized carbon coated Li$_2$FeSiO$_4$ nanoparticles of 20 nm size. The Li$_2$FeSiO$_4$/C sample showed excellent rate performance and cyclic stability with stable discharge capacities of 148 mAhg^{-1} at 25°C and 204 mAhg^{-1} at 55°C. Li$_2$MnSiO$_4$/C showed severe capacity fading especially at elevated temperatures. This poor capacity retention of Li$_2$MnSiO$_4$ is due to the presence of

Mn^{3+} ions, which causes J-T distortion and manganese dissolution as in case of spinel LiMn$_2$O$_4$.[42]

16.2.1.7 POLYANILINE

Conducting HClO$_4$-doped poly (aniline) (PANI) nanotubes and nanofibers prepared through a template route and a spray technique respectively exhibited higher electrical conductivity, large charge-discharge capacity, and better cycling capability than the commercial doped PANI powders. The electrochemical results showed the promising application of the doped PANI nanostructures in Li/PANI rechargeable batteries.[43] Polyaniline–LiV$_3$O$_8$ nanocomposites are also reported. The PANI formed a network of electrically conductive paths among the LiV$_3$O$_8$ particles, so the active LiV$_3$O$_8$ material can be fully utilized for Li extraction and insertion reactions and PANI–LiV$_3$O$_8$ nanocomposite exhibits much better electrochemical performance than bare LiV$_3$O$_8$. PANI incorporation in LiV$_3$O$_8$ could decrease the charge transfer resistance, increase the Li diffusion coefficient during the lithiation process, and improve the electrochemical reversibility.[44]

16.2.1.8 LiF/Fe NANOCOMPOSITE

LiF/Fe nanocomposite, prepared by a simple route of mechanical ball-milling of Li fluoride and iron using rigid TiN nanoparticles, has been demonstrated as a cathode material for Li-ion cells. Electrochemical measurements demonstrate that the LiF/Fe nanocomposite containing 50 wt% active materials of LiF and Fe can deliver a high reversible capacity of 568 mAhg^{-1} at 20 mAg^{-1}, approaching the theoretical capacity of the composite (600 mAhg^{-1}), and also show a strong power capability with a reversible capacity of ~300 mAhg^{-1} even at a very high rate of 500 mAg^{-1} at room temperature.[45]

16.2.2 ANODE

Graphite is the most widely used anode material in Li-ion cells. The maximum specific capacity achievable from graphite is limited to 372 mAh/g. Also, the rate capability of graphite is limited due to the slow solid-state diffusion of Li$^+$ ion within the electrode materials. Moreover, graphite undergoes volumetric expansion and contraction during the cell charging and discharging, respectively, which results in the peeling off of active materials from the current collector at a high rate, leading to capacity fading or cell shorting. The performance gets further affected under subzero as well as high operating temperatures. The Li-ion batteries for modern applications require anode materials with excellent electrochemical properties to achieve high energy density, power density, and cyclability.

A large number of anode materials are being developed with very high capacity (e.g., Si, Sn, metal oxides, etc.), improved rate capability, and cyclability (e.g., $Li_4Ti_5O_{12}$). However, an important difficulty with these high capacity anode materials is the high volumetric change during cycling which leads to capacity fading. Another difficulty with materials like $Li_4Ti_5O_{12}$ is their low electrical conductivity. Converting the material into nanostructures, (nanoparticles, nanorods, nanowires, etc.), forming nanocomposites with suitable materials, and coating with nanoparticles could solve these issues to some extent and provide anode materials with improved properties.

16.2.2.1 CARBON BASED ANODES

Carbon-based materials are the most widely used anode material for Li-ion batteries due to availability, stability in thermal, chemical, and electrochemical environment, low cost, and good Li intercalation and de-intercalation reversibility. Currently, most of the research on carbon based anodes focus on porous carbon, carbon nanotubes (CNTs), nanofibers, and graphene. The size reduction and the unique shape of these structures introduce novel properties that can considerably enhance the energy storage capacity in Li-ion batteries.[46]

16.2.2.1.1 Nano porous carbon

Nano-porous hard carbons were synthesized from pyrolyzed sucrose and their electrochemical performances for Li-ion batteries were evaluated. These nanoporous materials exhibited specific capacity close to 500 $mAhg^{-1}$ along with a good cycling life and rate capability. High rate capability has been demonstrated to be related to the fast Li diffusion ~4.11 × 10^{-5} cm^2s^{-1} for hierarchical nanoporous hard carbon.[47] Ordered mesoporous carbon

(CMK-3) has been considered as a possible candidate as high capacity anode for Li-ion batteries.[48] The ordered mesochannel and large surface area of CMK-3 shorten the distance for Li-ion diffusion, and its high conductivity favors electron transport.[49]

16.2.2.1.2 Carbon Nanotubes

CNTs have recently emerged as a potential candidate for energy storage devices. CNTs can be used as such or in combination with other anode active material. The maximum reversible theoretical capacity obtained for CNTs has been estimated to be 1116 mAhg^{-1} for SWCNTs in LiC_2 stoichiometry. This is by far the highest capacity among carbon based active materials, and it has been attributed to the intercalation of Li into stable sites located on the surface of pseudo graphitic layers and inside the central tube. However, the achievement of high coulombic efficiency with CNTs remains challenging because of the presence of large structure defects and high voltage hysteresis. When used together with other active anode materials, they exhibited better output than non CNT composite due to their superior electronic conductivity, good mechanical, and thermal stability, adsorption and transport properties.[46] Studies have been focused on the morphological features of CNTs, such as wall thickness, tube diameter, porosity, and shape (bamboo-like, quadrangular cross section, etc.) to improve the performance.[50,51]

CNT free-standing electrodes offer a constant capacity as a function of thickness which can significantly enhance the usable electrode capacity in a full battery, particularly in a high power battery design. Welna et al. synthesized vertically aligned MWCNTs on nickel substrate and demonstrated the performance as anode material for Li-ion battery. This material has shown high reversible specific capacities (up to 782 mAhg^{-1} at 57 mAg^{-1}). The intimate contact between each of the vertically aligned CNTs and the nickel metal current collector provides a low resistance pathway allowing for effective connectivity throughout the electrode.[52]

16.2.2.1.3 Graphene

Graphene is a promising anode for Li-ion batteries owing to its good electrical conductivity, excellent mechanical properties, high values of charge mobility, and high surface area.[46,53–56] When a single layer of graphene is considered, the performance will be inferior to graphite;[57] however, when we consider layers of graphene together, it can give higher capacities, leading either to 780 mAhg^{-1} or to 1116 mAhg.[53,58,59] In particular, the former assumes absorption of Li-ions on both faces of graphene (Li_2C_6 stoichiometry), while the latter assumes Li trapped at the benzene ring in a covalent bond (LiC_2 stoichiometry).[46] The addition of graphene to anode materials leads to superior electrical conductivity, high surface area, high surface-to-volume ratio, and ultra-thin thickness, which can shorten the diffusion distance of ions and the structural flexibility that paves the way for constructing flexible electrodes, and thermal and chemical stability which guarantee its durability in harsh environments.[60]

16.2.3   ALLOYING MATERIALS

Many materials (such as Si, Sn, Ge, etc.) are capable of accommodating Li by the formation of Li alloys (Li_xM) and show Li storage capacities much higher than those of carbonaceous materials.[5,61,62] The major drawbacks of these materials are the poor cycling life due to the high volume expansion/contraction and the larger irreversible capacity at the initial cycles. In order to overcome these issues, various approaches have been followed, namely the downsizing from micro to nanoscale particle and the fabrication of composites with both lithium active and inactive materials.[46]

16.2.3.1   SILICON BASED ANODES

Silicon has both the highest gravimetric capacity (4200 mAhg^{-1}, $Li_{22}Si_5$) and volumetric capacity (9786 mAhcm^{-3}) among the anode materials. In addition, the lithiation potential of silicon is almost close to graphite, that is, 0.4 V vs Li/Li$^+$. Si is the second most abundant element on earth, hence inexpensive and eco-friendly. Previous studies showed that the electrical contact between the active material and both the conductive carbon and current collector undergoes a drop due to the large volume expansion/contraction of the Si anode, leading to an irreversibility of lithium insertion/extraction. This leads to capacity fading and shorter cycling life.[46] To solve this problem, nanosized Si particles have been utilized and have achieved partial improvement by reducing the absolute volume change.[63] Different morphologies of Si nanostructures have been studied. Si nanocomposites have also been tried to improve the performance. Substantial improvement in cyclic stability and high rate capabilities could be achieved when the Si-based materials have (1) nanoporous structure to accommodate the volume variation during cycling; (2) fast electron conducting pathways to improve the poor electrical conductivity of Si; and (3) surface coating to form stable and thin SEI layer on the surface of Si.[64]

Liu et al. reported that rice husk, an abundant agricultural byproduct, can be converted directly into pure Si nanoparticles, which can be used as anode in Li-ion cells.[65] Nest-like Si nanospheres prepared by a modified solvothermal method exhibited a large specific capacity of 3052 mAhg^{-1} under a current density of 2000 mAg^{-1} (ca. 0.5 C). After cycling of up to 48 cycles under 2000 mAg^{-1}, the electrode modified by the nest-like Si nanospheres still showed a specific capacity of 1095 mAhg^{-1}.[66] An excellent capacity retention was noted for an interconnected Si hollow nanosphere electrode, synthesized via chemical vapor deposition of Si on silica particles followed by etching of SiO$_2$ by HF. The SEM of the Si hollow nanospheres at two different magnifications is shown in Figure 16.7. The hollow spherical structure was capable of delivering a high initial discharge capacity of 2725 mAhg^{-1} at a rate of 0.1 C. Even after 700 cycles, the Si hollow sphere electrode still retained 1420 mAhg^{-1} at a rate of 0.5 C.[67]

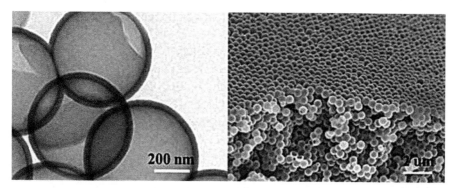

FIGURE 16.7 SEM images of Si hollow nanospheres at different magnifications. (Reprinted with permission from Yao, Y.; McDowell, M. T.; Ryu, I.; Wu, H.; Liu, N.; Hu, L.; Nix, W. D.; Cui, Y. Interconnected Silicon Hollow Nanospheres for Lithium-Ion Battery Anodes with Long Cycle Life. *Nano Lett.* **2011**, *11*, 2949–2954. © 2011 American Chemical Society.)

Si nanowire electrode synthesized via the vapor-liquid-solid process without any binder on stainless steel substrates using Au catalyst, exhibited a reversible capacity of 3193 mAhg^{-1} over 10 cycles at a C/20 rate.[68] With relaxed strain of nanowires, efficient electron transport, good electrical contact, and low fading during cycling, silicon nanowires have become a potential anode material. The Cui group reported that Si and Ge nanowires grown directly on a current collector exhibited stable cycle performances, because the nanowires were not pulverized or broken due to facile strain relaxation of the nanowire geometry.[69,70] They showed that the diameter of the nanowires increased without pulverization after several charge-discharge cycles.

2D Si thin film is another promising nanostructure with improved cycle stability and rate capabilities.[64] A 50 nm-thick Si film prepared from n-type Si was found to deliver a charge capacity over 3500 mAhg^{-1} and maintained during 200 cycles at a rate of 2 C, while a 150 nm-thick film showed around 2200 mAhg^{-1} during 200 cycles at 1 C in the electrolyte of 1 molL^{-1} LiClO$_4$ in propylene carbonate.[71]

Nanostructured electrodes can absorb large volume expansion/contraction, preserving the integrity of the electrode thereby improving the cycle performance. Nanostructured Si/C composite electrode showed a very high reversible capacity (about 1000 mAhg^{-1}) and good capacity retention (~99.8%) from the second cycle.[72] The nanosized composite electrode has accommodated the large volume change of the Si during charge and discharge processes, keeping the integrity of the electrode.[2] Carbon-coated Si nanocomposites produced by a spray-pyrolysis technique exhibited a capacity of 1489 mAhg^{-1} and high coulombic efficiency above 99.5%, even after 20 cycles.[73]

An effective method to synthesize nanosized C-coated Si composite was reported by Zhang et al. Polyacrylonitrile (PAN)-coated Si nanoparticles are formed by emulsion polymerization, and the PAN-coated precursor is heat-treated under argon to generate a Si/C core-shell nanocomposite. The conductive carbon shell envelops the Si nanoparticles and suppresses the aggregation of the Si nanoparticles during electrochemical cycling, subsequently ameliorating the capacity retention of this composite anode material.[74]

Ng et al. prepared Si/C nanocomposite using a spray-pyrolysis technique which showed a specific capacity of 1489 mAhg^{-1} after 20 cycles, evidencing the potential effect of pyrolyzed carbon coating on Si particles.[75] Another approach is that active silicon nanoparticles can be coated with inactive silica materials and conductive carbon together. Here, inactive materials play a key role in minimizing the mechanical stress induced by huge volume change in active silicon and thus prevents deterioration in electrochemical performance.[5]

Hu et al. reported the fabrication of flexible and conductive nanopaper aerogels with incorporated CNTs. The conductive nanopaper is made from aqueous dispersions with dispersed CNTs and cellulose nanofibers. Such aerogels are highly porous with open channels that allow the deposition of a thin-layer of silicon through a plasma-enhanced CVD (PECVD) method. The open channels allow for excellent ion accessibility to the surface of silicon. A stable capacity of 1200 mAhg^{-1} for 100 cycles in half-cells is achieved. Si-conductive nanopapers exhibited good cyclability. After 100 deep charge/discharge cycles, the discharge capacity still remains 83% compared to the second cycle and 77% compared to the first cycle. After 100 cycles, the capacity still remains above 1200 mAhg^{-1}, which is more than three times the theoretical capacity of graphite.[76]

Silvercoated 3D macroporous Si has been reported, in which silver nanoparticles formed an interconnected conductive network, providing electron pathways from the current collector to the whole surface area of the 3D porous Si particles. The silver-coated 3D macroporous Si delivered not only a reversible capacity of 1163 mAhg^{-1} at a rate of 0.2 C after 100 cycles, but an enhanced rate capability of 1930 mAhg^{-1} at a rate of 1 C.[77] Si–C composites prepared by dispersing nanocrystalline Si in carbon aerogel and subsequent carbonization exhibit a reversible Li storage capacity of 1450 mAhg^{-1}.[78]

Zhou et al. synthesized Si/TiSi$_2$ nanonet and demonstrated the performance as anode. The schematic and the TEM image of the Si/TiSi$_2$ nanonet are shown in Figure 16.8a,b, respectively. In this architecture, TiSi$_2$ served as a conductive backbone that enhanced the charge transport mechanism. The TiSi$_2$ backbone also played a critical role in improving the structural integrity of the anode, thereby facilitating a longer cycle life. Si nanoparticles attached to the TiSi$_2$ allowed for high capacities and high current densities. The anodes provided capacities of ~1000 mAhg^{-1} at a relatively high current density of 8.4 Ag^{-1}. The nanonet showed impressive capacities with a capacity fade as low as ~0.1% over 100 charge/discharge cycles (Fig.16.8c).[79]

Nanosized Si—Ni alloys also have been explored for Li-ion batteries. During Li insertion into the alloy electrodes, Si act as active center, which reacts with Li to form amorphous Li$_x$Si alloys, while the Si$_2$Ni in the sample plays the role of matrix as an inertial phase, which can buffer silicon volume expansion and facilitate charge transfer among silicon particles. A high Li storage capacity of 1304 mAhg^{-1} is observed for the Si–Ni alloy at Ni 9.0 wt% in nanoparticles with some reversibility.[80]

SiO is considered as an alternative choice to silicon with high theoretical capacity (> 1600 mAhg^{-1}). In addition, lithium oxygen co-ordination implies minimal volume change, and at the same time, lowers activation energy. The electrochemical reactions happening during the charge-discharge process determine the conversion of SiO into Si and lithium oxides and, possibly, the formation of Si–Li or, alternatively, the direct formation of Si–Li alloy and lithium silicates. Also, SiO undergoes consistent changes in the volume during the lithiation and de-lithiation processes. Moreover, both the electrical conductivity and the lithium insertion/de-insertion rates remain quite poor.[46] Core-shell Si/SiO nanocomposite displays better reversibility of Li insertion/extraction and higher coulombic efficiency than the virginal Si nanoparticles. Moreover, good capacity retention for the Si/SiO core-shell nanocomposite was achieved. The SiO shell acts as a barrier to prevent the aggregation of the Si nanoparticle and serves as a buffer as well to alleviate the volume expansion during the lithiation process. Consequently, the capacity retention of the core Si nanoparticles is improved.[81]

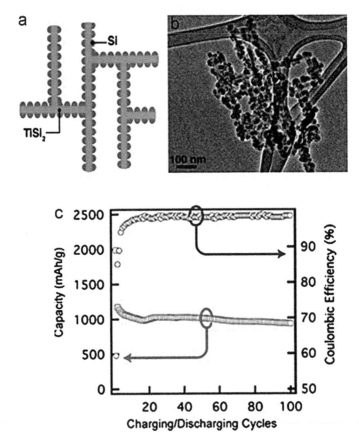

FIGURE 16.8 (a) Schematic and (b) TEM image of TiSi₂ nanonets coated with Si nanoparticles (c) cycling performance of the TiSi₂ nanonets coated with Si nanoparticles (Reprinted with permission from Zhou, S.; Liu, X.; Wang, D.; Si/TiSi₂ Hetero nanostructures as High-Capacity Anode Material for Li Ion Batteries. *Nano Lett.* **2010**, *10* (3), 860–863. © 2010 American Chemical Society.)

16.2.3.2   TIN BASED ANODES

Tin based materials have received considerable attention as anodes in Li-ion batteries due to their high theoretical capacity. However, the main disadvantage with Sn is the large volumetric change during cycling akin to that in the case of Si. Nanosized Sn electrodes exhibit much better cycling performance than bulkier Sn electrodes, due to a smaller absolute volume change.[82,83] Nanosized tin powder was prepared by laser-induced vapor deposition and evaluated as an anode material. The nano tin particles are spherical, and their size varied from 5 to 80 nm. The initial charge capacity is nearly same to the theoretical reversible capacity of lithium insertion into tin, resulting in $Li_{22}Sn_5$.[7]

Sn–C nanocomposites are also reported to give good electrochemical performance. Wang et al. synthesized Sn–C nanocomposite with metallic tin nanocrystals embedded into graphitic mesoporous carbon (GMC) walls via a simple one-step solid-liquid grinding/templating route. The Sn-GMC nanocomposite exhibits high initial coulombic efficiency, excellent cyclability, and rate performance. Highly dispersed metallic Sn nanocrystals with the sizes of 3–5 nm are well embedded in the highly conductive graphitic ordered mesoporous carbon walls, and the obtained Sn-GMC nanocomposite possesses hierarchical mesostructures with moderate surface area and large pore volume which are beneficial for the electric and ionic diffusion/transfer and thereby result in the improvement of rate performance and cycling capability.[84]

SnO_2 is another promising anode material for Li-ion batteries. The electrochemical reaction involves a first partially irreversible step, where SnO_2 is reduced into Sn and lithium oxides ($SnO_2+4Li \leftrightarrow Sn+2Li_2O$), the second step is the reversible Sn-lithium alloying/dealloying reaction ($Sn+4.4Li \leftrightarrow Li_{4.4}Sn$). The overall reaction involves 8.4Li for one SnO_2 formula unit. The corresponding theoretical capacity is 1491 mAhg^{-1} but it is reduced to 783 mAhg^{-1} when the second highly reversible step is reached. Hence, 783 mAhg^{-1} is commonly considered as the actual theoretical capacity. Furthermore, severe electrode degradation is observed because of the large volume change (> 200%) upon cycling.[46] A modified hydrothermal method was successfully developed to prepare tetragonal rutile tin oxide single crystals with diameters of 20–100 nm. At current densities of 0.4 mAcm^{-2} the reversible capacity still remained greater than 530 mAhg^{-1} after 20 cycles. The results showed that the as-prepared SnO_2 with higher crystallinity displayed a good electrochemical behavior.[85]

The SnO$_2$ nanoparticles based electrodes showed an initial discharge/charge capacity of 1518/916 mAhg^{-1}. They retained the charge capacity as high as 402 mAhg^{-1} at a current density of 156 mAg^{-1} after 40 cycles.[86] Kim et al. have shown that SnO$_2$ particles with 3 nm size have a superior capacity and cycling stability than 4 and 8 nm diameter particles (as illustrated in Fig.16.9) because they show less aggregation of Sn nanoparticles into large clusters since the nanoparticles are better dispersed in the Li$_2$O matrix.[87]

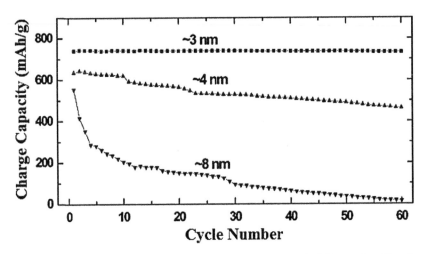

FIGURE 16.9 Cycling performance of SnO$_2$ nanoparticles with different size. (Reprinted with permission from Kim, C.; Noh, M.; Choi, M.; Cho, J.; Park, B. Critical Size of a Nano SnO2 Electrode for Li-Secondary Battery. *Chem. Mater.* **2005,** *17,* 3297–3301. © 2005 American Chemical Society.)

SnO$_2$/graphene nanocomposite prepared via an in situ chemical synthesis method exhibited first cycle charge and discharge capacities of 1775 and 862 mAhg^{-1}, respectively. Here, SnO$_2$ nanoparticles of 2–4 nm are uniformly attached on the graphene nanosheets matrix. The coulombic efficiency in the first cycle was around 48%; at the end of 10 cycles it retained 93% of the capacity. The discharge capacity dropped rapidly in the first cycle due to irreversible conversion of SnO$_2$ to Sn and Li$_2$O upon lithiation. In the subsequent charge/discharge cycles, Li$^+$ were reversibly inserted into Sn as Li$_x$Sn alloys. From the second cycle, the reversibility of the electrode was gradually improved on cycling. The SnO$_2$/graphene nanocomposite showed a reversible capacity of 665 mAhg^{-1} after 50 cycles and an excellent cycling performance, which was ascribed to the three-dimensional architecture of SnO$_2$/graphene nanocomposite.[88]

Meduri et al. presented a simple design of hybrid nanostructures involving Sn nanoclusters covered on SnO_2 nanowires as exceptionally stable anode materials with high reversible capacity. The SnO_2 nanowires were synthesized by reacting Sn metal powders directly in the gas phase with oxygen containing plasma. Pure SnO_2 nanowires were exposed to H_2 plasma in a microwave CVD reactor to synthesize the Sn nanocluster covered SnO_2 nanowires. Figure 16.10 shows the TEM image of Sn nanoclusters covered on SnO_2 nanowires, and the cycling performance of the Sn/SnO_2 hybrid nanostructure. The hybrid structure exhibited a reversible storage capacity of more than 800 mAhg^{-1} over 100 cycles. The capacity drop after the first few cycles was less than 1% per cycle.[89]

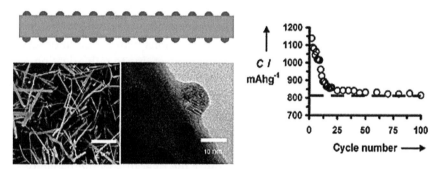

FIGURE 16.10   TEM image of Sn nanoclusters covered on SnO_2 nanowires and the cycling performance of the hybrid nanostructure. (Reprinted with permission from Meduri, P.; Pendyala, C.; Kumar, V.; Sumanasekera, G.U.; Sunkara, M.K.; Hybrid Tin Oxide Nanowires as Stable and High Capacity Anodes for Li-Ion Batteries. *Nano Lett.* **2009,** *9,* 612–616. © 2009 American Chemical Society.)

16.2.3.3   GERMANIUM

Germanium is also a promising anode material for lithium ion batteries owing to its high intrinsic electrical conductivity (104 times higher than silicon), higher capacity (1623 mAhg^{-1}) than graphite anode and a narrow band gap (0.67 eV). The practical usage of Ge as active electrode in LIBs is hindered by the dramatic volume change during lithium insertion/deinsertion.[46,90] Formations of $Li_{15}Ge_4$ and $Li_{22}Ge_5$ alloys induce a volume expansion of ~270%, thereby affecting the structural integrity of the anode and its cyclability. As in the case of other alloying anode materials, the electrochemical properties are improved by forming various nanostructures.

Park et al. fabricated novel hollow 0D and porous 3D Ge nanoparticles. The 3D Ge nanoparticles in particular showed impressive integrity after 100 cycles at a rate of 1 C. The excellent structural integrity also ensured that there was very little capacity fading associated with pulverization or loss of electrical contact of the germanium nanoparticles.[91]

Ge nanowire electrodes fabricated by using vapor-liquid-solid growth on metallic current collector substrates were found to have good performance. Schematic of the Ge nanowire electrode and the cycling performance are illustrated in Figure 16.11a,b, respectively. An initial discharge capacity of 1141 mAhg^{-1} was obtained which was stable over 20 cycles at the C/20 rate. High power rates were also observed up to 2 C with coulombic efficiency > 99%. The nanowire electrode design has several advantages. Mainly, there is good electrical contact between the current collector and every nanowire so that more active material can contribute to the capacity. For materials that undergo large volume changes during reactions with Li, often observation is that active material loses electronic contact due to pulverization and insufficient binding power of the polymer. With each nanowire contacting the current collector, this problem is avoided. The small diameter and 1D conductivity can allow for facile volume changes without pulverization and efficient charge transport, both for Li ions from the electrolyte and electrons from the current collector. Moreover, no binders or conducting carbon are needed, which add extra weight and lower the overall specific capacity of the battery.[70]

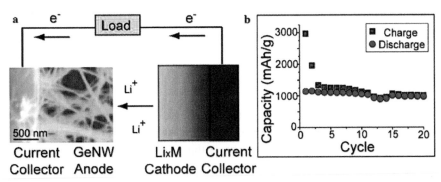

FIGURE 16.11 (a) Schematic of Ge nanowire electrode, (b) cycling performance of Ge nanowire electrode. (Reprinted with permission from Chan, C. K.; Zhang, X. F.; Cui, Y. High Capacity Li Ion Battery Anodes Using Ge Nanowires. *Nano Lett.* **2008,** *8,* 307–309.© 2008 American Chemical Society.)

16.2.4 TRANSITION METAL COMPOUNDS

As discussed earlier, the physical and chemical phenomena can be vastly different in a nanomaterial, compared to its bulk state as illustrated in the case of transition metal compounds. The transition metal compounds such as oxides, phosphides, sulphides, and nitrides (M_xN_y; $M = $ Fe, Co, Cu, Mn, and Ni and $N = $ O, P, S, and N) can be used in lithium ion batteries. The electrochemical reaction mechanism involving these compounds together with lithium, implies the reduction (oxidation) of the transition metal along with the composition (decomposition) of lithium compounds (Li_xN_y; here $N = $ O, P, S, and N). The electrochemical conversion reactions can be described as follows:[46]

$$M_xN_y + zLi^+ + ze^- \leftrightarrow Li_zN_y + xM \text{ (Here M = Fe, Co, Cu, Mn,}$$
$$\text{and Ni and N = O, P, S, and N).}$$

16.2.4.1 TI BASED MATERIALS

Ti based materials gained importance owing to the safety features, inexpensiveness, low toxicity, and low volume change (2–3%) on both lithium insertion and de-insertion, along with an excellent cycling performance. However, their main disadvantages are low theoretical capacity (175–330 mAhg^{-1}) and low electronic conductivity. Most extensively studied materials are TiO_2 and $Li_4Ti_5O_{12}$ (LTO). Nanostructured titanium oxides lead to better capacity, longer cycle life, and higher rate capability than the bulk materials.[46]

TiO_2 can host 1 mol of lithium per 1 mol of TiO_2 with a theoretical maximum capacity of 330 mAhg^{-1} and $LiTiO_2$ stoichiometry (almost double than LTO).[46] However, the exploitation of the entire capacity is a major experimental challenge. Titania has several allotropic forms; the most wellknown are rutile, anatase, and brookite.[92,93] Even though anatase titania has been considered the most electroactive form, other allotropes such as rutile and brookites are also widely studied for anode purposes.[94–96] Porous and dense TiO_2 nanospheres have been synthesized via a simple sol–gel process followed by subsequent hydrothermal or annealing treatments, respectively. Electrochemical evaluation in Li-ion batteries demonstrated that the porous TiO_2 samples possess a higher discharge capacity and better rate capability than the dense TiO_2 materials. Such considerable improvement of the electrochemical activity is believed to be caused by unique porous structures that favor electrochemical reactions.[97]

Fu et al. developed an effective method for the preparation of TiO_2/C core–shell nanocomposites. In the first step, a precursor, PAN coated nano-TiO_2 particles, was formed by emulsion polymerization. Then the precursor was heat-treated under argon atmosphere to achieve the nanocomposites. The conductive carbon shell enveloped TiO_2 nanoparticles suppressed the aggregation of nanoparticles during cycling. Meanwhile, it combined closely with the nanocores, significantly enhancing the kinetics of lithium intercalation and de-intercalation and diffusion coefficient of lithium ion. This provides a good approach to improve the cycling and kinetics of nano anode materials. In the case of TiO_2/C nanocomposite, after 10 cycles, the charge capacity still remained at 96.7% (i.e., 118 mAhg^{-1} titania) of its original capacity (i.e., 122 mAhg^{-1} titania), which is much higher than that of TiO_2.[98]

In the case of TiO_2 nanotubes, the charge/discharge capacities and rate performance are found to be dependent on the wall thickness. Maximum reversible capacity for Li-insertion in anatase TiO_2 (~330 mAhg^{-1}) has been achieved by reducing the tube wall thickness to 5 nm. Nanotubes with the wall thickness of 40 nm showed reversible capacity of ~170 mAhg^{-1} which is similar to the maximum theoretical capacity of the bulk anatase. Thus, with decrease in the wall thickness, rate performance of the nanotubes is significantly improved.[99]

Graphene oxides (GO) have been reported to be used as a support for the synthesis of anatase TiO_2 nanocrystals by hydrolysis of tetrabutyl titanate. After reduction, in situ formed TiO_2 nanoparticles were attached to graphene sheets and prevented the aggregation of the reduced GO sheets. Hybrid TiO_2/graphene/PVdF film was prepared by solvent evaporation of a PVdF/NMP solution containing TiO_2/graphene composites. The initial charge/discharge capacities of flexible electrodes were about 212 mAhg^{-1} and 202 mAhg^{-1}, respectively. Furthermore, flexible electrode showed an excellent cycling performance, which was attributed to the interconnected graphene conducting network.[100]

Lithium titanate, $Li_4Ti_5O_{12}$ (LTO) is another promising anode material for lithium ion batteries, owing to its good cyclability, rate capability, and safety.[101] Though there are many advantages with LTO, a major problem limiting its practical applications is its poor electrical conductivity (< 10^{-13} Scm^{-1}).[102] Moreover, the chemical diffusion coefficient is also low, about 2×10^{-8} cm^{2}s^{-1} at 25°C.[103,104] Formation of nanoparticles and forming nanocomposites with suitable materials could solve these problems to some extent. The synthesis of LTO ultra-fine fiber that exhibit good electrochemical performance can be synthesized by electrospinning route.[105] The SEM image and cycling performance of LTO ultrafine fibers, with a fiber diameter of 200–300 nm, prepared by a combination of sol–gel and electrospinning technique, is shown in Figure 16.12a,b, respectively. The performance of these fibers can be further improved by carbon coating, which is obtained by calcination of the electrospun precursor fiber in an inert atmosphere.[106]

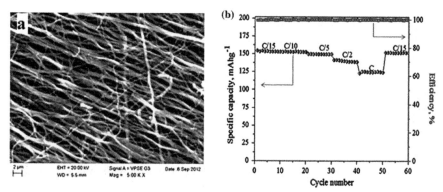

FIGURE 16.12 (a) SEM image and (b) cycling performance of lithium titanate ultrafine fibers. With kind permission from Ref. [105]. (Reprinted with permission from Sandhya, C. P.; John, B.; Gouri, C. Synthesis and Electrochemical Characterisation of Electrospun Lithium Titanate Ultrafine Fibres. *J. Mater. Sci.* **2013,** *48,* 5827–5832. © 2013 American Chemical Society.)

Ni et al. made a comparative study of the performance of nano-$Li_4Ti_5O_{12}$/ CNT and bulk $Li_4Ti_5O_{12}$/CNT. The $Li_4Ti_5O_{12}$ nanoparticles with a size of about 50nm were homogeneously anchored on CNTs which helps to prevent the agglomeration of $Li_4Ti_5O_{12}$ nanoparticles. The discharge capacities of bulk $Li_4Ti_5O_{12}$/CNT decrease steeply with increasing discharge rates, whereas in the case of nano- $Li_4Ti_5O_{12}$/CNT, the decrease is much slower at the same rate. The specific capacity of $Li_4Ti_5O_{12}$/ CNT nanocomposites (112 mAhg⁻¹) at 20C is higher than that obtained at the 5 C (106.5 mAhg⁻¹) for the bulk $Li_4Ti_5O_{12}$/CNT. $Li_4Ti_5O_{12}$/CNT composite also exhibited excellent cyclability with no noticeable decrease in performance over 100 cycles. The discharge capacity loss was 0.5% at 1 C and 2% at 5 C.[107]

16.2.4.2 MANGANESE BASED MATERIALS

Manganese oxide (MnO_2) has received substantial attention as promising anode material for Li-ion batteries due to their high theoretical specific capacities, low charge potential *vs.* Li/Li⁺, environmental benignity and natural abundance.[108] High reversible capacity was observed for MnO_2/CNT hybrid coaxial nanotubes compared to MnO_2 nanotubes, indicating the reduced structural changes during cycling. The first discharge capacity of 2170 mAhg^{-1} and a reversible capacity of ~500 mAhg^{-1} after 15 cycles were observed for MnO_2/CNT hybrid coaxial nanotubes. The nano-sized and porous structure of the MnO_2 allows fast ion diffusion, while the CNT facilitates electron transport to the MnO_2. In addition, CNTs act as additional lithium storage sites, which lead to an improved reversible capacity.[38] MnO_2 nanofibers, containing 20–25 nm oxide grains, prepared by electrospinning route exhibited more than 450 mAh/g in 50 cycles at 0.5 mA/cm.[2,109]

16.2.4.3   IRON BASED MATERIALS

Iron based oxides have been extensively studied for Li-ion batteries because of their low cost, non-toxicity and high natural abundance.[46] Iron oxides, both hematite (α-Fe_2O_3) and magnetite (Fe_3O_4), are capable of participating in reversible conversion reactions with lithium, providing a theoretical capacity of 1007 and 926 mAhg^{-1}, respectively.[110] Iron oxides tend to exhibit poor cycling performance due to low electrical conductivity, low diffusion of Li-ions, high volume expansion, and iron aggregation during charging and discharging.[46] Fe_2O_3 nanorods, with diameters of 60–80 nm and lengths of 300–500 nm, synthesized by hydrothermal method, exhibited improved capacity and cycling stability (763 mAhg^{-1} capacity after 30 cycles), compared to microcrystalline α-Fe_2O_3 powders.[111]

16.2.4.4   COBALT BASED MATERIALS

Co_3O_4 and CoO have received much attention as anode material for Li-ion batteries, owing to their high specific capacity, 890 mAhg^{-1} for Co_3O_4 and 715 mAhg^{-1} for CoO. However, they suffer from poor cycling stability because of severe particle aggregations and large volume changes. In order to improve the electrochemical performance, a variety of Co_3O_4 nanostructures, such as nanotubes, nanowires, nanorods, nanoneedles, nanosheets, nanocubes, hollow microspheres, and agglomerated nanoparticles have been studied extensively.[112]

Co_3O_4 nanowires possessed excellent high-rate capability compared to the commercial Co_3O_4 nanoparticles. Co_3O_4 nanowires with diameters in the range of 30–60 nm, delivered a discharge/charge capacity of 611 mAhg^{-1} and 598 mAhg^{-1} after fifty cycles at a current density of 0.11 Ag^{-1}, which was much higher than those of commercial Co_3O_4 nanoparticles. The excellent electrochemical performances could be attributed to the alleviation of the mechanical stress induced by the volume change during the repeated lithiation/delithiation processes for the 1D nanostructured feature of Co_3O_4 nanowires.[113]

A hierarchical porous MWCNTs/Co_3O_4 nanocomposite has been synthesized by Huang et al. with the help of a morphology-maintained annealing treatment of CNTs inserted metal organic frameworks (MOFs). The SEM images of the nanocomposites is shown in Figure 16.13. The MWCNTs/Co_3O_4 integrates the high theoretical capacity of Co_3O_4 and excellent conductivity, as well as strong mechanical/chemical stability of MWCNTs. The nanocomposite displayed a high reversible capacity of 813 mAhg^{-1} at a current density of 100 mAg^{-1} after 100 charge-discharge cycles. Even at 1000 mAg^{-1}, a stable capacity as high as 514 mAhg^{-1} could be maintained. The improved reversible capacity, excellent cycling stability, and good rate capability of MWCNTs/Co_3O_4 can be attributed to the hierarchical porous structure and the synergistic effect between Co_3O_4 and MWCNTs.[114]

FIGURE 16.13 SEM images of the as-synthesized MWCNTs/Co_3O_4. (Reprinted with permission from Huang, G.; Zhang, F.; Du, X.; Qin,Y.; Yin,D.; Wang, L. Metal Organic Frameworks Route to in Situ Insertion of Multiwalled Carbon Nanotubes in Co_3O_4 Polyhedra as Anode Materials for Lithium-Ion Batteries. *ACS Nano.* **2015**, *9* (2), 1592–1599. © 2015 American Chemical Society.)

CoO based nanocomposites are also reported. Zhang et al. developed a novel ordered mesoporous carbon hybrid composite, CoO/CMK-3, by an infusing method using $Co(NO_3)_2 \cdot 6H_2O$ as the cobalt source. In this composite, the CoO nanoparticles are loaded in the channels of mesoporous carbon. CoO/CMK-3 composites have higher reversible capacities (more than 700 mAhg^{-1}) and better cycle performance in comparison with the pure mesoporous carbon (CMK-3). This substantial improvement of electrochemical performance may be attributed to the synergistic effects in the CoO/CMK-3 composites.[48]

CoO/Cu nanowire arrays prepared through electrodeposition and sputtering, delivered long cycle life and enhanced power performance than CoO/Cu films. The large accessible surface area and improved electronic/ionic conductivity of the nanostructured electrodes may be responsible for the improved performance. The nanostructured hybrid CoO/Cu electrode delivered a capacity of 970 mAhg^{-1} at a current of 0.3 C after 200 discharge/charge cycles. Even at 10 C (7160 mAg^{-1}), the electrode delivered a stable capacity of 530 mAhg^{-1}. Upon decreasing the rate from 10 C to 0.3 C, nearly 100% of the initial capacity at 0.3 C (about 800 mAhg^{-1}) can be recovered.[115]CoCO$_3$ submicrocube/Graphene nanosheet (GNS) composites, prepared through a solvothermal route, delivered a reversible capacity of 930 mAhg^{-1}, 107% higher than the theoretical value based on traditional lithium storage mechanisms.[116]

CoS$_2$ has attracted significant attention in the past few years as a suitable anode material for Li-ion batteries due to its high electrical conductivity, good thermal stability, and low cost. A yolk–shell structure with a distinctive core@void@shell configuration has received much attention. The core of the yolk–shell particle will improve the energy density of the powder by increasing the weight fraction of the electrochemically active component. The unique interstitial void space between the yolk and the shell can also accommodate the volume variation of particles during lithium insertion and extraction. The as-synthesized yolk–shell CoS$_2$@NG hybrid displayed excellent cycling performance with high specific capacity of 882 mAhg^{-1} at a current density of 100 mAg^{-1} after 150 cycles, and excellent rate properties with a capacity of 655 mAhg^{-1}, even at a high current rate of 5 C.[117]

16.2.4.5   NICKEL BASED MATERIALS

NiO is another transition metal based anode for Li-ion batteries. For the NiO electrodes, the severe capacity degradation should be attributed to the large volume expansion–contraction leading to the pulverization and degradation of the electrode.[118] NiO nanotubes that are 60 μm long with a 200 nm outer diameter and wall thickness of 20–30 nm was prepared by Needham et al. by a template synthesis technique. Electrochemical studies indicated a large irreversible capacity loss. First cycle capacity was about 600 mAhg^{-1}, which was reduced to less than 200 mAhg^{-1} in 20 cycles. The template synthesis techniques can be used as an effective method for producing experimental quantities of high aspect ratio NiO nanotubes.[119]

Ma et al. synthesized a variety of mesoporous nickel oxide (NiO) nano-structures by annealing the corresponding β-Ni(OH)$_2$ precursor nanostructures by manipulating the reaction parameters (e.g., alkaline sources, reactant concentration). The precursor β-Ni(OH)$_2$ with various morphologies from nanoplates to assembled structures such as stacked nanoplates and flower-like superstructures were prepared by employing different alkaline sources. NiO nanomaterials are then obtained by annealing the corresponding precursors at high temperature. Here, the size and shape of the NiO nanostructures are consistent with the β-Ni(OH)$_2$ nanostructures. All the samples displayed higher capacities than theoretical value (about 718 mAhg^{-1}).[118]

Xia et al. reported the synthesis of hierarchically porous NiO/C micro-spheres via a facile biotemplating method using natural porous lotus pollen grains as both the carbon source and the template. The resultant NiO nanopar-ticles are in the shape of nanowalls growing directly on the surface of the carbonized pollen, forming good adhesion, and better electrical contact between NiO and the porous biocarbon, which enhances the kinetics of NiO during cycling under multiple current rates. The specific capacities of the porous NiO/C nanocomposites after every 10 cycles at 0.1, 0.5, 1, and 3 Ag^{-1} are about 698, 608, 454, and 352 mAhg^{-1}.[120]

16.2.4.6 COPPER BASED MATERIALS

Cu$_2$O nanocubes have been prepared by Park et al. using a one-pot polyol process. As-synthesized Cu$_2$O nanocubes were converted into CuO hollow cubes, hollow spheres, and urchin-like particles by controlling the amount of aqueous ammonia solutions added in air. The urchin-like electrodes retained high capacities of above 560 mAhg^{-1} up to 50 cycles. The good cycling performance of urchin-like particles might be attributed to the structural robustness and the stability of the large single-crystalline domains.[121]

16.2.4.7 ZINC BASED MATERIALS

Different morphologies of ZnO nanomaterials have been evaluated as anode in Li-ion batteries. Liu et al. grew ZnO nano-needles on Nickel substrate, and evaluated it as an anode material. The material exhibited a first cycle discharge capacity of 1219 mAhg^{-1}.[122] ZnO/C nanocomposites delivered a stable reversible capacity of 610 mAh g^{-1} at 100 mA g^{-1} even after 50 cycles.[123] The microspheres assembled with distorted nanosheets, hexag-onal nanorods, and radial assembly of nanorods of ZnO were successfully prepared by the hydrothermal reaction. The capacity retention of the microspheres and radial assembled nanorods are higher than that of hexagonal nanorods. This may be due to their inner spacing of specific structure patterns that can accommodate and restrain the volume changes during cycling. The ZnO microspheres calcinated at 600°C show the best performance with a specific capacity of 1328.2 mAhg^{-1} for the first cycle and 662.8 mAhg^{-1} for the 50th cycle at 0.1C.[124] Mesoporous ZnO nanosheets having micron-sized length/width and nanosized thickness are constructed by many interconnected nanostrips and form a mesoporous net-like structure. The mesoporous ZnO nanosheets exhibited an initial charge capacity of 640 mAhg^{-1} and a 50th cycle charge capacity of 420mAhg^{-1}. Both the capacity and cycling performance are enhanced, compared with those of common solid ZnO particles.[125]

16.2.5   TRANSITION METAL SULPHIDES

Layered transition metal sulphides such as MoS_2, WS_2, VS_2 etc. are structurally similar to that of graphite. They have analogous layered structures that are made up of covalently bonded 2D sandwiched SM—S layers, which are held together by weak Van der Waals force.[126] These transition metal sulphides allow foreign atoms or molecules to be introduced between the layers through intercalation. This property has been explored for their use as anode material for Li-ion batteries.[127] Since lithium ions can intercalate into small holes/channels, nanostructured transition metal sulphides with plenty of defects can deliver higher lithium intercalation capacity.[128]

Among the transition metal sulphides, MoS_2 has been the most widely studied because of its high theoretical capacity and inexpensive components of molybdenum and sulfur. However, its practical application as anode of LIBs is hindered by its low reversible capacity and poor cyclic life due to its poor electronic conductivity, massive structural reorganization, and volumetric changes during cycling and the loss of sulfur in the form of polysulfides. Two main strategies, that is, modification of MoS_2 morphology and preparation of nanocomposites have been attempted to overcome these problems.[129]

Layered MoS_2 nanostructures with hierarchical morphology and nanoscale size are reported to be much superior to its bulk counterpart in Li storage, owing to multiple reasons. First, the initial 2D nanostructures of the MoS_2 can lead to the formation of the unique nanostructures with an enhanced interlayer spacing and enlarged surface area by the restacking process, thus favoring Li$^+$ intercalation and stress accommodation. Secondly, large electrode/electrolyte interface area of nanostructured MoS$_2$ can offer more active sites for conversion reactions. Thirdly, the porous structure can provide voids to alleviate severe volume changes during cycling, and additionally mesoporous MoS$_2$ composed of smaller grain size may shorten Li$^+$ diffusion pathway and thus enhance rate capability.[126]

Feng et al. reported a rheological phase reaction method for preparing MoS$_2$ nanoflakes. The MoS$_2$ nanoflake electrode exhibited higher specific capacity, with very high cycling stability, compared to MoS$_2$ nanoparticle electrode. The reversible capacity remains at 840 mAhg^{-1} after 20 cycles, which is 84% of the initial reversible capacity. There are four possibilities for lithium intercalation in the MoS$_2$ nanoflake electrodes, namely lithium ion intercalation into nanoflake clusters, defect sites in nanoflakes, intratubal sites (the hollow core) through the open end and into the MoS$_2$ layer sites to form Li$_x$MoS$_2$. All these four possibilities contribute to the high lithium insertion capacity of MoS$_2$ nanoflake electrodes.[128] Hwang et al. prepared MoS$_2$ nanoplates, with increased interlayer distance of 0.69 nm, which can accommodate much more intercalated Li$^+$ ions than the bulk counterpart before decomposing to Mo and Li$_2$S. As a result, excellent cycling stability and extremely high rate capability were achieved. Capacity of 907 mAhg^{-1} was maintained after 50 cycles at 1 C rate (1.062 Ag^{-1}), with a retention of 98%. Reversible capacity of 700 mAhg^{-1} can still be achieved at even extremely high rate of 50 C (53.1 Ag^{-1}), and 554 mAhg^{-1} after 20 cycles (Fig.16.14). HR-TEM images (Fig.16.14) confirmed that the nanostructure of MoS$_2$ active materials might be inherited after conversion reaction, and the converted M and Li$_2$S were uniformly dispersed in the matrix contributing to elevated rate capacity and stability.[130]

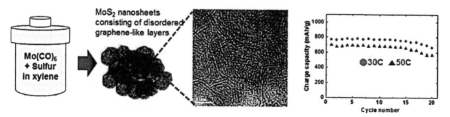

FIGURE 16.14 MoS$_2$ nanosheets consisting of disordered graphene-like layers and the cycling performance in coin-type lithium half-cell between 0 and 3 V at high charge rates of 30 and 50 C (1 C = 1.062 Ag^{-1}) at room temperature. (Reprinted with permission from Hwang, H.; Kim, H.; Cho, J.; MoS$_2$ Nanoplates Consisting of Disordered Graphene-Like Layers for High Rate Lithium Battery Anode Materials. *Nano Lett.* **2011**, *11*, 4826–4830. © 2011 American Chemical Society.)

Restacked MoS_2 nanosheets prepared from chemical lithiation and exfoliation have been evaluated as anode for LIBs. The restacked MoS_2 showed much better cycling stability and retained a capacity of 750 mAhg^{-1} after 50 cycles compared to the bulk MoS_2 which showed a decrease in capacity from 800 mAhg^{-1} to 226 mAhg^{-1} after 50 cycles. The restacked MoS_2 nanosheets possessed an enlarged inter-sheet spacing compared to the raw MoS_2, which lowered the barrier for Li intercalation, to give better ionic conductivity.[131] The massive production of single and multi-layer MoS_2 nanosheets from the top-down solution-phase exfoliation or the bottom-up chemical synthesis may enable their widespread application for batteries.[132]

MoS_2 nanoflowers synthesized by ionic liquid assisted hydrothermal reaction exhibited a reversible capacity of about 900 mAhg^{-1}, however, the capacity decreased to 520 mAhg^{-1} after 100 cycles due to the aggregation and pulverization of the active particles during cycling.[133]

In order to improve the cycle performance of MoS_2, an effective strategy is to prepare nanocomposites with carbon. The carbon can act as a barrier to alleviate volume expansion of active materials and suppress their aggregation and pulverization during cycling. In addition, the carbon can improve the conductivity of the active materials. Li et al. synthesized MoS_2/C nanocomposites, by one-pot hydrothermal route, which consisted of 2D nanosheet MoS_2 crystal and amorphous carbon. The MoS_2 was uniformly dispersed in the amorphous carbon. The annealed MoS_2/C composites exhibited much higher reversible capacity and better cyclability than the MoS_2 sample. The initial reversible capacity was 1065 mAhg^{-1} and the capacity of 1011 mAhg^{-1} was retained after 120 cycles. The amorphous carbon in the nanocomposites stabilized the electrode structures and also kept the active materials electrically connected, thus significantly improving the cycle performances of the MoS_2/C nanocomposite electrodes.[127]

Gao et al. reported composites of MoS_2 and MWNT. The studies show that MoS_2/MWNT composite with molar ratio of 1:4 exhibits the highest specific reversible capacity of 1112.5 mAhg^{-1} at a current of 100 mAg^{-1} after 90 cycles. The composites showed an improved cycling performance and an excellent rate capability over pristine MoS_2. The excellent electrochemical properties of MoS_2/MWNT composites are attributed to the synergistic effects of layered MoS_2 and MWNT.[134]

A facile process is demonstrated for the synthesis of layered SiCN-MoS_2 structure via pyrolysis of polysilazane functionalized MoS_2 flakes. SiCN-MoS_2 showed the classical three-stage reaction with improved cycling stability and capacity retention than neat MoS_2. SiCN-MoS_2 in the form of self-supporting paper electrode (at 6 mg/cm^2) exhibited even better performance, regaining initial charge capacity of approximately 530 mAhg^{-1} when the current density returned to 100 mAg^{-1} after continuous cycling at 2400 mAg^{-1} (192 mAhg^{-1}). MoS$_2$ cycled electrode showed mud-cracks, film delamination whereas SiCN-MoS$_2$ electrodes were intact and covered with a uniform solid electrolyte interphase coating.[135] Flexible and free-standing MoS$_2$/GS films are constructed for use as free-standing and binder-free electrodes for Li-ion batteries. The optimized hybrid film with graphene content of 32% exhibits superior rate capability (994, 880, and 598 mAhg^{-1} at 0.5, 1, and 5 Ag^{-1}, respectively) and outstanding cycling performance at high rates (retaining 100.6% of the initial capacity after 1000 cycles at 1000 mAg^{-1}).[136]

(2D) layered MoS$_2$/graphene and MoS$_2$/XC-72 composites are also evaluated as anode in Li-ion cells. Due to the outstanding properties of graphene and the synergistic interaction between 2D MoS$_2$ and graphene nanosheets, the 2D MoS$_2$/graphene composite exhibits a very high reversible capacity of 1060 mAhg^{-1} with excellent cycle stability and significantly enhanced rate capability compared with pristine MoS$_2$ and the MoS$_2$/XC-72 composite. The satisfying electrochemical performance could be attributed to a strong suite of physical properties of graphene, and the synergistic interaction between 2D layered MoS$_2$ and graphene nanosheets.[137]

Han et al. developed a ternary MoS$_2$/SiO$_2$/graphene hybrid (MSGs) by incorporating amorphous SiO$_2$ with MoS$_2$ and graphene nanosheets. The inert silica in the hybrids can effectively mediate the volume expansion of MoS$_2$ during lithiation and delithiation cyclings, when they serve as the anode material in Li-ion cells. As a result, MSGs with 40 wt% of SiO$_2$ and 42 wt% of MoS$_2$ manifest the capacity stabilized at 1060 mAhg^{-1} for more than 100 cycles at the current density of 0.1 Ag^{-1}. Moreover, this electrode maintains a specific capacity of 580 mAhg^{-1} at an ultrahigh current density of 8 Ag^{-1}.[138]

MoS$_2$/poly (3,4-ethylenedioxythiophene): poly (styrenesulfonate) (PEDOT: PSS) composite (MoS$_2$/P) prepared through a facial and environmental friendly dip-coating method showed an enhanced electric conductivity of 1.0×10^{-1} Scm^{-1}, about 5 times of that (2.2×10^{-2} Scm^{-1}) of the pristine MoS$_2$. MoS$_2$/P composite electrode exhibited significantly improved electrochemical performances such as cyclability and rate capability. The composite electrode delivered a reversible capacity of 712 mAhg^{-1} at a current density of 50 mAg^{-1} and retained 81% capacity after 100 cycles. At a higher current density of 200/300 mAg^{-1}, it still retained a capacity of 439/363 mAhg^{-1}, respectively, as compared to 191/140 mAhg^{-1} for the pristine MoS$_2$.[129]

Tungsten disulphide (WS$_2$), whose physical and chemical properties are more or less similar to that of MoS$_2$, also has been studied extensively as an anode active material in Li-ion batteries.[132]Feng et al. synthesized WS_2 nano-flakes and found that the WS_2 nanoflakes exhibited high reversible capacity (800 mAhg^{-1}).[139] WS_2 nanotubes synthesized by sintering amorphous WS_2 at high temperature under flowing hydrogen show a lithium insertion capacity of about 915 mAhg^{-1} (corresponding to 8.6 mol lithium per mole of WS_2 nanotubes), which is much higher than for WS_2 powders (where the lithium insertion capacity was only 0.6 mol Li$^+$ per mole of crystalline WS_2). After the first cycle, the WS_2 nanotube electrode exhibited stable cycling behavior over a wide voltage range of 0.1–3.1V versus Li/Li$^+$.[140]

Few layered WS_2–graphene nanosheet composites prepared by a simple and scalable hydrothermal reaction and a subsequent freeze-drying method exhibited good cycling stability and outstanding high-rate capability. The reversible capacity remains 647 mAhg^{-1} after 80 cycles at a current density of 0.35 Ag^{-1}. Comparable capacities of 541 and 296 mAhg^{-1} can still be maintained when cycling at even higher current densities of 7 and 14Ag^{-1} (7 and 14 mAcm^{-2}), respectively.[141]

A graphene-like few-layer tungsten sulfide (WS_2) supported on reduced graphene oxide (RGO) with WS_2:RGO = 80:20 and 70:30 exhibited good enhanced electrochemical performance and excellent rate capability performance when used as anode materials for Li-ion batteries. The WS_2–RGO composite delivered a capacity of 400–450 mAhg^{-1} after 50 cycles when cycled at a current density of 100 mAg^{-1}. At 4000 mAg^{-1}, the composites showed a stable capacity of approximately 180–240 mAhg^{-1}, respectively. The noteworthy electrochemical performance of the composite is not additive, rather it is synergistic in the sense that the electrochemical performance is much superior compared to both WS_2 and RGO. As the observed lithiation/delithiation for WS_2–RGO is at a voltage of 1.0 V (E0.1 V for graphite, Li$^+$/Li), the Li-ion battery with WS_2–RGO is expected to possess high interface stability, safety, and management of electrical energy is expected to be more efficient and economic.[142]

Vanadiun sulphides have been rarely investigated as the anode for Li-ion batteries as compared to MoS_2 or WS_2.[126] $Li_{1+x}VSe_2-yS_y$ and $Li_{1+x}VS_2$ were tested as anode materials by intercalating Li-ions in their tetrahedral sites, showing a capacity lower than 200 mAhg^{-1}.[143,144] The synthesis of VS_4 using a simple, facile hydrothermal method with a graphene oxide (GO) template and the characterization of the resulting material was reported. The SEM image of the VS_4/rGO composite is shown in Figure 16.15. Initial discharge and charge capacities of the VS4/rGO were found to be 1669 and 1105 mAhg^{-1}, respectively, and good cycling performance, with the retention of a high charge capacity of 954 mAhg^{-1} after 100 cycles at 0.1 C, was observed.[145] In another study, a graphene-attached VS_4 composite prepared by a simple hydrothermal method is studied in terms of its lithium reaction mechanism and high rate capability. The nanocomposite exhibited a good cycling stability and an impressive high-rate capability for lithium storage, delivering a comparable capacity of 630 and 314 mAhg[-1], even at high rates of 10 and 20 (C = 10 and 20 Ag[-1], or 10 and 20 mAcm[-2]), respectively.[146]

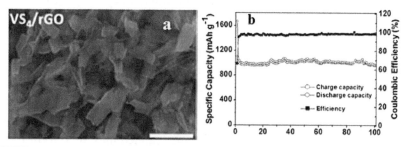

FIGURE 16.15 (a) SEM image and (b) cycling performance of VS_4/rGO. (Reprinted with permission from Rout, C. S.; Kim, B. H.; Xu, X.; Yang, J.; Jeong, H. Y.; Odkhuu, D.; Park, N.; Cho, J.; Shin, H. S. Synthesis and Characterization of Patronite form of Vanadium Sulphides on Graphitic Layer. *J. Am. Chem. Soc.* **2013**, *135,* 8720–8725. © 2013 American Chemical Society.)

Ultrathin ZrS_2 nano discs, whose TEM images are shown in Figure 16.16, prepared by using a colloidal route exhibited reversible discharge capacity of 650mAhg[-1] when cycled between 0.02 and 1.2 V at a current density of 69 mAg[-1]. The capacity of the ZrS_2 nano discs decreased slightly after 50 cycles, while the capacity of the bulk ZrS_2 dropped dramatically to less than 200 mAhg[-1]. The rate capability was also improved considerably in the ZrS_2 nano discs, showing the capacity retention of 80% when the current density increased from 69 to 552 mAg[-1].[147]

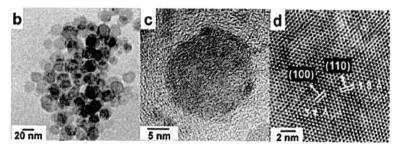

FIGURE 16.16 (b) Low- and (c) HR-TEM images of 20 nm single-crystalline ZrS_2 nanodiscs (d) HR-TEM image of a nanodisc showing (1 0 0) and (1 1 0) lattice fringes. (Reprinted with permission from Jang, J.T.; Jeong, S.; Seo, J.W.; Kim, M. C.; Sim, E.; Oh, Y.; Nam, S.; Park, B.; Cheon, J.; Ultrathin Zirconium Disulfide Nanodiscs. *J. Am. Chem. Soc.* **2011**, *133,* 7636–7639. © 2011 American Chemical Society.)

16.2.6   INTERMETALLIC ANODES

Various intermetallic compounds like Cu_6Sn_5, Sb_2Ti, Sb_2V, Sn_2Co, Sn_2Mn, Sn_2Fe, Al_2Cu, Sn_2Ru, Ge_2Fe, Mg_2Si, Mg_2Ni, and Mg_2Ge have been proposed as anode material in Li-ion cells.[148] The cycling stability of intermetallic anodes can be significantly improved by decreasing the particle size due to the reduction of volume changes and enhancement of Li-alloying kinetics.[149]

Zhang et al. synthesized nanoscale $FeSn_2$ intermetallic compound by two different facile methods, namely, chemical reduction process and solvothermal method. The $FeSn_2$ intermetallic compound synthesized by solvothermal method showed lower initial discharge capacity and better cyclability than that prepared by chemical reduction process due to the better crystallinity. The nanoscale $FeSn_2$ powders both delivered a high reversible discharge capacity (500 mAhg^{-1}). During the cycling, $FeSn_2$ firstly decomposed to Fe and Sn then lithium ions are intercalated into Sn to form Li–Sn alloys. The reaction of Li^+ and Sn was responsible for the reversible capacity in cycling process. The electrochemical reaction of $FeSn_2$ with Li can be expressed as follows:

$$FeSn_2 + 8.8Li^+ + 8.8e^- \longrightarrow 2Li_{4.4}Sn + Fe \qquad (16.1)$$

$$Li_{4.4}Sn \leftrightarrow Sn + 4.4Li^+ + 4.4e^- \qquad (16.2)$$

$FeSn_2$ could reversibly react with $8.8Li^+$ per formula unit, leading to a reversible capacity as high as 804 mAhg^{-1}.[150]

Nanocrystalline intermetallic NiSi alloy prepared by high energy ball milling showed a high lithium storage capacity of 1180 mAhg^{-1} in the initial discharge, in which Si acted as an active element to combine with Li to form Li_xSi. This reaction was partially reversible and its capacity declined with each cycle.[151]

16.2.7   ELECTROLYTES

In the case of electrolytes for Li-ion batteries, the major development is in the use of nanomaterial additive to improve the performance. The addition of nanosized fillers to electrolytes modifies their physical and electrochemical properties. Various types of nanoparticles including ceramics and clay were used as additive to electrolytes commonly used in Li-ion batteries such as liquid electrolytes, polymer electrolytes, and plastic crystals.[152]

The dispersions of nanoparticles in liquid electrolyte are known as soggy sand or colloidal electrolyte.[152] The name "soggy sand" is derived from the fact that, in the regime of high conductivities, the composite possesses a soft matter consistency similar to that of "soggy" sand.[153] The conductivity of the liquid electrolyte is improved by the addition of nanosized particles.[154] The composite conductivity has been found to be dependent on several parameters viz. surface acidity of oxides, volume fraction and particle size, solvent dielectric constant, salt concentration, and temperature.[153] The addition of silica nanoparticles improved the conductivity of liquid electrolytes by few to ten folds.[154] Dispersion of oxide particles varying in surface acidity–basicity, for example, SiO_2, TiO_2, Al_2O_3 (size $\approx$300 nm) in 0.1 M $LiClO_4$ in MeOH (εr = 32.6) resulted in a remarkable variation in overall composite conductivity. Apart from ionic conductivity, the mechanical properties are also improved. The high effective viscosity of the "soggy sand" electrolytes endows them with improved mechanical strength (in terms of partial shapeability) compared to liquid electrolytes in the assembly of an electrochemical cell.[153]

The addition of nanoparticles to polymer electrolytes significantly improved the properties. The most widely used polymer electrolyte is polyethylene oxide (PEO) mixed with Li salt. The main disadvantages with this system are poor conductivity at temperatures below 70°C, low ion transference number (~0.3), and reactivity toward lithium metal. Polymer nanocomposites obtained by dispersing ceramics, clays, and mesoporous materials exhibit better ion transfer number, ionic conductivities (up to ten folds), mechanical properties, and electrolyte–electrode interface compared to the host polymer.[152] For example, nanosized TiO_2 powders can effectively improve the conductivity of PEO-$LiClO_4$ by a factor of two, both at room temperature and at elevated temperature.[155] This is because nanoparticles with large surface area can prevent the PEO chain reorganization at room temperature and thus enhance the ionic conductivity. Nanosized TiO_2 powders are highly polar with many hydrophilic groups on the surface of the grains, such as $-COOH$ and $-OH$. These highly polar groups can react with the anions and the PEO segments, and thus reduce the ion aggregation, increase the fraction of free lithium cations and enhance the ionic conductivity.[73] The incorporation of the inert filler reduces the crystallinity of the polymer host and acts as "solid plasticizer" capable of enhancing the transport properties.[156] Also, fast ion conduction can proceed at the highly conductive interface layer between the PEO matrix and the nanoparticles, according to the effective medium theory.[157]

Nanosized inorganic fillers can also be used with gel polymer electrolytes made of solid polymer and liquid electrolytes. The addition of TiO_2 to poly (methyl methacrylate) gel made of $LiClO_4$, EC, PC, and PMMA in a ratio 4.5:46.5:19:30 has improved the ionic conductivity and Li-ion transference number.[158] Ionic conductivity and interfacial stability of a polymer gel electrolyte made of $PMMA/LiV_3O_8$ saturated with a solution of 1 M $LiClO_4$-PC/DEC (1:1) were superior to the same gel in the absence of LiV_3O_8. Room temperature ionic conductivity of $PMMA/LiV_3O_8$ gel nanocomposite (1.8×10^{-3} Scm^{-1}) was about four times higher than PMMA gel in the absence of LiV_3O_8 (5.1×10^{-4} Scm^{-1}). This improvement could be attributed to the fact that the negative $V_3O_8^-$ layers have high dielectric constant ($\varepsilon = 4.5$), which help dissolve more of $LiClO_4$ thereby increasing the conductivity.[159]

Organic ionic plastic crystals are a new type of solid state conducting molecular plastic materials consisting entirely of ions. Doping these materials with inorganic filler such as lithium salts and oxides (e.g., SiO_2) results in significant increase in conductivity. This enhancement in conductivity was attributed to the induced dissociation of ionic aggregates at the interface and increased concentration of charge carriers in a space-charge layer around the filler surface.[152]

16.3   CHALLENGES IN NANOMATERIALS FOR LITHIUM ION BATTERIES

Nanomaterials have the potential to improve the energy density and power density of Li-ion cell. So far, we have discussed the potentials and advantages of nanomaterials. However, there are a few challenges/disadvantages in the use of nanomaterials for Li-ion batteries. The challenges/disadvantages of nanomaterials are summarized below:

1. The high specific surface area of nanomaterials makes the electrochemical reaction more active; however, the high surface area increases parasitic side reactions between the electrode and the electrolyte and hence leads to large irreversible capacity. This can also lead to capacity fading. Also, the high surface area could induce poor crystallinity, resulting in fewer storage sites for Li$^+$ ions.[6] In the case of nanosized $LiCoO_2$, it was observed that the coulombic efficiencies and reversible capacities decreased to ca. 26% and 65 mAhg^{-1}, respectively as the size of $LiCoO_2$ is decreased from bulk to 6 nm.[160] Bulkier $LiCoO_2$ usually exhibits > 95% and > 140 mAhg^{-1},

respectively. Therefore, greater focus should be given to obtain more stable surfaces and better crystallinity of nanomaterials for their practical utilization in batteries.[6]

2. Nanomaterials have low tap density. Therefore, the use of nanomaterials leads to reduction in volumetric energy density.[40]

3. Nanomaterials tend to form agglomerates during the electrode fabrication process, making it difficult to uniformly disperse them in the electrodes.[5] Also, nanoparticles tend to aggregate during prolong charge/discharge cycling, ultimately reducing the calendar life of the cell.[40]

4. Nanomaterials are difficult to synthesize.[40] Nanomaterials may incur high fabrication cost due to complex synthetic processes. Purification of materials is also a challenge. Consistent and cost-efficient production may be difficult. Therefore, a major challenge will be to develop simple synthesis methods for large-scale production of nanostructured active materials.[2]

5. Safety aspects of many of the nanomaterials are unknown. There is both foreseen and unexplored environmental, health, and safety risks associated with the manufacture, use, recycling, and disposal of nanoscale Li-ion batteries. The handling characteristics of nanomaterials are sub-optimal.

16.4   NANOMATERIALS IN COMMERCIAL LITHIUM ION CELLS

Nanomaterials are used in commercial Li-ion batteries. Nanosized carbon is used as conductive diluent in cathode of many commercial lithium ion cells. Apart from that some commercial cells use nanosized active materials, the details of which are given below:

Nano Lithium Iron Phosphate is used as the cathode material in A123's high-performance Nanophosphate lithium iron phosphate ($LiFePO_4$) cells. These cells deliver high power and energy density combined with excellent safety performance and extensive life cycling in a lighter and more compact package. These cells have low capacity loss and impedance growth over time. The advantages of the cell include high power, higher usable energy, excellent safety, and extended cycle life. At low rates these cells can deliver thousands of cycles at 100% depth of discharge (DOD). Even when cycled at 10 C discharge rates, these cells deliver in excess of 1000 full DOD cycles.[161] Saphion Li-ion technology by Valence Technology, Inc. also use nanophosphate technology.[162]

In the Nano Safe cells produced by Altairnano, a patented nano LTO material is used as the negative electrode.[163] Toshiba's Supercharge Ion battery (SCiB) also use nanostructured LTO as anode material. SCiB batteries can be charged in as little as 10 min and have excellent thermal performance, reducing or eliminating the need for battery cooling. The LTO chemistry contained in SCiB is not susceptible to thermal runaway or lithium metal plating, providing exceptional battery safety characteristics.[164] Nexeon Li-ion cells use a Si based nanomaterial as anode.[165]

16.5 SUMMARY

Nanotechnology has revolutionarized Li-ion cell technology akin to other fields. A lot of research is being pursued worldwide to develop nanomaterials which could find application in Li-ion cells. A plethora of cathode and anode materials with high capacity and rate capability are being developed. A few materials have been commercialized and a few are under various stages of commercialization. Different architectures of nanomaterials like nanoparticles, nanotubes, nanorods, and nanosheets are synthesized and explored as electrode-active materials that can improve the energy density as well as power density of lithium ion cells. However, the challenges like high irreversible capacity (or low efficiency), difficulty in bulk production of nanomaterials, etc. need to be solved to make a large scale entry of these materials in market. Toward fulfilling the power requirement for EV and modern devices, the development and commercialization of new and promising nanomaterials is definitely going to play a major and crucial role in the coming years.

KEYWORDS

- lithium ion batteries
- phytic acid
- nanomaterials
- anode
- cathode

REFERENCES

1. Mukherjee, R.; Krishnan, R.; Lu, T. M.; Koratkar, N. Nanostructured Electrodes for High-Power Lithium Ion Batteries. *Nano Energy* **2012,** *1,* 518–533.

2. Jiang, C.; Holon, E.; Zhou, H. Nanomaterials for Lithium Ion Batteries. *Nano Today* **2006,** *1* (4), 28–33.

3. Tang, W.; Wang, X. J.; Hou, Y. Y.; Li, L. L.; Sun, H.; Zhu, Y. S.; Bai, Y.; Wu, Y. P.; Zhu, K.; Ree, T. V. Nano $LiMn_2O_4$ as Cathode Material of High Rate Capability for Lithium Ion Batteries. *J. Power Sour.* **2012,** *198,* 308–311.

4. Goodenough, J. B.; Park, K. S. The Li-ion rechargeable battery: A Perspective. *J. Am. Chem. Soc.* **2013,** 135 (4), 1167–1176.

5. Song, M. K.; Park, S.; Alamgir, F. M.; Cho, J.; Liu, M. Nanostructured Electrodes for Lithium-Ion and Lithium-Air Batteries: The Latest Developments, Challenges, and Perspectives. *Mater. Sci. Eng.* **2011,** *72,* 203–252.

6. Lee, K. T.; Cho, J. Roles of Nanosize in Lithium Reactive Nanomaterials for Lithium Ion Batteries. *Nano Today.* **2011,** *6,* 28–41.

7. Zhang, T.; Fu, L. J.; Gao, J.; Wu, Y. P.; Holze, R.; Wu, H. Q. Nanosized Tin Anode Prepared by Laser-Induced Vapour Deposition for Lithium Ion Battery. *J. Power Sources.* **2007,** *174,* 770–773.

8. Jamnik, J.; Maier, J. Nanocrystallinity Effects in Lithium Battery Material, Aspects of Nano-Ionics–Part IV. *Phys. Chem. Chem. Phys.* **2003,** *5,* 5215–5220.

9. Zhukovskii, Y. F.; Balaya, P.; Kotomin, E. A.; Maier, J. Evidence for Interfacial-Storage Anomaly in Nanocomposites for Lithium Batteries from First Principle Simulations. *Phys. Rev. Lett.* **2006,** *96,* 058302.

10. Balaya, P.; Bhattacharyya, A. J.; Jamnik, J.; Zhukovskii, Y. F.; Kotomin, E. A.; Maier, J. Nano-Ionics in the Context of Lithium Batteries. *J. Power Sour.* **2006,** *159,* 171–178.

11. Zhang, X.; Hou, Y.; He, W.; Yang, G.; Cui, J.; Liu, S.; Song, X.; Huang, Z. Fabricating High Performance Lithium-Ion Batteries Using Bionanotechnology. *Nanoscale* **2015,** *7,* 3356–3372.

12. Schodek, D. L.; Ferreira, P.; Ashby, M. F. *Nanomaterials, Nanotechnologies and Design: An Introduction for Engineers and Architects;* Butterworth-Heinemann: Oxford, 2009.

13. Amatucci, G. G.; Tarascon, J. M.; Klein, L. C. Cobalt Dissolution in $LiCoO_2$-Based Non-Aqueous Rechargeable Batteries. *Solid State. Ion.* **1996,** *83* (1–2), 167.

14. Chen, H.; Qiu, X.; Zhu, W.; Hagenmuller, P. Synthesis and High Rate Properties of Nanoparticled Lithium Cobalt Oxides as the Cathode Material for Lithium-Ion Battery. *Electrochem. Commun.* **2002,** *4* (6), 488–49.

15. Qian, X.; Cheng, X.; Wang, Z.; Huang, X.; Guo, R.; Mao, D.; Chang, C.; Song, W. The Preparation of $LiCoO_2$ Nanoplates via a Hydrothermal Process and the Investigation of Their Electrochemical Behavior at High Rates. *Nanotechnology.* **2009,** *20,* 115608.

16. Li, G.; Zhang, J. Synthesis of nano-sized lithium cobalt oxide via a sol–gel method. *Appl. Surface Sci.* **2012,** *258* (19), 7612–7616.

17. Luo, S.; Wang, K.; Wang, J.; Jiang, K.; Li, Q.; Fan, S. Binder-Free $LiCoO_2$/Carbon Nanotube Cathodes for High-Performance Lithium Ion Batteries. *Adv. Mater.* **2012,** *24,* 2294–2298.

18. Li, X.; Cheng, F. Y.; Guo, B.; Chen, J. Template-Synthesized $LiCoO_2$, $LiMn_2O_4$, and $LiNi_{0.8}Co_{0.2}O_2$ Nanotubes as the Cathode Materials of Lithium Ion Batteries. *J. Phys. Chem. B.* **2005,** *109,* 14017–14024.

19. Hao, Q.; Ma, H.; Ju, Z.; Li, G.; Li, X.; Xu, L.; Qian, Y. Nano-CuO Coated $LiCoO_2$: Synthesis, Improved Cycling Stability and Good Performance at High Rates. *Electrochim. Acta.* **2011,** *56* (25), 9027–9031.

20. Cho, J.; Kim, Y. W.; Kim, B.; Lee, J. G.; Park, B. A Breakdown in the Safety of Lithium Secondary Batteries by Coating the Cathode with $AlPO_4$ Nanoparticles. *Ang. Chem.* **2003,** *42* (14), 1618–1621.

21. Lee, D. J.; Scrosati, B.; Sun, Y. K. $Ni_3(PO_4)_2$-Coated $Li[Ni_{0.8}Co_{0.15}Al_{0.05}]O_2$ Lithium Battery Electrode with Improved Cycling Performance at 55°C. *J. Power Sources.* **2011,** *196,* 7742–7746.

22. Kim, Y.; Cho, J. Lithium-Reactive $Co_3(PO_4)_2$ Nanoparticle Coating on High-capacity $LiNi_{0.8}Co_{0.16}Al_{0.04}O_2$ Cathode Material for Lithium Rechargeable Batteries. *J. Electrochem. Soc.* **2007,** *154* (6), 495–499.

23. Kim, D. K.; Muralidharan, P.; Lee, H. W.; Ruffo, R.; Yang, Y.; Chan, C. K.; Peng, H.; Huggins, R. A.; Cui, Y. Spinel $LiMn_2O_4$ Nanorods as Lithium Ion Battery Cathodes. *Nano Lett.* **2008,** *8* (11), 3948–3952.

24. Hosono, E.; Kudo, T.; Honma, I.; Matsuda, H.; Zhou, H. S. Synthesis of Single Crystalline Spinel $LiMn_2O_4$ Nanowires for a Lithium Ion Battery with High Power Density. *Nano Lett.* **2009,** *9,* 1045–1051.

25. Shaju, K. M.; Bruce, P. G. A Stoichiometric Nano-$LiMn_2O_4$ Spinel Electrode Exhibiting High Power and Stable Cycling. *Chem. Mater.* 2008, *20* (17), 5557–5562.

26. Ding, Y. L.; Xie, J.; Cao, G. S.; Zhu, T. J.; Yu, H. M.; Zhao, X. B. Single-Crystalline $LiMn_2O_4$ Nanotubes Synthesized Via Template-Engaged Reaction as Cathodes for High-Power Lithium Ion Batteries. *Adv. Funct. Mater.* **2011,** *21,* 348–355.

27. Sun, Y. K.; Hong, K. J.; Prakash, J. The Effect of ZnO Coating on Electrochemical Cycling Behavior of Spinel $LiMn_2O_4$ Cathode Materials at Elevated Temperature. *J. Electrochem. Soc.* **2003,** *150,* 970–972.

28. Lim, S.; Yoon, C. S.; Cho, J. Synthesis of Nanowire and Hollow $LiFePO_4$ Cathodes for High-Performance Lithium Batteries. *Chem. Mater.* **2008,** *20* (14), 4560–4564.

29. Murugan, A. V.; Muraliganth, T.; Manthiram, A. Rapid Microwave-Solvothermal Synthesis of Phospho-Olivine Nanorods and their Coating with a Mixed Conducting Polymer for Lithium Ion Batteries. *Electrochem. Commun.* **2008,** *10,* 903–906.

30. Manthiram, A.; Murugan, A. V.; Sarkar, A.; Muraliganth, T. Nanostructured Electrode Materials for Electrochemical Energy Storage and Conversion. *Energy Environ. Sci.* **2008,** *1,* 621–638.

31. Sides, C. R.; Croce, F.; Young, V. Y.; Martin, C. R.; Scrosati, B. A High-Rate, Nanocomposite $LiFePO_4$ Carbon Cathode. *Electrochem. Solid-State Lett.* **2005,** *8* (9), 484–487.

32. Kang, B.; Ceder, G. Battery Materials for Ultrafast Charging and Discharging. *Nature* **2009,** *458,* 190–193.

33. Wang, G.; Liu, H.; Liu, J.; Qiao, S.; Lu, G.; M.; Munroe, P.; Ahn. H. Mesoporous $LiFePO_4$/C Nanocomposite Cathode Materials for High Power Lithium Ion Batteries with Superior Performance. *Adv. Mater.* **2010,** *22,* 4944–4948.

34. Jin, B.; Jin, E. M.; Park, K. H.; Gu, H. B. Electrochemical Properties of $LiFePO_4$-Multiwalled Carbon Nanotubes Composite Cathode Materials for Lithium Polymer Battery. *Electrochem. Commun.* **2008,** *10,* 1537–1540.

35. Wang, Y. G.; Wang, Y. R.; Hosono, E. J.; Wang, K. X.; Zhou, H. S. The Design of a $LiFePO_4$/Carbon Nanocomposite With a Core–Shell Structure and Its Synthesis by an In Situ Polymerization Restriction Method. *Angew. Chem. Int. Edit.* **2008,** *47,* 7461–7465.

36. Liu, Y.; Zhang, W.; Zhu, Y.; Luo, Y.; Xu, Y.; Brown, A.; Culver, J. N.; Lundgren, C. A.; Xu, K.; Wang, Y.; Wang, C. Architecturing Hierarchical Function Layers on Self-Assembled Viral Templates as 3D Nano-Array Electrodes for Integrated Li-Ion Micro-batteries. *Nano Lett.* **2013,** *13,* 293–300.

37. Su, J.; Wu, X. L.; Yang, C. P.; Lee, J. S.; Kim, J.; Guo, Y. G.; Self-Assembled LiFePO$_4$/C Nano/Microspheres by Using Phytic Acid as Phosphorus Source. *J. Phys. Chem. C.* **2012,** *116,* 5019–5024.

38. Hu, Y. S.; Guo, Y. G.; Dominko, R.; Gaberscek, M.; Jamnik, J.; Maier, J. Improved Electrode Performance of Porous LiFePO$_4$ Using RuO$_2$ as an Oxidic Nanoscale Inter-connect. *Adv. Mater.* **2007,** *19,* 1963–1966.

39. Du, X. Y.; He, W.; Zhang, X. D.; Yue, Y. Z.; Liu, H.; Zhang, X. G.; Min, D. D.; Ge, X. X.; Du, Y. Enhancing the Electrochemical Performance of Lithium Ion Batteries Using Mesoporous Li$_3$V$_2$(PO$_4$)$_3$/C Microspheres. *J. Mater. Chem.* **2012,** *22,* 5960.

40. Sen, U. K.; Sarkar, S.; Veluri, P. S.; Singh S.; Mitra, S. Nano Fimensionality: A Way towards Better Li-Ion Storage. *Nanosci. Nanotech. Asia.* **2013,** *3,* 21–35.

41. Dominko, R. Li$_2$MSiO$_4$ (M= Fe and/or Mn) Cathode Materials. *J. Power Sources,* **2008,** *184,* 462–468.

42. Muraliganth, T.; Stroukoff, K. R.; Manthiram, A. Microwave- Solvothermal Synthesis of Nanostructured Li$_2$MSiO$_4$/C (M = Mn and Fe) Cathodes for Lithium-Ion Batteries. *Chem. Mater.* **2010,** *22,* 5754–5761.

43. Cheng, F.; Tang, W.; Li, C.; Chen, J.; Liu, H.; Shen, P.; Dou, S. Conducting Poly(aniline) Nanotubes and Nanofibres: Controlled Synthesis and Application in Lithium/poly(aniline) Rechargeable Batteries. *Chem. Eur. J.* **2006,** *12,* 3082–3088.

44. Guo, H.; Liu, L.; Wei, Q.; Shu, H.; Yang, X.; Yang, Z.; Zhou, M.; Tan, J.; Yan, Z.; Wang, X. Electrochemical Characterization of Polyaniline–LiV$_3$O$_8$ Nanocomposite Cathode Material for Lithium Ion Batteries. *J. Power Sources.* **2012,** *217,* 54–58.

45. Li, T.; Chen, Z. X.; Ai, X. P.; Cao, Y. L.; Yang, H. X.; LiF/Fe Nanocomposite as a Lithium-Rich and High Capacity Conversion Cathode Material for Li-Ion Batteries. *J. Power Sources.* **2012,** *217,* 54–58.

46. Goriparti, S.; Miele, E.; Angelis, F. D.; Fabrizio, E. D.; Zaccaria, R. P.; Capiglia, C. Review on Recent Progress of Nanostructured Anode Materials for Li-Ion Batteries. *J. Power Sources.* **2014,** *257,* 421–443.

47. Yang, J.; Zhou, X. Y.; Li, J.; Zou, Y. L.; Tang, J. J. Study of Nano-Porous Hard Carbons as Anode Materials for Lithium Ion Batteries. *Mater. Chem. Phys.* **2012,** *135,* 445–450.

48. Zhang, H.; Tao, H.; Jiang, Y.; Jiao, Z.; Wu, M.; Zhao, B.; Ordered CoO/CMK-3 Nano-composites as the Anode Materials for Lithium-Ion Batteries. *J. Power Sour.* **2010,** *195,* 2950–2955.

49. Jun, S.; Joo, S. H.; Ryoo, R.; Kruk, M.; Jaroniec, M.; Liu, Z.; Ohsuna, T.; Terasaki, O. Synthesis of New, Nanoporous Carbon with Hexagonally Ordered Mesostructure. *J. Am. Chem. Soc.* **2000,** *122,* 10712–10713.

50. Lv, R.; Zou, L.; Gui, X.; Kang, F.; Zhu, Y.; Zhu, H.; Wei, J.; Gu, J.; Wang, K.; Wu, D. High-Yield Bamboo-Shaped Carbon Nanotubes from Cresol for Electrochemical Appli-cation. *Chem. Commun.* **2008,** *17,* 2046–2048.

51. Zhou, J.; Song, H.; Fu, B.; Wu, B.; Chen, X. Synthesis and High-Rate Capability of Quadrangular Carbon Nanotubes with One Open End as Anode Materials for Lithium-Ion Batteries. *J. Mater. Chem.* **2010,** *20,* 2794–2800.

52. Welna, D. T.; Qu, L.; Barney E.; Taylor, C.; Daid, L.; Durstock, M. F. Vertically Aligned Carbon Nanotube Electrodes for Lithium-Ion Batteries. *J. Power Sour.* **2011,** *196,* 1455–1460.

53. Hou, J.; Shao, Y.; Ellis, M.W.; Moore, R. B.; Yi, B. Graphene-Based Electrochemical Energy Conversion and Storage: Fuel Cells, Supercapacitors and Lithium Ion Batteries. *Phys. Chem. Chem. Phys.* **2011,** *13,* 15384–15402.

54. Cui, G.; Gu, L.; Zhi, L.; Kaskhedikar, N.; Aken, P. A.; Mullen, K.; *J. Maier,* A. Germanium-Carbon Nanocompsite Material for Lithium Batteries. *Adv. Mater.* **2008,** *20,* 3079–3083.

55. Liang, M.; Zhi, L. Graphene-Based Electrode Materials for Rechargeable Lithium Batteries. *J. Mater. Chem.* **2009,** *19,* 5871–5878.

56. Brownson, D. A. C.; Kampouris, D. K.; Banks, C. E. An Overview of Graphene in Energy Production and Storage Applications. *J. Power Sour.* **2011,** *196,* 4873–4885.

57. Liu, Y.; Artyukhov, V. I.; Liu, M.; Harutyunyan, A. R.; Yakobson, B. I. Feasibility of Lithium Storage on Graphene and its Derivatives. *J. Phys. Chem. Lett.* **2013,** *4,* 1737–1742.

58. Hwang, H. J., Koo, J.; Park, M.; Park, N.; Kwon, Y.; Lee, H. Multilayer Graphynes for Lithium Ion Battery Anode. *J. Phys. Chem. C.* **2013,** *117,* 6919–6923.

59. Pan, D.; Wang, S.; Zhao, B.; Wu, M.; Zhang, H.; Wang, Y.; Jiao, Z. Li Storage Properties of Disordered Graphene Nanosheets. *Chem. Mater.* **2009,** *21,* 3136–3142.

60. Zhu, J.; Duan, R.; Zhang, S.; Jiang, N.; Zhang, Y.; Zhu, J; The Application of Graphene in Lithium Ion Battery Electrode Materials. *Springer Plus.* **2014,** *3* (585), 1–8.

61. Garcia, M. F.; Arias, A. M.; Hanson, J. C.; Rodriguez, *J. A.* Nanostructured Oxides in Chemistry: Characterization and Properties. *Chem. Rev.* **2004,** *104,* 4063–4104.

62. Nazri, G.A.; Pistoia, G. *Lithium Batteries Science and Technology*; 1st ed.; Springer: New York, 2003.

63. Zhao, N.; Fu, L.; Yang, L.; Zhang, T.; Wang, G.; Wu, Y.; Ree, T. V. Nanostructured Anode Materials for Li-Ion Batteries. *Pure Appl. Chem.* **2008,** *80* (11), 2283–2295.

64. Yin, Y. X.; Wan, L. J.; Guo, Y. G. Silicon-Based Nanomaterials for Lithium-Ion Batteries. *Chin. Sci. Bull.* **2012,** *57,* 4104–4110.

65. Liu, N.; Huo, K.; McDowell, M. T.; Zhao, J.; Cui, Y. Rice Husks as a Sustainable Source of Nanostructured Silicon for High Performance Li-Ion Battery Anodes. *Sci. Rep.* **2013,** *3,* 1919, 1–7.

66. Ma, H.; Cheng, F.; Chen, J.Y.; Zhao, J.Z.; Li, C.S.; Tao, Z.L.; Liang, J. Nest-like Silicon Nanospheres for High-Capacity Lithium Storage. *Adv. Mater.* **2007,** *19,* 4067–4070.

67. Yao, Y.; McDowell, M. T.; Ryu, I.; Wu, H.; Liu, N.; Hu, L.; Nix, W. D.; Cui, Y. Interconnected Silicon Hollow Nanospheres for Lithium-Ion Battery Anodes with Long Cycle Life. *Nano Lett.* **2011,** *11,* 2949–2954.

68. Chan, C. K.; Peng, H.; Liu, G.; McIlwrath, K.; Zhang, X. F.; Huggins, R. A.; Cui, Y. High-Performance Lithium Battery Anodes Using Silicon Nanowires. *Nat. Nanotechnol.* **2008,** *3,* 31–35.

69. Chan, C.K.; Peng, H.L.; Liu, G.; McIlwrath, K.; Zhang, X.F.; Huggins, R.A.; Cui, Y.. High-performance Lithium Battery Anodes Using Silicon Nanowires. *Nat. Nanotech.* **2008,** *3,* 31–35.

70. Chan, C. K.; Zhang, X. F.; Cui, Y. High Capacity Li Ion Battery Anodes Using Ge Nanowires. *Nano Lett.* **2008,** *8,* 307–309.

71. Ohara, S.; Suzuki, J.; Sekine, K.; Takamura, T. A Thin Film Silicon Anode for Li-Ion Batteries Having a Very Large Specific Capacity and Long Cycle Life. *J. Power Sour.* **2004**, *136*, 303–306.

72. Holzapfel, M.; Buqa, H.; Scheifele, W.; Novák P.; Petrat, F. M. A New type of Nano-sized Silicon/carbon Composite Electrode for Reversible Lithium Insertion. *Chem. Commun.* **2005**, *0*, 1566–1568.

73. Liu, H. K.; Wang, G. X.; Guo, Z. P.; Wang, J. Z.; Konstantinov, K. The Impact of Nano-materials on Li-ion Rechargeable Batteries. *J. New. Mat. Electr. Sys.* **2007**, *10*, 101–104.

74. Zhang, T.; Fu, L. J.; Gao, J.; Yang, L. C.; Wu, Y. P.; Wu, H. Q. Core-shell Si/C Nano-composite as Anode Material for Lithium Ion Batteries. *Pure Appl. Chem.* **2006**, *78*, 1889–1896.

75. Ng, S. H.; Wang, J. Z.; Wexler, D.; Konstantinov, K.; Guo, Z. P.; Liu, H. K. Highly Reversible Lithium Storage in Spheroidal Carbon Coated Silicon Nanocomposites as Anodes for Lithium-Ion Batteries. *Angew. Chem. Int. Ed.* **2006**, *45*, 6896–6899.

76. Hu, L.; Liu, N.; Eskilsson, M.; Zheng, G.; McDonough, J.; Wagberg, L.; Cui, Y.; Silicon-Conductive Nanopaper for Li-Ion Batteries. *Nano Energy.* **2013**, *2* (1), 138–145.

77. Yu, Y.; Gu, L.; Zhu, C.; Tsukimoto, T.; Aken, P. A. V.; Maier, J. Reversible Storage of Lithium in Silver-Coated Three-Dimensional Macroporous Silicon. *Adv. Mater.* **2010**, *22*, 2247–2250.

78. Wang, G. X.; Ahn, J. H.; Yao, J.; Bewlay, S.; Liu, H. K. Nanostructured Si–C Composite Anodes for Lithium-Ion Batteries. *Electrochem. Commun.* **2004**, *6*, 689–692.

79. Zhou, S.; Liu, X.; Wang, D.; Si/TiSi$_2$ Hetero nanostructures as High-Capacity Anode Material for Li Ion Batteries. *Nano Lett.* **2010**, *10* (3), 860–863.

80. Wang, Z.; Tian, W. H.; Liu, X. H.; Lia, Y.; Li, X. G. Nanosized Si–Ni Alloys Anode Prepared by Hydrogen Plasma–Metal Reaction for Secondary Lithium Batteries. *Mater. Chem. Phys.* **2006**, *100*, 92–97.

81. Zhang, T.; Gao, J.; Zhang, H. P.; Yang, L. C.; Wu, Y. P.; Wu, H. Q. Preparation and Electrochemical Properties of Core-Shell Si/SiO Nanocomposite as Anode Material for Lithium Ion Batteries. *Electrochem. Commun.* **2007**, *9*, 886–890.

82. Yang, J.; Winter, M.; Besenhard, J. O. Small Particle Size Multiphase Li-Alloy Anodes for Lithium-Ion Batteries. *Solid State. Ionics.* **1996**, *90*, 281–287.

83. Yang, J.; Wachtler, M.; Winter, M.; Besenhard, J. O. Sub-Microcrystalline Sn and Sn - SnSb Powders as Lithium Storage Materials for Lithium-Ion Batteries. *Electrochem. Solid. State Lett.* **1999**, *2* (4), 161–163.

84. Wang, Y.; Li, B.; Zhang, C.; Tao, H.; Kang, S.; Jiang, S.; Li, X. Simple Synthesis of metallic Sn nanocrystals embedded in graphitic ordered mesoporous carbon walls as Superior Anode Materials for Lithium Ion Batteries. *J. Power Sources.* **2012**, *219*, 89–93.

85. Liang, Y.; Fan, J.; Xia, X.; Jia, Z. Synthesis and Characterisation of SnO$_2$ Nano-Single Crystals as Anode Materials for Lithium-Ion Batteries. *Mater. Lett.* **2007**, *61*, 4370–4373.

86. Xia, G.; Li, N.; Li, D.; Liu, R.; Xiao, N.; Tian, D. Molten-Salt Decomposition Synthesis of SnO$_2$ Nanoparticles as Anode Materials for Lithium Ion Batteries. *Mater. Let.* **2011**, *65*, 3377–3379.

87. Kim, C.; Noh, M.; Choi, M.; Cho, J.; Park, B. Critical Size of a Nano SnO$_2$ Electrode for Li-Secondary Battery. *Chem. Mater.* **2005**, *17*, 3297–3301.

88. Du, Z.; Yin, X.; Zhang, M.; Hao, Q.; Wang, Y.; Wang, T. In Situ Synthesis of SnO$_2$/Graphene Nanocomposite and Their Application as Anode Material for Lithium Ion Battery. *Mater. Lett.* **2010**, *64*, 2076–2079.

89. Meduri, P.; Pendyala, C.; Kumar, V.; Sumanasekera, G.U.; Sunkara, M.K.; Hybrid Tin Oxide Nanowires as Stable and High Capacity Anodes for Li-Ion Batteries. *Nano Lett.* **2009,** *9,* 612–616.

90. Park, C. M.; Kim, J. H.; Kim, H.; Sohn, H. J. Li-Alloy Based Anode Materials for Li Secondary Batteries. *Chem. Soc. Rev.* **2010,** *39,* 3115–3141.

91. Park, M. H.; Kim, K.; Kim, J.; Cho, J. Flexible Dimensional Control of High-Capaciy Li-Ion Battery Anodes: From 0D to 3D porous Germanium Nanoparticle Assemblies. *Adv. Mater.* **2010,** *22,* 415–418.

92. Dambournet, D.; Belharouak, I.; Amine, K. Tailored Preparation Methods of TiO_2 Anatase, Rutile, Brookite: Mechanism of Formation and Electrochemical Properties. *Chem. Mater.* **2010,** *22,* 1173–1179.

93. Su, X.; Wu, Q.; Zhan, X.; Wu, J.; Wei, S.; Guo, Z. Advanced Titania Nanostructures and Composites for Lithium Ion Battery. *J. Mater. Sci.* **2012,** *47,* 2519–2534.

94. Deng, D.; Kim, M. G.; Lee, J. Y.; Cho, J. Green Energy Storage Materials: Nanostructured TiO_2 and Sn-Based Anodes for Lithium-Ion Batteries. *Energy Environ. Sci.* **2009,** *2,* 818–837.

95. Armstrong, A. R.; Armstrong, G.; Canales, J.; García, R.; Bruce, P. G. Lithium-Ion Intercalation into TiO2-B Nanowires. *Adv. Mater.* **2005,** *17,* 862–865.

96. Jiang, C.; Honma, I.; Kudo, T.; Zhou, H. Nanocrystalline Rutile TiO_2 Electrode for High-Capacity and High-Rate Lithium Storage Batteries and Energy Storage. *Electrochem. Solid-State Lett.* **2007,** *10,* 127–129.

97. Wang, H. E.; Cheng, H.; Liu, C.; Chen, X.; Jiang, Q.; Lu, Z.; Li, Y. Y.; Chung, C. Y.; Zhang, W.; Zapien, J. A.; Martinu, L.; Igor Bello, I. Facile Synthesis and Electrochemical Characterization of Porous and Dense TiO_2 Nanospheres for Lithium-Ion Battery Applications. *J. Power Sources.* **2011,** *196,* 6394–6399.

98. Fu, L. J.; Liu, H.; Zhang, H. P.; Li, C.; Zhang, T.; Wu, Y. P.; Wu, H. Q. Novel TiO_2/C Nanocomposites for Anode Materials of Lithium Ion Batteries. *J. Power Sources.* **2006,** *159,* 219–222.

99. Panda, S. K.; Yoon, Y.; Jung, H. S.; Yoon, W. S.; Shin, H. Nanoscale Size Effect of Titania (anatase) Nanotubes with Uniform Wall Thickness as High Performance Anode for Lithium-Ion Secondary Battery. *J. Power Sources.* **2012,** *204,* 162–167.

100. Ren, H. M.; Ding, Y. H.; Chang, F. H.; He, X.; Feng, J. Q.; Wang, C. F.; Jiang, Y.; Zhang, P. Flexible Free-Standing TiO_2/Graphene/PVdF Films as Anode Materials for Lithium-Ion Batteries. *App. Surf. Sci.* **2012,** *263,* 54–57.

101. Sandhya, C. P.; John, B.; Gouri, C. Lithium Titanate as Anode Material for Lithium-Ion Cells: A Review. *Ionics.* **2014,** *20,* 601–620.

102. Guerfi, A.; Charest, P.; Kinishita, K.; Perrier. M.; Zaghib, K. Nano Electronically Conductive Titanium-Spinel as Lithium Ion Storage Negative Electrode. *J. Power Sources.* **2004,** *126* (1/2), 163–168.

103. Zaghib, K.; Simoneau, M.; Armand, M.; Gauthier, M. Electrochemical Study of $Li_4Ti_5O_{12}$ as Negative Electrode for Li Ion Polymer Rechargeable Batteries. *J. Power Sources.* **1999,** *81* (82), 300–305.

104. Porotnikov, N. V.; Chaban, N. G.; Petrov, K. I. Synthesis and Investigation of Electrical Conductivity of Complex Oxides in the System Li_2O-ZnO-TiO_2. *Neorg Mater.* **1982,** *18* (6), 1066.

105. Sandhya, C. P.; John, B.; Gouri, C. Synthesis and Electrochemical Characterisation of Electrospun Lithium Titanate Ultrafine Fibres. *J. Mater. Sci.* **2013,** *48,* 5827–5832.

106. Wang, J.; Shen, L.; Li, H.; Ding, B.; Nie, P.; Dou, H.; Zhang, X.; Mesoporous $Li_4Ti_5O_{12}$/ Carbon Nanofibers for High-Rate Lithium-Ion Batteries. *J. Alloys Compd.* **2014**, *587*, 171–176.

107. Ni, H.; Fan, L. Z. Nano-$Li_4Ti_5O_{12}$ Anchored on Carbon Nanotubes by Liquid Phase Deposition as Anode Material for High Rate Lithium-Ion Batteries. *J. Power Sour.* **2012**, *214*, 195–199.

108. Deng, Y.; Wan, L.; Xie, Y.; Qin, X.; Chen , G. Recent Advances in Mn-based Oxides as Anode Materials for Lithium Ion Batteries. *RSC Adv.* **2014**, *4*, 23914–23935.

109. Fan, Q.; Whittingham, M. S. Electrospun Manganese Oxide Nanofibers as Anodes for Lithium-Ion Batteries. *Electrochem. SolidState Lett.* **2007**, *10* (3), 48–51.

110. Xu, J.S.; Zhu, Y.J. Monodisperse Fe_3O_4 and γ-Fe_2O_3 Magnetic Mesoporous Microspheres as Anode Materials for Lithium-Ion Batteries. *ACS Appl. Mater. Inter.* **2102**, *4*, 4752–4757.

111. Liu, H.; Wang, G.; Park, J.; Wang, J.; Liu, H.; Zhang, C.; Electrochemical Performance of α-Fe_2O_3 Nanorods as Anode Material for Lithium-Ion Cells. *Electrochim. Acta.* **2009**, *54*, 1733–1736.

112. Wen, W.; Wu, J.M.; Tu, J.P. A Novel Solution Combustion Synthesis of Cobalt Oxide Nanoparticles as Negative-Electrode Materials for Lithium Ion Batteries. *J. Alloys Comp.* **2012**, *513*, 592–596.

113. Yao, X.; Xing Xin, X.; Zhang, Y.; Wang, J.; Liu, Z.; Xu, X. Co_3O_4 Nanowires as High Capacity Anode Materials for Lithium Ion Batteries. *J. Alloys Comp.* **2012**, *521*, 95–100.

114. Huang, G.; Zhang, F.; Du, X.; Qin,Y.; Yin,D.; Wang, L. Metal Organic Frameworks Route to in Situ Insertion of Multiwalled Carbon Nanotubes in Co_3O_4 Polyhedra as Anode Materials for Lithium-Ion Batteries. *ACS Nano.* **2015**, *9* (2), 1592–1599.

115. Qi, Y.; Du, N.; Zhang, H.; Wang, J.; Yang, Y.; Yang, D. Nanostructured Hybrid Cobalt Oxide/Copper Electrodes of Lithium-Ion Batteries with Reversible High-Rate Capabilities. *J. Alloys Comp.* **2012**, *521*, 83–89.

116. Su, L.; Zhou, Z.; Qin, X.; Tang, Q.; Wu, D.; Shen, P. $CoCO_3$ Submicrocube/Graphene Composites with High Lithium Storage Capability. *Nanoenergy.* **2013**, *2* (2), 276–282.

117. Qiu, W.; Jiao, J.; Xia, J.; Zhong, H.; Chen, L. A Self-Standing and Flexible Electrode of Yolk–Shell CoS_2 Spheres Encapsulated with Nitrogen-Doped Graphene for High-Performance Lithium-Ion Batteries. *Chem. A. Eur. J.* **2015**, *21*, 1–10.

118. Ma, J.; Yang, J.; Jiao, L.; Mao, Y.; Wang, T.; Duan, X.; Lian, J.; Zheng, W. NiO Nanomaterials: Controlled Fabrication, Formation Mechanism and the Application in Lithium-Ion Battery. *Cryst. Eng. Comm.* **2012**, *14*, 453–459.

119. Needham, S. A.; Wang, G. X.; Liu, H. K. Synthesis of NiO Nanotubes for Use as Negative Electrodes in Lithium Ion Batteries. *J. Power Sources.* **2006**, *159*, 254–257.

120. Xia, Y.; Zhang, W. K.; Xiao, Z.; Huang, H.; Zeng, H. J.; Chen, X. R.; Chen, F.; Gan, Y. P.; Tao, X. Y. Biotemplated Fabrication of Hierarchically Porous NiO/C Composite from Lotus Pollen Grains for Lithium-Ion Batteries. *J. Mater. Chem.* **2012**, *22*, 9209–9215.

121. Park, J. C.; Kim, J.; Kwon, H.; Song, H. Gram-Scale Synthesis of Cu_2O Nanocubes and Subsequent Oxidation to CuO Hollow Nanostructures for Lithium Ion Battery Anode Materials. *Adv. Mater.* **2009**, *21*, 803–807.

122. Liu, J.; Li, Y.; Huang, X. Proceedings of the 2nd IEEE International Nanoelectronics Conference (INEC 2008), March 24–27, 1998.

123. Li, P.; Liu, Y.; Liu, J.; Li, Z.; Wu, G.; Wu, M.; Facile Synthesis of ZnO/Mesoporous Carbon Nanocomposites as High-Performance Anode for Lithium-Ion Battery. *Chem. Eng. J.* **2015**, *271*, 173–179.

124. Xiao, L.; Mei, D.; Cao, M.; Qu, D.; Deng, B. Effects of Structural Patterns and Degree of Crystallinity on the Performance of Nanostructured ZnO as Anode Material for Lithium-Ion Batteries. *J. Alloys Comp.* **2015,** *627,* 455–462.

125. Huang, X. H.; Guo, R. Q.; Wu, J. B.; Zhang, P. Mesoporous ZnO Nanosheets for Lithium Ion Batteries. *Mate. Lett.* **2014,** *122,* 82–85.

126. Xu, X.; Liu, W.; Kim, Y.; Cho, J. Nanostructured Transition Metal Sulfides for Lithium Ion Batteries: Progress and Challenges. *Nano Today.* **2014,** *9,* 604–630.

127. Li., H.; Ma, L.; Chen, W. X.; Wang, J. M. Synthesis of MoS$_2$/C nanocomposites by hydrothermal route used as Li-ion intercalation electrode materials. *Mat. Lett.* **2009,** *63,* 1363–1365.

128. Feng, C.; Ma, J.; Li, H.; Zeng, R.; Guo, Z.; Liu, H. Synthesis of Molybdenum Disulfide (MoS$_2$) for Lithium Ion Battery Applications. *Mater. Res. Bull.* **2009,** *44,* 1811–1815.

129. Zhao, X.; Mai, Y.; Luo, H.; Tang, D.; Lee, B.; Huang, C.; Zhang, L. Nano-MoS$_2$/poly (3,4-ethylenedioxythiophene): Poly (styrenesulfonate) Composite Prepared by a Facial Dip-Coating Process for Li-Ion Battery Anode. *Appl. Surf. Sci.* **2014,** *288,* 736–741.

130. Hwang, H.; Kim, H.; Cho, J.; MoS$_2$ Nanoplates Consisting of Disordered Graphene-Like Layers for High Rate Lithium Battery Anode Materials. *Nano Lett.* **2011,** *11,* 4826–4830.

131. Du, G.; Guo, Z.; Wang, S.; Zeng, R.; Chen, Z.; Liu, H. Superior Stability and High Capacity of Restacked Molybdenum Disulfide as Anode Material for Lithium Ion Batteries. *Chem. Commun.* **2010,** *46,* 1106–1108.

132. Late, D. J.; Rout, C.S.; Chakravarty D.; Ratha, S. Emerging Energy Applications of Two-Dimensional Layered Materials. *Canad. Chem. Trans.* **2015,** *3,* 2, 118–157.

133. Li, H.; Li, W.; Ma, L.; Chen, W.; Wang, J. Electrochemical Lithiation/Delithiation Performances of 3D Flowerlike MoS$_2$ Powders Prepared by Ionic Liquid Assisted Hydrothermal Route. *J. Alloys. Compd.* 2009, *47,* 442–447.

134. Gao, P; Yang, Z.; Liu, G.; Qiao, Q. Facile Synthesis of MoS$_2$/MWNT Anode Material for High-Performance Lithium-Ion Batteries. *Ceram. Inter.* **2015,** *41,* 1921–1925.

135. David, L.; Bhandavat, R.; Barrera, U.; Singh, G. Polymer-Derived Ceramic Functionalized MoS$_2$ Composite Paper as a Stable Lithium-Ion Battery Electrode. *Sci. Rep.* **2015,** *5,* 1–7.

136. Wang, R.; Xu, C.; Sunn, J.; Liu, Y.; Gao, L.; Yao, H.; Lin, C. Heat-induced formation of Porous and Free-Standing MoS$_2$/GS Hybrid Electrodes for Binder-Free and Ultra Long-life Lithium-Ion Batteries. *Nano Energy.* **2014,** *8,* 183–195.

137. Zhou, X.; Wang, Z.; Chen, W.; Ma, L.; Chen, D.; Lee, J. Y. Facile Synthesis and Electrochemical Properties of two Dimensional Layered MoS$_2$/Graphene Composite for Reversible Lithium Storage. *J. Power Sour.* **2014,** *251,* 264–268.

138. Han, S.; Zhao, Y.; Tang, Y.; Tan, F.; Huang, Y.; Feng, X.; Wu, D., Ternary MoS$_2$/SiO$_2$/Graphene Hybrids for High-Performance Lithium Storage. *Carbon* **2015,** *81,* 203–209.

139. Feng, C.; Huang, L.; Guo, Z.; Liu, H. Synthesis of Tungsten Disulfide (WS$_2$) Nanoflakes for Lithium ion Battery Application. *Electrochem Commun.* 2007, *9,* 119–122.

140. Wang, G. X.; Bewlay, S.; Yao, J.; Liu, H. K.; Dou, S. X. Tungsten Disulfide Nanotubes for Lithium Storage. *Electrochem. Solid State Lett.* **2004,** *7* (10), A321–A323.

141. Xu, X.; Rout, C. S.; Yang, J.; Cao, R.; Oh, P.; Shin, H. S.; Cho, J. Freeze-Dried WS$_2$ Composite with Low Content of Graphene as High-Rate Lithium Storage Materials. *J. Mater. Chem. A.* **2013,** *1,* 14548–14554.

142. Shiva, K.; Matte, H. S. S. R.; Rajendra, H. B.; Bhattacharyya, A. J.; Rao, C. N. R. Employing Synergistic Interactions between Few-layer WS$_2$ and Reduced Graphene

Oxide to Improve Lithium Storage, Cyclability and Rate Capability of Li-Ion Batteries. *Nano Energy.* **2013,** *2,* 787–793.

143. Guzman, R.; Lavela, P.; Morales, J.; Tirado, J. L. VSe_2-yS_y Electrodes in Lithium and Lithium-ion Cells. *J. Appl. Electrochem.* **1997,** *27,* 1207—1211.

144. Kim, Y.; Park, K.S.; Song, S.H.; Han, J.; Goodenough, J. B. Access to M^{3+}/M^{2+} Redox Couples in Layered $LiMS_2$ Sulfides (M=Ti,V,Cr) as Anodes for Li-Ion Battery. *J. Electrochem. Soc.* **2009,** *156,* A703–A708.

145. Rout, C. S.; Kim, B. H.; Xu, X.; Yang, J.; Jeong, H. Y.; Odkhuu, D.; Park, N.; Cho, J.; Shin, H. S. Synthesis and Characterization of Patronite form of Vanadium Sulphides on Graphitic Layer. *J. Am. Chem. Soc.* **2013,** *135,* 8720–8725.

146. Xu, X.; Jeong, S.; Rout, C. S.; Oh, P.; Ko, M.; Kim, H.; Kim, M. G.; Cao, R.; Shin, H. S.; Cho, J. Lithium Reaction Mechanism and High Rate Capability of VS_4–Graphene Nanocomposite as an Anode Material for Lithium Batteries. *J. Mater. Chem. A.* **2014,** *2,* 10847–10853.

147. Jang, J.T.; Jeong, S.; Seo, J.W.; Kim, M. C.; Sim, E.; Oh, Y.; Nam, S.; Park, B.; Cheon, J.; Ultrathin Zirconium Disulfide Nanodiscs. *J. Am. Chem. Soc.* **2011,** *133,* 7636–7639.

148. Li, H.; Shi, L.; Wang, Q.; Chen, L.; Huang, X. Nano-Alloy Anode for Lithium Ion Batteries. *Solid State. Ionics.* **2002,** *148,* 247–258.

149. Yang, J.; Takada, Y.; Imanishi, N.; Yamamoto, O. Ultrafine Sn and $SnSb_{0.14}$ Powders for Lithium Storage Matrices in Lithium-Ion Batteries. *J. Electrochem. Soc.* **1999,** *146,* 4009–4013.

150. Zhang, C. Q.; Tu, J. P.; Huang, X. H.; Yuan, Y. F.; Wang, S. F.; Mao, F. Preparation and Electrochemical Performances of Nanoscale $FeSn_2$ as Anode Material for Lithium Ion Batteries. *J. Alloys Comp.* **2008,** *457,* 81–85.

151. Wang, G. X.; Sun, L.; Bradhurst, D. H.; Zhong, S.; Dou, S. X.; Liu. H. K. Nanocrystalline NiSi Alloy as an Anode Material for Lithium-Ion Batteries. *J. Alloys Comp.* **2000,** *306,* 249–252.

152. Yaser A. L.; Isobel, D. Nanotechnology for Lithium-ion batteries; Springer: New York, 2013.

153. Balaya, P.; Bhattacharyya, A. J.; Jamnik, J.; Yu, F.; Zhukovskii, E. A.; Kotomin, Maier, J. Nano-Ionics in the Context of Lithium Batteries. *J. Power Sour.* **2006,** *159,* 171–178.

154. Bhattacharyya, A. J.; Patel, M.; Das, S. K.; Soft Matter Lithium Salt Electrolytes: Ion Conduction and Application to Rechargeable Batteries. *Monatsh. Chem.* **2009,** *140* (9), 1001–1010.

155. Ahn, J. H.; Wang, G. X.; Liu, H. K.; Dou, S. K. Nanoparticle Dispersed PEO Polymer Electrolytes for Li Batteries. *J. Power Sources.* **2003,** *119–12,* 422–426.

156. Stephan, A. M.; Nahm, K. S. Review on Composite Polymer Electrolytes for Lithium Batteries. *Polymer* **2006,** *47,* 5952–5964.

157. Bhattacharyya, A. J.; Tarafdar, S.; Middya, T. R. Effective Medium Theory for Ionic Conductivity in Polycrystalline Solid Electrolytes. *Solid State Ionics.* **1997,** *95,* 283–288.

158. Adebahr, J.; Best, A. S.; Byrne, N.; Jacobsson, P.; MacFarlane, D. R.; Forsyth, M. Ion Transport in Polymer Electrolytes Containing Nanoparticulate TiO_2: The Influence of Polymer Morphology. *Phys. Chem. Chem. Phys.* **2003,** *5,* 720–725.

159. Deka, M.; Kumar, A. Enhanced Ionic Conductivity in Novel Nanocomposite Gel Polymer Electrolyte Based on Intercalation of PMMA into Layered LiV_3O_8. *J. Solid State Electrochem.* **2010,** *14* (9), 1649–1656.

160. Okubo, M.; Hosono, E.; Kim, J.; Enomoto, M.; Kojima, N.; Kudo, T.; Zhou, H. S.; Honma, I. Nanoszie Effect on High-Rate Li-Ion Intercalation on $LiCoO_2$ Electrode. *J. Am. Chem. Soc.* **2007,** *129,* 7444–7452.
161. http://www.a123systems.com/lithium-iron-phosphate-battery.htm
162. Bergmann, C. P.; Andrade, M. J. d. *Nanostructured Materials for Engineering Applications;* Springer Science and Business Media: Berlin, 2011.
163. http://www.altairnano.com/products/performance/
164. https://www.toshiba.com/tic/industrial/rechargable-battery
165. http://www.nexeon.co.uk/technology-2/

INDEX

Z

Printed and bound by CPI Group (UK) Ltd, Croydon, CR0 4YY

23/10/2024

01777705-0008